Building Cross-Platform Mobile and Web Apps for Engineers and Scientists

AN ACTIVE LEARNING APPROACH

Pawan Lingras
Saint Mary's University, Halifax, Nova Scotia

with

Matt Triff
Canada Border Services Agency, Ottawa, Ontario

Rucha Lingras
Statistics Canada, Ottawa, Ontario

Australia • Brazil • Mexico • Singapore • United Kingdom • United States

***Building Cross-Platform Mobile and Web Apps for Engineers and Scientists: An Active Learning Approach*, First Edition**

Pawan Lingras with Matt Triff and Rucha Lingras

Product Director, Global Engineering: Timothy L. Anderson

Senior Content Developer: Mona Zeftel

Associate Media Content Developer: Ashley Kaupert

Product Assistant: Alexander Sham

Marketing Manager: Kristin Stine

Director, Content and Media Production: Sharon L. Smith

Content Project Manager: D. Jean Buttrom

Production Service: RPK Editorial Services, Inc.

Copyeditor: Lori Martinsek

Proofreader: Harlan James

Indexer: Shelly Gerger-Knechtl

Compositor: SPi-Global

Senior Art Director: Michelle Kunkler

Cover and Internal Designer: Harasymczuk Design

Cover Image: Sergey Nivens/Shutterstock.com

Internal Image: Excellent backgrounds/Shutterstock.com

Intellectual Property
Analyst: Christine Myaskovsky
Project Manager: Sarah Shainwald

Text and Image Permissions Researcher: Kristiina Paul

Senior Manufacturing Planner: Doug Wilke

© 2017 Cengage Learning®

WCN: 01-100-101

ALL RIGHTS RESERVED. No part of this work covered by the copyright herein may be reproduced, transmitted, stored, or used in any form or by any means graphic, electronic, or mechanical, including but not limited to photocopying, recording, scanning, digitizing, taping, Web distribution, information networks, or information storage and retrieval systems, except as permitted under Section 107 or 108 of the 1976 United States Copyright Act, without the prior written permission of the publisher.

For product information and technology assistance, contact us at **Cengage Learning Customer & Sales Support, 1-800-354-9706**.

For permission to use material from this text or product, submit all requests online at **www.cengage.com/permissions**. Further permissions questions can be emailed to **permissionrequest@cengage.com**.

Library of Congress Control Number: 2015949700

ISBN: 978-1-305-10596-6

Cengage Learning
20 Channel Center Street
Boston MA 02210
USA

Cengage Learning is a leading provider of customized learning solutions with employees residing in nearly 40 different countries and sales in more than 125 countries around the world. Find your local representative at **www.cengage.com**.

Cengage Learning products are represented in Canada by Nelson Education Ltd.

To learn more about Cengage Learning Solutions, visit **www.cengage.com/engineering**.

Purchase any of our products at your local college store or at our preferred online store **www.cengagebrain.com**.

Unless otherwise noted, all items © Cengage Learning

Printed in the United States of America
Print Number: 01 Print Year: 2015

CONTENTS

PREFACE

Past app development process tended to be specific to a platform in the native application development environment such as Objective-C or Swift for Apple iOS, or Java for Google's Android. While some apps still need to be developed in the native development environment, the emergence of HTML5 and JavaScript-based jQuery and Node.js makes it possible to develop apps that can run on most communication devices. These technologies help businesses with limited resources to keep up with the recent shift to mobile technologies. Businesses still need to support the desktops/laptops/netbooks platforms. However, mobile devices such as tablets, smartphones, smart TVs, and phablets increasingly serve our personal and home information technology needs, and as such are equally and in some cases more important. In order to address a fragmented computing, information, and communication device market, we need to adopt a mobile-first or mobile-friendly development strategy. The major web browsers for desktop/laptop/netbook computing such as Internet Explorer, Firefox, Chrome, Safari, and Opera that run on Windows, Mac, Linux, and Chrome OS now also include versions for smartphones, smart TVs, and other mobile devices that run on Android by Google, iOS by Apple, Windows RT by Microsoft, as well as several other competing mobile platforms. The cross-platform versions of these browsers makes it easier to create cross-platform apps.

This book guides readers through the process of building apps from a variety of domains including science, engineering, and business using an active learning approach. The technologies that are introduced work with all major mobile and web platforms and are applicable in any domain. The book presents an app-centric development methodology using the ubiquitous HTML5, JavaScript, jQuery, Node.js and JSON. Instead of describing various features of these languages and demonstrating their usage in an app, we identify important features that should be part of an app and then introduce the necessary language constructs. We present the material in an organized and readable manner, which makes the book appropriate as a course textbook, yet at the same time it can be used as a reference for app development. Instructors will find it useful for teaching either a first course on mobile and web programming, or for a sequel to a web programming course.

Structure

Every chapter focuses on an important app feature that is illustrated by building an app in a science, engineering, or economics/finance/business domain. The apps are incrementally developed. The development process is illustrated by figures that include the code and corresponding screenshots. Students can follow the app building process by typing in the code in parallel. Each chapter ends with exercises that encourage students to make changes to the code and see the resulting changes in app behavior, as well as hands-on app-projects from different application domains. These app-projects can be used as part of students' portfolios after the course, providing excellent professional preparation.

The book consists of thirteen chapters and makes no assumptions about the knowledge of any specific technology. The book begins with an introduction to web page creation and deployment using HTML5 for specifying the content and structure of web pages and CSS3 for describing the presentation details. The focus quickly shifts to web programming with the use of JavaScript. After getting basic knowledge of these three important underlying technologies, we explore mobile app development with the help of the jQuery Mobile platform that facilitates both the presentation and programming of our web pages for mobile devices. Along with various features of the jQuery Mobile platform, other libraries such as Bootstrap are used to develop non-trivial applications. Once students have a handle on client-side mobile computing, the book leads them to server-side computing with the help of another variant of JavaScript called Node.js. In addition to computing, students explore different aspects of data storage including local storage on devices, NoSQL databases, and the conventional SQL databases. Chapter 12 introduces templates that can be used for creating a consistent interface for all pages in an app. The book concludes with an illustration of technologies to convert the web-based apps to native apps. Depending on the length of a course and the background of the students, it will be possible to select a subset of the chapters. For example, students at an introductory level may choose to learn HTML5, CSS3, jQuery Mobile, bootstrap, local storage, Node.js, and only one of the database management systems (chapters 1-9) and defer the templates and native app creation for a later time. Students with prior knowledge of HTML, CSS, and JavaScript can quickly browse through the initial chapters (chapters 1-3) and begin an in-depth study of the mobile computing and data storage.

Given the breadth of the technologies covered in the book, a student may need to refer to additional material for mastering any one of these technologies. However, the app-centric view of the book leads to a reasonable understanding of the cores of these technologies for rapid prototyping of the apps.

MindTap Online Course and Reader

This textbook is also available online through Cengage Learning's MindTap, a personalized learning program. Students who purchase the MindTap have access to the book's multimedia-rich electronic Reader and are able to complete homework and assessment material online, on their desktops, laptops, or iPads. The new MindTap Mobile app makes it easy for students to study anywhere, anytime. Instructors who use a Learning Management System (such as Blackboard, Canvas, or Moodle) for tracking course content, assignments, and grading, can seamlessly access the MindTap suite of content and assessments for this course.

With MindTap, instructors can:

- Personalize the Learning Path to match the course syllabus by rearranging content or appending original material to the online content
- Connect a Learning Management System portal to the online course and Reader
- Highlight and annotate the digital textbook, then share their notes with students.
- Customize online assessments and assignments
- Track student engagement, progress and comprehension
- Promote student success through interactivity, multimedia, and exercises

Additionally, students can listen to the text through ReadSpeaker, take notes in the digital Reader, study from and create their own Flashcards, highlight content for easy reference, and check their understanding of the material through practice quizzes and automatically-graded homework.

Acknowledgments

Two apps that span over multiple chapters in this book started as student projects. We would like to acknowledge Drew Rajaraman for providing the conceptual design of the app to monitor Thyroid Cancer patients. The first version of the app was developed by Rafael Paravia. The graphs and gadgets in the app were developed by Siddharth Jain and Srushti Bharadwaj. The final version presented in the book was developed by Matt Triff. The Explorador app was developed for a competition by a team consisting of Matt Triff, Andrew Valencik, Siddharth Jain, and Rucha Lingras. We thank all of them for their contributions.

We thank these reviewers who provided valuable feedback: Donald Ekong, Mercer University; Chengqi Guo, James Madison University; Ling Rothrock, Penn State University; and Jun Zheng, New Mexico Institute of Mining and Technology.

We wish to acknowledge and thank our Global Engineering team at Cengage Learning for their dedication to this new book:

Timothy Anderson, Product Director; Mona Zeftel, Senior Content Developer; D. Jean Buttrom, Content Project Manager; Kristin Stine, Marketing Manager; Elizabeth Murphy, Engagement Specialist; Ashley Kaupert, Associate Media Content Developer; Teresa Versaggi and Alexander Sham, Product Assistants; and Rose Kernan of RPK Editorial Services, Inc.. They have skillfully guided every aspect of this text's development and production to successful completion.

Pawan Lingras
Matt Triff
Rucha Lingras

ABOUT THE AUTHORS

Pawan Lingras is a Professor of Mathematics and Computing Science and the Director of the M.Sc. in Computing and Data Analytics program at Saint Mary's University, Halifax. He has authored more than 200 research papers in various international journals and conferences. He has also co-authored three textbooks, co-edited two books, and nine volumes of research papers. He has been a Natural Sciences and Engineering Research Council (NSERC) Discovery Grant recipient for 25 years, as well as other commercialization funding programs including NSERC Engage, NRC/IRAP, Nova Scotia Productivity and Innovation Fund, and Mitacs. His research and commercialization projects cover data mining, big data, retail mining, wearable technology, Internet of Things, sensor data analysis, web usage mining, image processing, engineering process mining, clustering/profiling, prediction, automated classification, association mining and optimization using evolutionary algorithms. He is an active organizer and reviewer for a number of international conferences and journals.

Pawan is a recipient of the Father William A. Stewart, S.J., Medal for Excellence in Teaching by the Alumni Association and the Faculty of Science Teaching Excellence Award by the Student Association, both at Saint Mary's University. He was the coach of the Saint Mary's CS Huskies programming team from 1999-2010, during which the team won eight Atlantic Canadian Championships, four runner-up finishes, and one trip to the World Finals (Hawaii, 2002). He continues to mentor students in a variety of app development competitions.

Pawan is a graduate of IIT Bombay with graduate studies from the University of Regina. His co-authors include professionals from Canada, India, China, Tunisia, USA, UK, Germany, Norway, and Chile. He was the director of a business administration and computing program in The Gambia, West Africa. He served as a University Grant Council (UGC) funded Scholar-in-Residence at SRTM University, Nanded, and as a visiting professor at IIT Gandhinagar, both in India. During these two visits in 2011-12, he extensively traveled through India, giving more than 40 invited talks. He has served as a Shastri Indo-Canadian Institute Scholar. He has also participated in a science tour of Germany and most recently served as a visiting professor at the Munich University of Applied Sciences.

Matt Triff is a veteran of many hackathons and has won awards at the LinkedIn Hackday TO in Toronto and the Data Mining Competition at the Joint Rough Set Symposium, 2013 in Halifax. He has software, system administration and development experience with IBM Canada, POS Bio-Sciences and OpenCare. He graduated from the University of Saskatchewan with High Honours in Computer Science and works as a programmer for the Canada Border Services Agency. His research and development projects span web, mobile, social networks, data analytics, computer vision, image processing, and video analysis.

Rucha Lingras has database, software development and project management experience with Clearwater Seafoods, IBM Canada, ExxonMobil, and Deloitte. She is currently working as a programmer analyst with Statistics Canada. She graduated from Saint Mary's University with a double major in Mathematics and Computing Science. She has been a recipient of a number of academic awards including the Saint Mary's University Presidential Scholarship, an NSERC research award, and the Governor General's Academic Medal.

CHAPTER 1

Introduction

This book will guide readers through the process of building apps from a variety of domains including science, engineering, and business using an active learning approach. The technologies that will be introduced will work with all major mobile and web platforms and will be applicable in any domain.

1.1 World of mobile computing

In recent years we have witnessed yet another shift in the computing paradigm. While we continue to use desktop/laptop/netbooks for professional computing, our personal and home information technologies needs are served by tablets, smartphones, smart TVs, wearable devices, and phablets (phones with large enough screens to be almost a tablet). As a result we have a fragmented computing, information, and communication device market. The major web browsers for desktop/laptop/netbook computing include Internet Explorer, Firefox, Chrome, Safari, and Opera running on Windows, Mac, Linux, and Chrome OS for serving both personal and professional needs. Tablets, smartphones, smart TVs, and other mobile devices tend to run on Android by Google, iOS by Apple, Windows RT, and several other competing platforms including Blackberry, Firefox, and Ubuntu/Linux. While each mobile device has a web browser, the primary way to access information on these devices is through dedicated applications. These dedicated applications are specific to a platform in the native application development environment. Figures 1.1 to 1.3 show a number of devices that run on Google's Android operating system including phones, tablets, and wearable watches that can be programmed using Java. Apple's iPhone and iPad shown in Figures 1.4 and 1.5 are programmed using Objective-C for Apple iOS. Microsoft's Windows phones (Figure 1.6) and tablets (Figure 1.7) are best programmed using their .Net framework and languages such as C# or Visual Basic. While a significant percentage of the apps still need to be developed in the native development environment, the emergence of HTML5 and JavaScript make it possible to develop apps that can be run across all mobile platforms through the web.

The objective of this book is to present an app-centric development methodology using the ubiquitous HTML5, JavaScript, jQuery, Node.js, and JSON. Instead of describing various features of these languages and demonstrating their usage in an app, we identify important features that should be part of an app and then introduce the necessary language features and constructs. We present the material in an organized and readable manner, which makes the book appropriate as a course textbook, while at the same time it can be used as reference for app development.

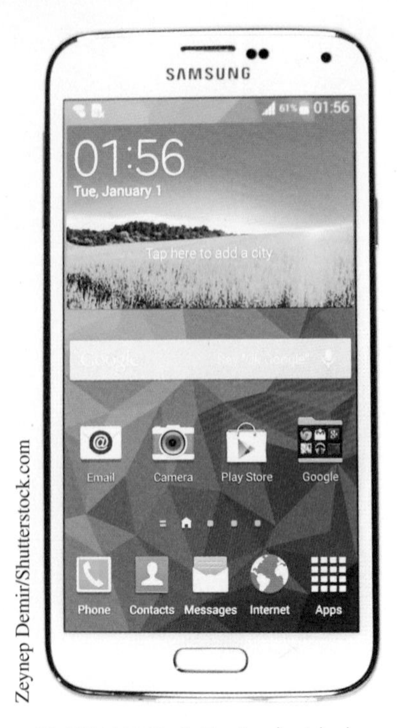

Zeynep Demir/Shutterstock.com

■ **FIGURE 1.1** Android phone

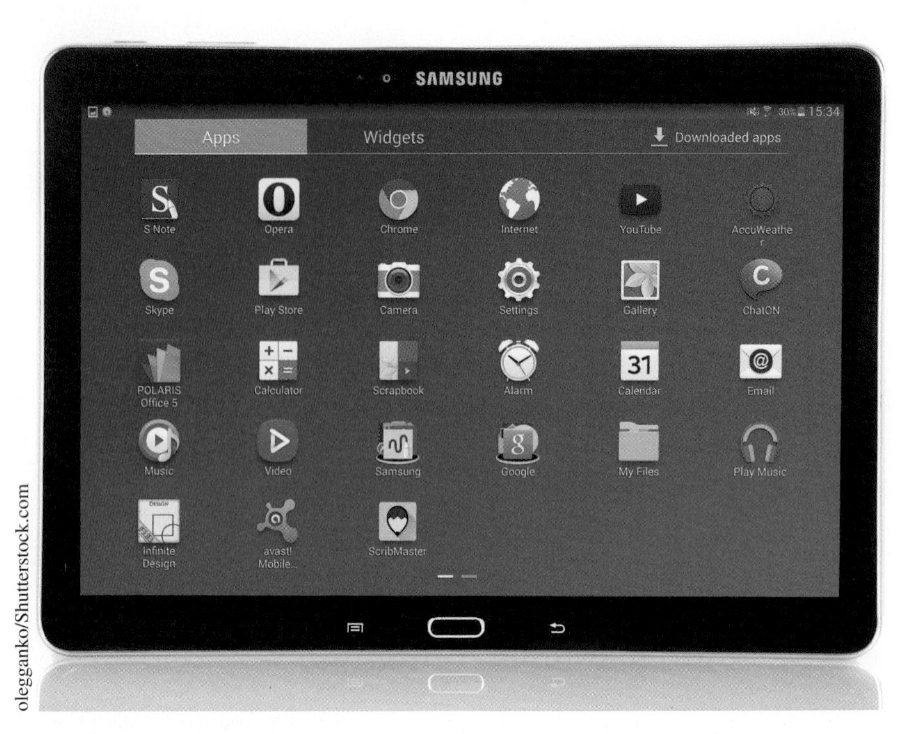

olegganko/Shutterstock.com

■ **FIGURE 1.2** Android tablet

Ivan Garcia/Shutterstock.com

■ **FIGURE 1.3** Android wearable device

Bloomua/Shutterstock.com

■ **FIGURE 1.4** Apple's iPhone

Denys Prykhodov/Shutterstock.com

■ **FIGURE 1.5** Apple's iPad

Bloomua/Shutterstock.com

■ **FIGURE 1.6** Windows phone

JeKh/Shutterstock.com

■ **FIGURE 1.7** Windows tablet

Every chapter will focus on an important app feature that will be illustrated by building an app for either science, engineering, or the economics/finance/business domain. The apps will be incrementally developed. The development process will be illustrated with the help of figures that include the code and corresponding screenshots. The readers will be able to follow the app-building process by typing in the code in parallel. Each chapter will end with exercises that encourage readers to make changes to the code and see the resulting changes in app behavior, as well as app projects from different application domains.

The book can serve as a textbook or a how-to book for self-learners. Teachers will find it useful for teaching either a first course on mobile and web programming or for a sequel to a web programming course.

In the following section, we will take a quick tour of various apps that will be developed in this book.

1.2 Tour of the apps developed in the book

Chapter 2. Developing, installing, and testing first app

Some readers may be familiar with the basic HTML and CSS, and may know how to deploy a website. These readers can quickly browse through Chapter 2. It will contain information such as how to manage the file access control on a UNIX/Linux server and incorporating videos (a feature that is new to HTML5). The chapter shows

- How to create a simple HTML5-based app
- How to put the app on the web for general access through the Internet

- How to enhance the presentation with CSS3
- How to add multimedia to the web page, including images and videos

Figure 1.8 shows a web page with an image, a video, and a table with presentation that is enhanced using CSS3, which is accessed from a UNIX server.

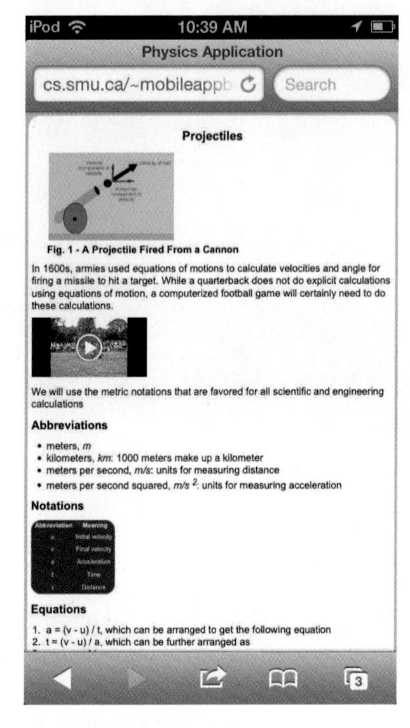

■ **FIGURE 1.8** First HTML5 and CSS3 web app

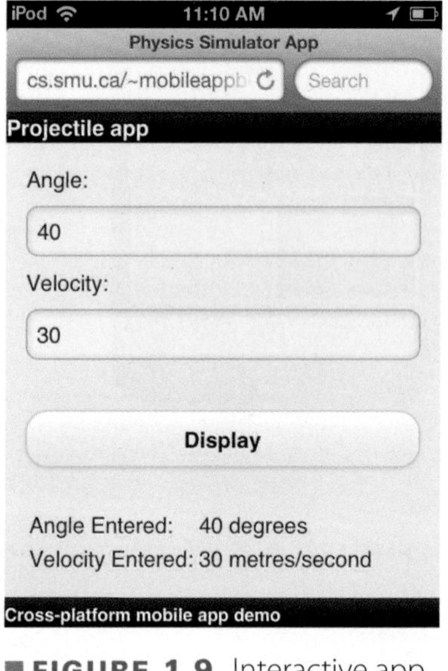

■ **FIGURE 1.9** Interactive app with textual input

Chapter 3. Making apps more interactive through data input

This chapter shows how to change the layout of the output depending on the types and sizes of screens. We will also discuss how to switch between textual and graphical output. Furthermore, the chapter will show how we can add more interaction/input so that a user can get more refined output if he or she desires. More specifically, we will learn more about

- JavaScript
- jQuery for mobile devices
- How to embed JavaScript in a web page
- How to use an external JavaScript file on a web page
- How to use CSS3 elements designed for mobile devices
- How to accept input through a web page
- How to use different types of input widgets
- How to link multiple mobile pages in an app

Figure 1.9 shows a physics Projectile app that accepts textual input through an HTML5 app. An app that essentially performs the same computations using inputs from a slider bar is shown in Figure 1.10. Another interesting feature of the app in Figure 1.10 is the fact that it integrates two web pages using tabs.

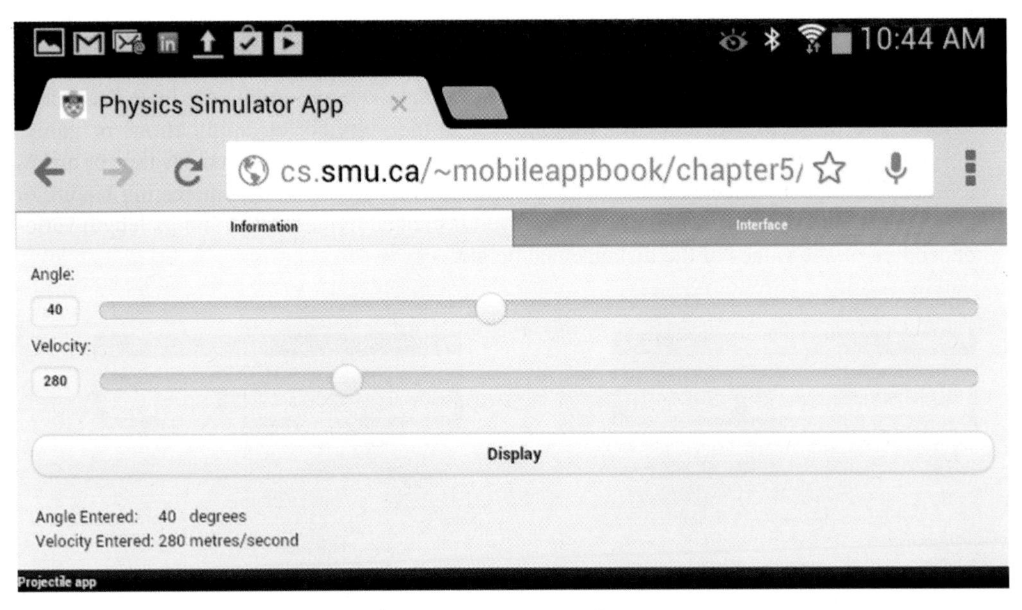

FIGURE 1.10 Slider bars for input and tab for different pages

Chapter 4. Making apps do significant computing

In this chapter, we use the ever popular temperature converter app to learn the basics of JavaScript computations. We then take the computations up a notch through more complex calculations to calculate distance and height of a projectile shot at an angle. Up to this point, the readers will be working with several individual data items. Oftentimes, our apps will deal with a collection, sequence, or array of data. The collection of items adds a new dimension to computing that will be explored in this chapter. In the process we will be learning a number of interesting aspects of computing including

- How to validate input
- How to do computations based on input
- How to do conditional computing
- How to do iterative computing
- How to use arrays in JavaScript

Figure 1.11 shows a temperature converter app that accepts temperature, uses radio buttons to let the user input whether the temperature needs to be converted to Fahrenheit or Celsius,

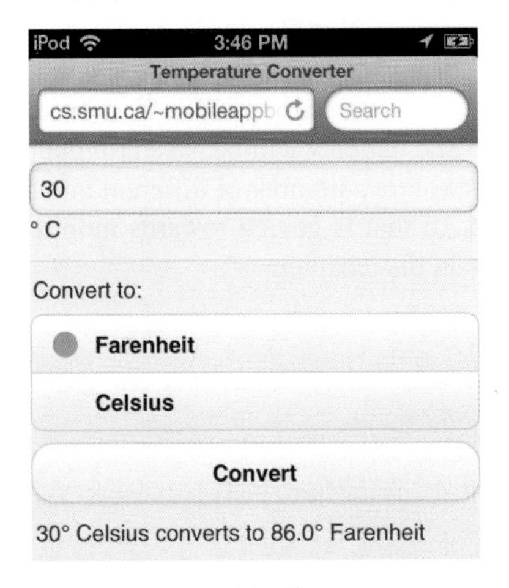

FIGURE 1.11 Temperature converter app

computes, and displays the conversion. Figure 1.12 shows an app that uses more complex mathematics including trigonometry to calculate the maximum distance traveled and maximum height reached by a projectile. The app also ensures that the angle and height have reasonable values. The projectile travels over time, and in some engineering applications or games programming, we may need to track the progress. Figure 1.13 uses three JavaScript arrays to track the time, distance, and height of the projectile's travel. Another interesting feature of this app is the fact that the table that is displayed is created dynamically, and its length varies depending on the values of the distance and height.

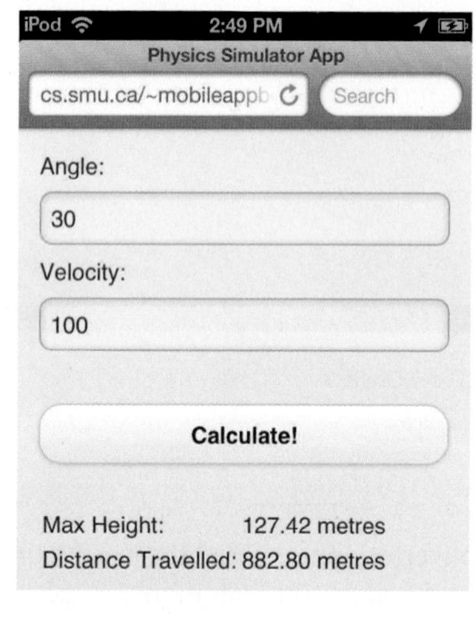

FIGURE 1.12 Computations of projectile's height and distance

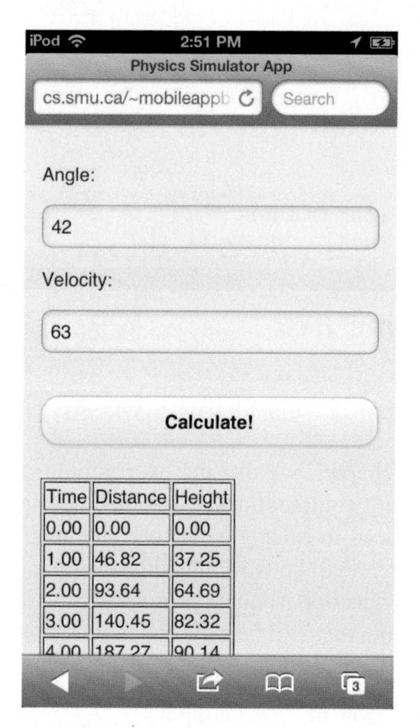

FIGURE 1.13 Tracking projectile's height and distance over time

Chapter 5. A menu-driven app to monitor important indicators

We now study an app that has many applications in engineering and sciences. Many times one has to track some important indicators such as pressure or temperature of an instrument. Oftentimes, the people who track these indicators are not experts and do not know what to do for different readings from the instrument. In this app, we use a real-world example that monitors blood test results of thyroid cancer patients. First, we create a numeric pad to enter a password for security purposes as shown in Figure 1.14. The patients will be asked to input their basic information as shown in Figure 1.15, where we explore a number of different input gadgets. We also show a menu-driven system in Figure 1.16 that is geared towards mobile devices. In particular, we will learn the following concepts in this chapter:

- How to secure an app with a password
- How to create a numeric pad for input
- How to navigate using menus

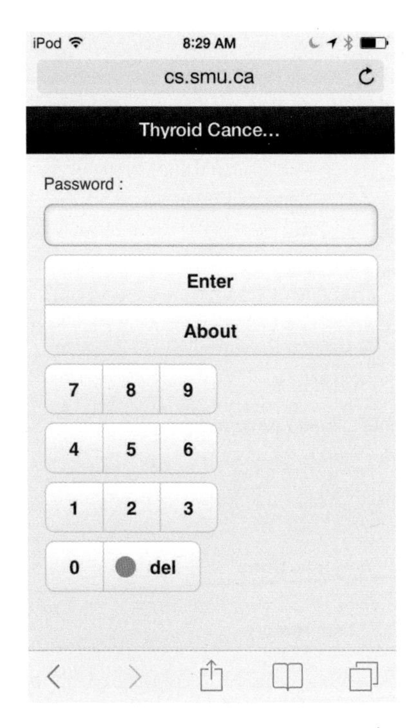

FIGURE 1.14 Numeric pad for password entry

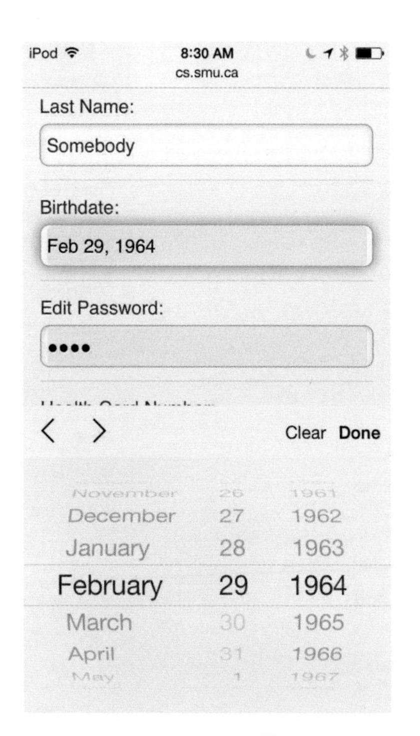

FIGURE 1.15 Different input gadgets for entering personal information

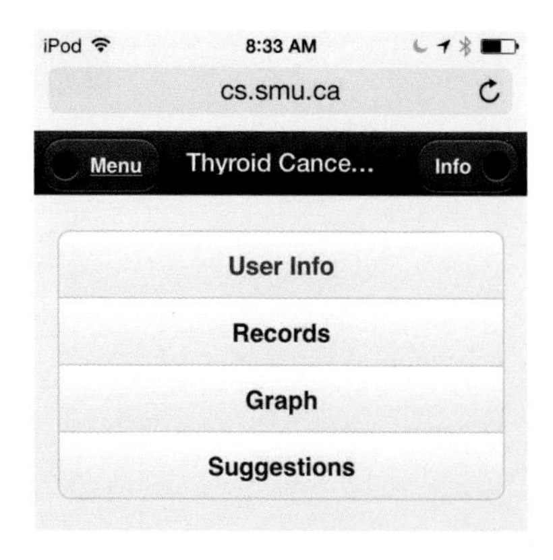

FIGURE 1.16 Menu-driven system geared towards mobile devices

Chapter 6. Data storage and retrieval

In Chapter 5, we set up the structure of the app to monitor blood tests of thyroid cancer patients. In this chapter, we are going to study all the intricate programming details involved in managing a data management system. We are going to securely store the information on our device, so that it is available even when we have no Internet connection. We will show how to read, write, and modify data stored locally on a device. We will discuss JSON (JavaScript Object Notation), which is syntax for storing and exchanging text information. The following is a list of questions that will be answered in this chapter:

- What is local storage?
- How can we store and retrieve data locally on the device?
- How can we check a password?
- How can we make sure that the user has accepted the disclaimer?
- How can we accept and manage user profiles?
- How can we store and manage an array of records?

Figure 1.17 shows the screen that can be used to add a record. Figure 1.18 shows the tabular history of records along with the personal information that is retrieved from the local storage of the device. Some of the subtle aspects of a data management system include variable length tables as well adding buttons that will help delete or modify specific records as can be seen in Figure 1.18.

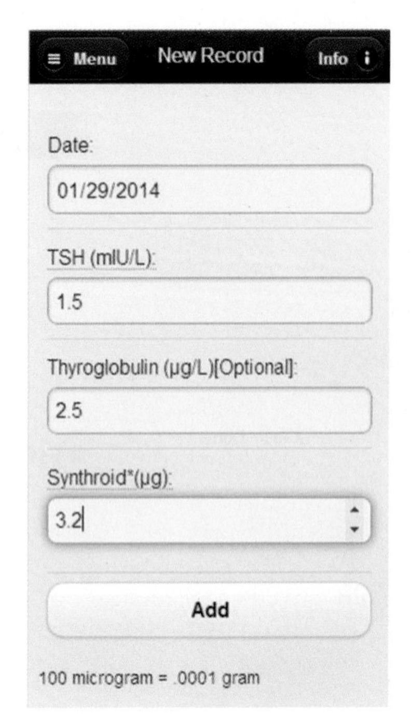

■ **FIGURE 1.17** Adding a record to the local storage

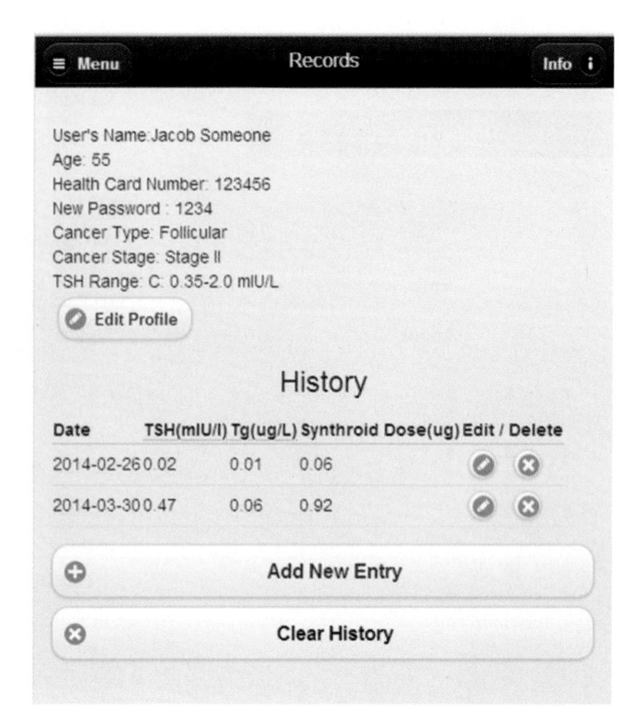

■ **FIGURE 1.18** Displaying personal information and records from the local storage

Chapter 7. Graphics on HTML5 canvas

Canvas is a new addition in HTML5. It allows us to create drawings on a web page. We will study the basics of canvas drawings as shown in Figure 1.19. Most drawings need complex programming. Fortunately, there are a number of packages available for us to create interesting graphics such as RGraph (http://www.rgraph.net/). The gauge meter that is created as part of the advice screen of the Thyroid app shown in Figure 1.20 uses RGraph. We also explore the bootstrap package from Twitter developers along with the RGraph package as shown with the graphs that are drawn for the thyroid blood test records in Figure 1.21. We also see how we can recycle code from previous version of the Projectile app and the graphing in the Thyroid app to create a graphical version of the Projectile app, as shown in Figure 1.22. Finally, we will study how to launch our browser-based apps from the home screen as well as with no Internet connection. The summary of skills presented in Chapter 7 is as follows:

- How to work with canvas
- How to draw various shapes
- How to display some of the graphical gadgets
- How to draw graphs
- How to create an icon for the app on the home screen
- How to run an app locally with no Internet connection

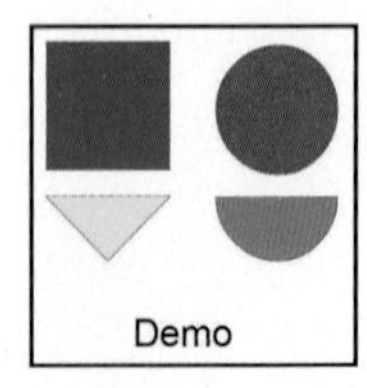

■ **FIGURE 1.19** Exploring the drawing canvas

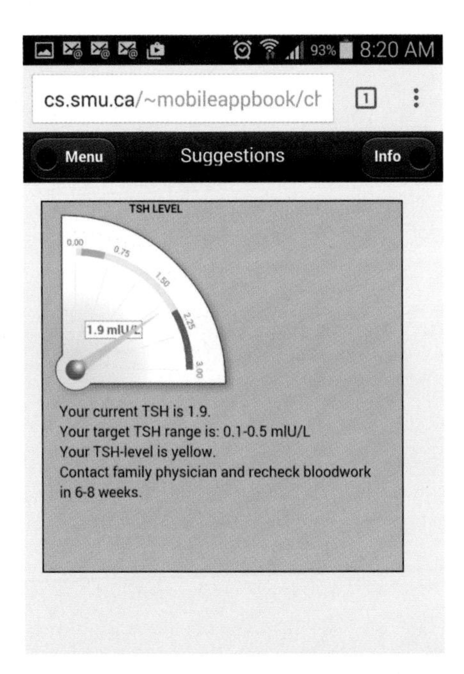

FIGURE 1.20 Displaying the advice with the help of a gauge meter

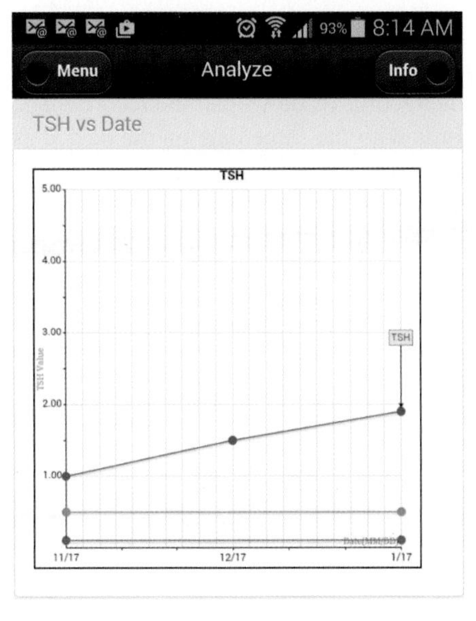

FIGURE 1.21 Graphical display of records from the local storage

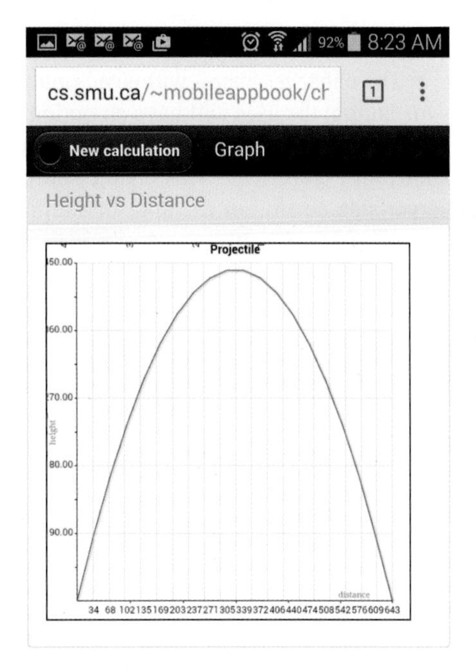

FIGURE 1.22 Graphical display of projectile track

Chapter 8. Using servers for sharing and storing information

So far all our computations and storage were on the local device. In the following chapters, we will see how we can run these apps even when the device is not connected to the Internet. However, one cannot completely rely on the mobile device for longer-term storage. In this chapter, we can sync information from our Thyroid app on a server. We will also use the server for emailing tables and charts. Figure 1.23 shows the modified login screen for the server-based Thyroid app. We will ask the user to login with an email and password. Users will have to create an account for themselves using the Create New User button the first time they use the app.

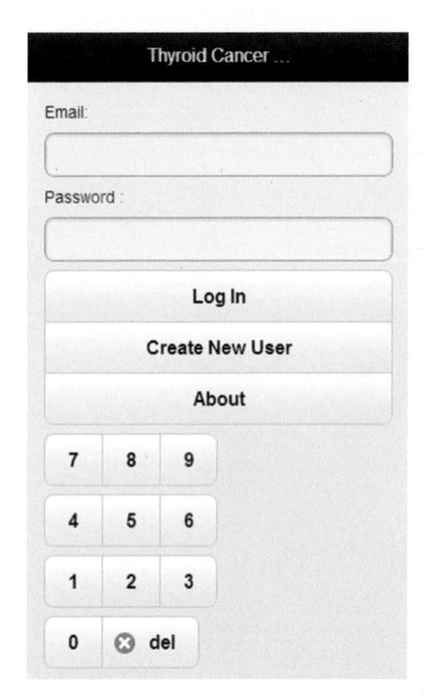

■ **FIGURE 1.23** Modified login screen for the server-based Thyroid app

The enhanced menu with sync option is shown in Figure 1.24. When a user presses the sync button, the user will have an option to upload the records or download the records. The user can choose whether or not to overwrite the records with the same date on the device as shown in Figure 1.25. This chapter introduces readers to the web server component of the server-side version of the app. In addition, we take our first look at Node.js, a server-side version of JavaScript.

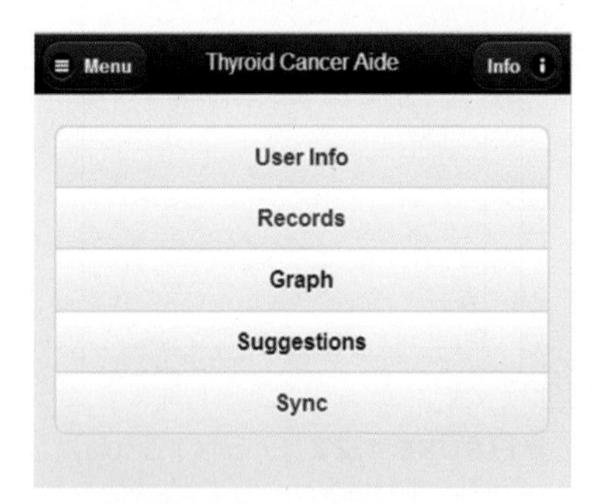

■ **FIGURE 1.24** Modified menu screen for the server-based Thyroid app with sync option

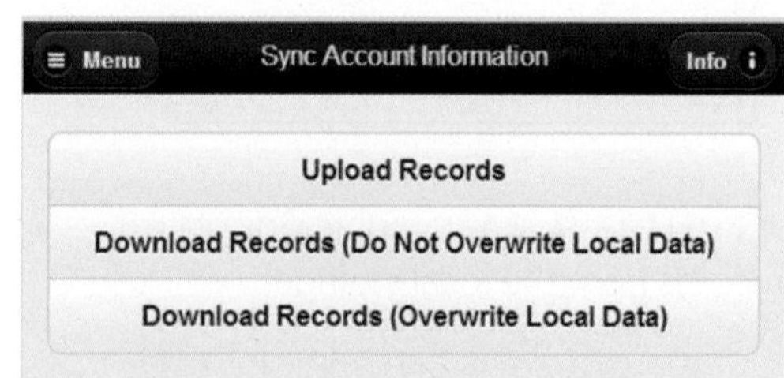

■ **FIGURE 1.25** Sync screen for the server-based Thyroid app

Chapter 9. Using MongoDB server for sharing and storing information

This chapter provides the data management part of the server-based Thyroid app. It will be based on a JSON-based non-relational database model. Some of the readers who are familiar with tabular view of the data will find these models a little less structured. We will discuss a popular implementation of non-relational databases called MongoDB. The chapter will include an introduction to MongoDB with simple illustrative examples such as the one shown in Figure 1.26, which shows how one can nest data in a non-relational model.

```
{
   {
     author : ";Pawan",
     books :
     [
        { title : ";Web mining", year : 2007 },
        { title : ";Web programming", year : 2012 },
        { title : ";Mobile app development", year : 2015 }
     ]
   }

   {
     author : ";Walter",
     books :
     [
       { title : ";Intro programming", year : 2000 },
       { title : ";Data structures", year : 2002 }
     ]
   }
}
```

FIGURE 1.26 A nested library collection in NoSQL

Chapter 10. Using a relational database server for sharing information

This chapter provides the same facilities as Chapter 9. However, it will be based on a relational database model. It is possible that some instructors may choose to use only one of the two database models. Therefore, this chapter will be self-contained—that is, the readers do not have to read Chapter 9 to understand Chapter 10. Relational databases are and will continue to play a major role in the management of data. They provide a more structured view of the database as shown in Figures 1.27 and 1.28 for our server-based Thyroid app. The table shown in Figure 1.27 stores information about a thyroid cancer patient as well as his test records in a single table providing a structured view of the data. However, the basic information for the user is repeated in the database for every test record. We will discuss the normal forms that will help us reduce the repetition in the data as shown by the two tables that appear in Figure 1.28.

email	Password	First name	Last Name	Date of birth	Health Insurance Card	Cancer type	Cancer stage	TSH range	Agreed to legal	Date of test	TSH	Tg	Synthroid
j@cs.smu.ca	1953	J.	Some	1953/02/03	123456	Medular	I	0.1–0.3	True	2011/12/05	0.20	0.6	1.7
j@cs.smu.ca	1953	J.	Some	1953/02/03	123456	Medular	I	0.1–0.3	True	2011/12/05	0.20	0.9	0.7
k@cs.smu.ca	1950	K.	Other	1950/02/23	183456	Folicular	II	0.01–0.1	True	2012/12/05	0.20	0.4	1.1
j@cs.smu.ca	1953	J.	Some	1953/02/03	123456	Medular	I	0.1–0.3	True	2012/11/15	0.20	0.5	1.8
k@cs.smu.ca	1950	K.	Other	1950/02/23	183456	Folicular	II	0.01–0.1	True	2010/01/25	0.02	0.2	1.9
j@cs.smu.ca	1953	J.	Some	1953/02/03	123456	Medular	I	0.1–0.3	True	2013/11/02	0.20	0.16	1.7

FIGURE 1.27 A non-normalized thyroid patient relational database

email	Password	First name	Last Name	Date of birth	Health Insurance Card	Cancer type	Cancer stage	Agreed to legal	TSH range
j@cs.smu.ca	1953	J.	Some	1953/02/03	123456	Medular	I	True	0.1-0.3
j@cs.smu.ca	1953	J.	Some	1953/02/03	123456	Medular	I	True	0.1-0.3
k@cs.smu.ca	1950	K.	Other	1950/02/23	183456	Folicular	II	True	0.01-0.1
j@cs.smu.ca	1953	J.	Some	1953/02/03	123456	Medular	I	True	0.1-0.3
k@cs.smu.ca	1950	K.	Other	1950/02/23	183456	Folicular	II	True	0.01-0.1
j@cs.smu.ca	1953	J.	Some	1953/02/03	123456	Medular	I	True	0.1-0.3

user	Date of test	TSH	Tg	Synthroid
j@cs.smu.ca	2011/12/05	0.20	0.6	1.7
j@cs.smu.ca	2011/12/05	0.20	0.9	0.7
k@cs.smu.ca	2012/12/05	0.20	0.4	1.1
j@cs.smu.ca	2012/11/15	0.20	0.5	1.8
k@cs.smu.ca	2010/01/25	0.02	0.2	1.9
j@cs.smu.ca	2013/11/02	0.20	0.16	1.7

FIGURE 1.28 A thyroid patient relational database in second normal form

Chapter 11. Using web templates

Web templates are becoming increasingly popular for using different data sources to present a large amount of information on similar-looking web pages. This chapter demonstrates one of the popular web template systems called *handlebars* to build an app called Explorador that helps us explore parks in a metro region. We will explore other packages that will allow us to implement features such as the sliding menu shown in Figure 1.29 and maps with pop-out screens shown in Figure 1.30.

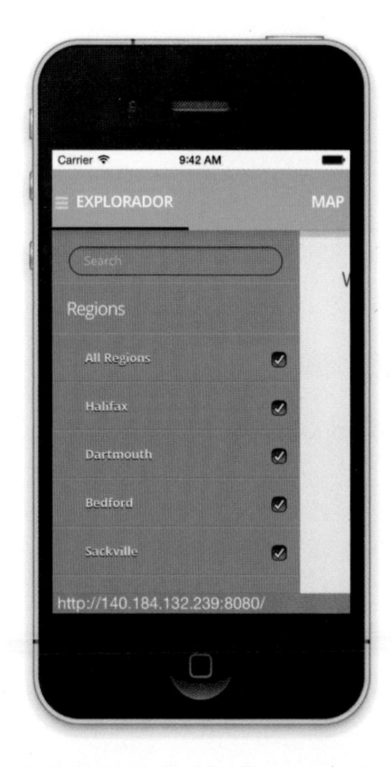

■ **FIGURE 1.29** Screenshot of sliding menu in Explorador

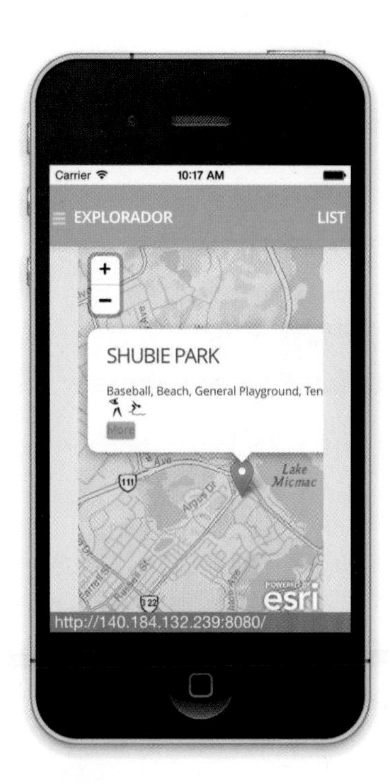

■ **FIGURE 1.30** Screenshot of detailed results shown on a map in Explorador

Chapter 12. Working with image databases, maps, and location tracking

Most modern mobile devices provide sensors to determine location identification services. This chapter will show how to obtain the location information and utilize it to enhance the user experience with the help of maps. We will also extend our understanding of databases for storing multimedia information.

Chapter 13. Cross-platform and native app development and testing

The apps developed in this book are designed to run on a number of platforms without any modifications. In order to exploit all the capabilities of a particular device, one needs to develop an app for that specific device. This chapter will discuss and showcase the hardware and software development tools necessary for native app development. We will also demonstrate how some of the apps we have developed in the previous chapter can be converted to the native platforms including iOS and Android. Figure 1.31 shows the Explorador for Android device that was made available on the Google Play store.

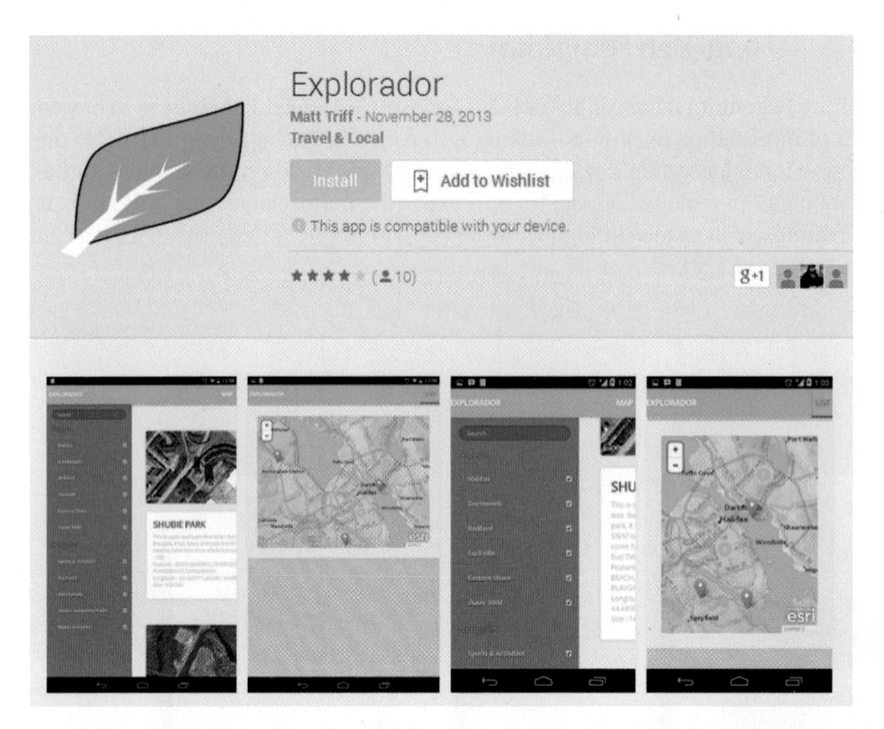

FIGURE 1.31 Explorador app for Android devices on Google Play

CHAPTER 2

Developing, installing, and testing first app

WHAT WE WILL LEARN IN THIS CHAPTER

1. How to create a simple HTML5-based app
2. How to put the app on the web for general access through the web
3. How to enhance the presentation with CSS3
4. How to add multimedia to the web page including images and videos

The end of the chapter has a Quick Facts section that lists all the technical terms discussed here.

2.1 Choice of programming platform

As mentioned earlier, different mobile devices use different programming languages for developing apps. Let us quickly recap the difference between hardware and operating system. The examples of hardware include all the mobile, computing, or communication devices such as iPhone, iPad, Samsung Galaxy phone and tablet, Nokia phones, Microsoft surface tablet, laptops, and desktops. All of these pieces of hardware run software called operating system (OS), which facilitates our interaction with the devices. The three major operating systems for mobile devices are

1. iOS by Apple
2. Windows 8 by Microsoft
3. Android by Google

All the mobile Apple devices including iPhone, iPod touch, iPad Mini, and iPad all run iOS. While iOS continues to be a dominant operating system, it only runs on Apple devices. Most of the other major hardware makers of mobile devices use either Android or Windows 8. The Android operating system has the major share of mobile devices, followed by Apple's iOS, with Windows 8 running distant third.

In order to make the best use of all the features of Apple devices, one needs to write apps in an integrated development environment (IDE) called Xcode. Xcode consists of a suite of tools for developing software for iOS. An app written in Objective C for iOS cannot run on other mobile devices such as those that run the Android operating system developed by Google or Windows RT or Windows Phone developed by Microsoft. The native apps for Android are written in Java, and those for Windows mobile devices are written in .Net framework. These native apps written for a particular platform cannot be run on another platform. Sometimes you need to use a variety of features of a mobile device and optimize the performance of your app. In that case, you must write your apps in the development environment for that platform. In many cases, such an intensive use of mobile device hardware is not necessary. In that case, we can write a single app that works on all major platforms. The recent fragmentation of the mobile device market has made the importance of such a cross-platform development a need of the hour. The developer community has responded with tools to make these cross-platform apps reasonably efficient and versatile. The primary focus of this book is to develop apps using these ubiquitous tools, which include HTML5, CSS3, jQuery mobile, and other related technologies. We will discuss an experiment with various issues in making sure that our apps run on all the major mobile platforms and devices. We will also see how to use the native app development environments, if only to transport our cross-platform apps to native apps for the three major mobile operating systems.

Fortunately, the study of cross-platform development begins with the well-established HTML—HyperText Markup Language. Those who are familiar with HTML may want to quickly browse through this chapter to ensure that they are familiar with some of the newer features in HTML5.

The history of HTML dates back to 1980s when Tim Berners-Lee started development of a system for sharing documents among fellow physicists. The first HTML specifications were proposed in 1991 by Sir Berners-Lee, when HTML was competing with a number of other communication protocols on the Internet such as gopher. Marc Andreessen shares some of the credit for the popularity of HTML as he developed the very first multimedia browser for the Internet based on the HTML protocol. HTML standards have evolved over time. There was a time when it was believed that HTML 4.01 would be the last version of HTML, which would make way for XHTML1.1. Yet here we are studying HTML5.

HTML5 is a response to the changing nature of the Internet, which includes devices of many sizes, shapes, abilities, and power. It is an attempt to create a single interoperable and more logically designed markup language that includes useful extensions to make it a viable candidate for cross-platform web and mobile applications. New elements in HTML5 include `<video>`, `<audio>`, and `<canvas>`. It also supports scalable vector graphics (SVG) for higher quality images and mathematical formulas using MathML. These features make it possible to incorporate a richer set of multimedia content on the web, which in the past required proprietary browser plugins and application programming interfaces (APIs). HTML5 also has new elements that help us add commonly present semantics to the documents such as `<section>`, `<article>`, `<header>`, and `<nav>`. The HTML5 further rationalizes and redefines some of the existing features from HTML and XHTML along with a set of procedures to deal with erroneous documents.

In this chapter, we are going to review some of the basic HTML5 syntax and create simple static web pages. We will discuss and demonstrate how to upload these web pages to a web server. We will view these web pages through all the key browsers on desktop/notebook computers as well as mobile devices. In the first chapter we have seen a preview of a wide variety of apps. For the first three apps we will use a projectile tracking app based on physics concepts.

2.2 How to create a simple HTML5 web page

Figure 2.1 shows the skeleton of our first HTML5 page. An HTML5 document is marked up with the help of tags. Tags are instructions to the web browser regarding the display and behavior of a web page. An HTML5 tag is enclosed in two angled braces (they are really less than and greater than signs) <. . .>.

The very first tag we see in Figure 2.1 is the `<!DOCTYPE html>` tag. It tells the browser that we are working with an HTML document. Many of the tags come in pairs—an open tag and a close tag. The `<!DOCTYPE>` tag is an exception. It does not have a close tag.

The next tag `<html>`, which indicates the beginning of an HTML document, has a corresponding close tag `</html>` that indicates the end of that HTML document. Every HTML document has two parts, the `<head>` and the `<body>`. We see those two tags in the document with their close tag counterparts `</head>` and `</body>`.

None of the information specified in the `<head>` section actually appears on the web page. It usually consists of various tags that are used to set up the web page. The pair `<title>` and `</title>` is such an example, which contains the title that will appear in the title bar at the

```
<!DOCTYPE html>
<html>
  <head>
    <title>Physics Application</title>
  </head>
  <body>

    <!--
     ...
     ...
    The body of the document goes here shown on the following pages
     ...
     ...
     -->

  </body>
</html>
```

FIGURE 2.1 Bird's-eye view of the HTML5 code of our first web page
Follow: ch02/physicsProjectileApp0/version1/physicsProjectileApp.html

top of the web browser. The screenshot of the complete page, shown in Figure 2.2, shows the title of the web page to be "Physics Application".

The `<body>` section enclosed in `<body>` and `</body>` contains the actual document that will appear in the web browser. We can see the detailed `<body>` section in Figure 2.3. In Figure 2.1, we have commented out the contents of the body using the comment tag `<!-- -->`. Anything that appears between `<!--` and `-->` is not displayed on the web page. Other usages of the comment tag will be discussed later.

Figure 2.3 shows the rest of the HTML5 code that goes into the `<body>` section of our first web page. We have already seen the screenshot of the entire web page in Figure 2.2. Let us look at the new HTML5 tags used in the `<body>` section. Before going into the details of the body, we want to direct your attention to the indentations in the HTML5 file. It is customary and strongly recommended that all the contents that appear between a given pair of open and close tags be indented by a few spaces. Therefore, we are indenting everything between the pair of `<body>...</body>` tags. We also recommend the use of spaces for the indentation instead of tabs as different text editors display the tabs differently, which can lead to less than reader-friendly HTML5 code.

At the very top we have a pair of tags, `<h1>...</h1>`. This tag is used to specify a header, which usually appears in the browser in bold and large font. There are many levels of headers, from 1 to 6, and they are specified as `<h1>...</h1>`, `<h2>...</h2>`, `<h3>...</h3>` up to level 6. The size and prominence of the header goes down from level 1 to level 6.

One of the first things we notice when we compare Figure 2.2 and Figure 2.3 is that the physical line breaks in Figure 2.2 do not match with those in the HTML5 code. The browsers uses physical line breaks based on the HTML5 tags. Therefore, we see a line break after the "Projectiles", when the browser encountered the end of the header, such as `</h1>`.

The other line breaks in this document essentially come from two tags, `<p>...</p>` and `<br>`. The pair of tags `<p>...</p>` indicates the beginning and end of a paragraph. There will be a natural line break at the end of a paragraph. If we wanted to have a line break in the middle of a paragraph or outside a paragraph we use the tag `<br>`. Readers who are familiar with XHTML will notice that HTML5 has gone back to the old standards and no longer recommends `<br/>` for indicating a line break.

It is strongly recommended that all the text should be in some pair of tags to indicate what type of formatting should be used by the browser for displaying the text. However, the users may notice that we have left the last paragraph in the file starting from "a = (v – u) / t" floating,

Projectiles

In 1600s, armies used equations of motions to calculate velocities and angle for firing a missile to hit a target. While a quarterback does not do explicit calculations using equations of motion, a computerized football game will certainly need to do these calculations.
We will use the metric notations that are favored for all scientific and engineering calculations

The following abbreviations will be used:

meters, m
kilometers, km: 1000 meters make up a kilometer
meters per second, m/s: units for measuring distance
meters per second squared, m/s^2: units for measuring acceleration

The following physics notation will be used:

Initial velocity, u
Final velocity, v
Acceleration, a
Time, t
Distance, s

Therefore, some of the useful equations of motion for us are:

$a = (v - u) / t$, which can be arranged to get the following equation
$t = (v - u) / a$, which can be further arranged as
$v = u + a * t$
$s = u * t + 0.5 * a * t^2$, another useful equation of motion

FIGURE 2.2 Screenshot of our first web page
Follow: ch02/physicsProjectileApp0/version1/physicsProjectileApp.html

```
<body>
    <h1>Projectiles</h1>

    <p>In 1600s, armies used equations of motions
     to calculate velocities and angle for firing
     a missile to hit a target.
     While a quarterback does not do explicit
     calculations using equations of motion, a
     computerized football game will certainly
     need to do these calculations.
     <br>
     We will use the metric notations that are
     favored for all scientific and engineering
     calculations
     <br><br>
     The following abbreviations will be used:
    </p>

      meters, m<br>
      kilometers, km: 1000 meters
```

```
        make up a kilometer<br>
      meters per second, m/s: units for
        measuring distance<br>
      meters per second squared, m/s
        <sup>2</sup>: units for measuring
        acceleration<br>

    <p>
      The following physics notation will be used:
    </p>

      Initial velocity, u<br>
      Final velocity, v<br>
      Acceleration, a<br>
      Time, t<br>
      Distance, s<br>

    <p>Therefore, some of the useful equations of
      motion for us are:
    </p>
      a = (v - u) / t, which can be
        arranged to get the following
        equation<br>
      t = (v - u) / a, which can be
        further arranged as<br>
      v = u + a * t<br>
      s = u * t + 0.5 * a * t<sup>2</sup>
        , another useful equation of
        motion<br>
</body>
```

FIGURE 2.3 Detailed HTML5 code of our first web page
Follow: ch02/physicsProjectileApp0/version1/physicsProjectileApp.html

with no explicit indication that it is part of a paragraph. Most browsers will be accepting of these omissions and will assume that it is part of a new paragraph.

Our first HTML5 document is rather simple and uses the typical HTML5 tags for formatting. In this book, we will explore a number of HTML5 tags in a reasonable amount of detail to ensure that the students can continue. However, this book is not meant to be a comprehensive guide to creating HTML5 web pages. There are a number of resources including books and web resources to get a more thorough introduction to HTML5. One of the best resources is http://www.w3schools.com, which provides a comprehensive introduction to HTML5 and other related technologies. The focus of this textbook is on developing mobile apps for engineering and scientific applications. We will cover only the necessary HTML5 features for the apps that we want to develop in this book. For example, in our HTML5 app, we are going to use equations of motions, which require the use of exponents. Therefore, our first web page uses an additional tag called `<sup>` and `</sup>`, which is used to make the enclosed text appear as a superscript. We have used it to denote squares of entities such as ‘s’ and ‘t’ in the web page. We will delay the discussion on physics until we are ready to program our app.

So far we have not really told the readers where to find or how to create or what do with the HTML5 file that we have discussed. While there are a number of software products that allow you to develop a web page using a “What You See Is What You Get” or WYSIWYG approach, the HTML5 code created by such software can be rather unwieldy and difficult to modify. We will use a simple text editor such as Notepad on Windows to work with our HTML5 documents. There are a number of smarter text editors that can provide additional features such as syntax highlighting or viewing the corresponding web page as you modify the HTML5 code. Everyone has their own favorite text editors, and we will let the readers decide their favorite editors. Everything we are going to discuss here can be done using an editor such as Notepad on Windows, TextEdit on Mac, or gedit on Linux.

2.3 How to put an HTML5 web page on the Internet

The book website has a directory called "web", which will contain all the documents discussed in this book. They are further organized by chapters. From this point onwards, we will simply mention the "web directory or folder" to refer to this repository. In our web directory, there is a folder called ch02. It has another folder called physicsProjectileApp0, which tells us that we are working on an app by the name "physicsProjectileApp0". It has another folder called version1, since we are going to enhance the app in stages with additional features. This Russian doll structure may seem rather cumbersome at the moment. However, as we progress, this structure will make it easier to keep track of various apps and their different versions.

We recommend that you copy the code to your computer's hard disk so that you can modify the code and see the effects of your modifications. You can always get the original copy from the USB drive or the book website. In order to view the first version of the physics Projectile app, navigate to the folder or directory:

ch02/physicsProjectileApp0/version1

and look for a file called physicsProjectileApp.html. Depending on the operating system that you use, the ".html" may not be visible to you. If you double click on the document, there is a good chance that the file will be opened in your favorite/default web browser. You can right click on the file (in Windows and Linux), and you will get an option to open the file with other programs. We will assume Windows as the default operating system for the rest of the discussion. Please use Notepad for opening the file. You will see the HTML5 text that we have discussed so far. Please feel free to make changes to the file, save it to your hard drive, and observe the effects of changes in the web browser by reloading/refreshing the web page.

Now that we know how to display and edit the web page from your local hard drive, we should consider putting it on the World Wide Web (WWW) or simply the web. Putting the app code on the web is the easiest way to download and distribute it to multiple mobile devices through wireless communication. We will see later on how the web page can be stored locally on the device.

In order to put the web page on the Internet and make it accessible through the World Wide Web, we need to contact an Internet Service Provider (ISP). Many ISPs usually provide you an ability to create your own website. The details may vary from ISP to ISP. In an educational setting, the system administrator at your institution may give you an account on the institutional server. We will use a server at Saint Mary's University as an example.

Our system administrator—let us call him Andrew—has given us an account on a server with the domain name called cs.smu.ca. The domain name is translated into an IP address. An IP address is used to refer to any computer on the Internet. IP stands for Internet Protocol. Andrew gave us a username "mobilebook" along with a password. He also told us that we should put all our web pages in a directory/folder called public_html. Once we have done that we can access these web pages by specifying the URL: http://cs.smu.ca/~mobilebook

URL stands for Uniform Resource Locator, which is a syntax for specifying resources on the Internet such as web pages. Our URL begins with http://. It stands for hypertext transfer protocol, the protocol that is used to serve web pages.

Depending on the web server you are using and the tools made available to you, you may or may not be able to upload files the way we are going to describe in this chapter. However, the tools we are going to use are ubiquitous and can be made to work with most setups.

A web server is typically an Internet-accessible computer that is specially configured to serve and receive documents at very high rates of transmission. The web servers run software that is also helpfully called web server. The software runs a number of parallel processes that facilitate the communication. Two prominent examples of software that serves web pages are Apache (www.apache.org) and Microsoft Exchange Server. The Apache server is an open source project that supports most major operating system platforms including Linux and Windows. Microsoft Exchange Server runs only on Windows servers and is a commercial product. Most of the discussion in this book will be based on the Apache server and the Linux operating system. However, it is applicable to other web servers and operating systems as well. Moreover, web servers play a small role in our mobile app development, as the actual development occurs on your personal computer. Most of our app development discussion will be based on the Windows operating system, which is still the most dominant operating system platform for personal computers, and it is well supported by all the software development platforms.

Let us recap our setup. We are doing all our development on a personal computer running a Windows operating system. We have access to a Linux server that is running the Apache software. Our system administrator, Andrew, has created an account for us with username "mobilebook". We are told to put our web pages in a folder called public_html. They will be accessible to the rest of the Internet using the URL: http://cs.smu.ca/~mobilebook. Empowered by this information, we are ready to put our first HTML5 web page on the web for accessing it from any web-enabled device.

We now need to transfer files from our personal computer to the web server cs.smu.ca. File transfer protocol (ftp) was used for a number of decades for transferring files. Now there are more secure transmission protocols called secure-ftp and scp (secure copy protocol). We are going to use a public domain software called WinSCP that helps us use either of the two secure file transfer protocols on a Windows desktop. The software can be downloaded from http://winscp.net. Once you have installed the software, you can launch the software. You will get a chance to open a new connection. WinSCP provides a number of options to make your transfers convenient, including drag and drop, remembering your previously accessed directory, and syncing folders on the server and your desktop. We will not explore all the options. Let us look at some of the basic operations necessary for file uploading to the web server using WinSCP. As mentioned before, even if you use a different software for transferring files, the basic principles described here will still apply.

Figure 2.4 shows the screenshot of the login screen for WinSCP. At the very least, we need to enter the "Host name" and "User name". You do not have to enter the password, as WinSCP

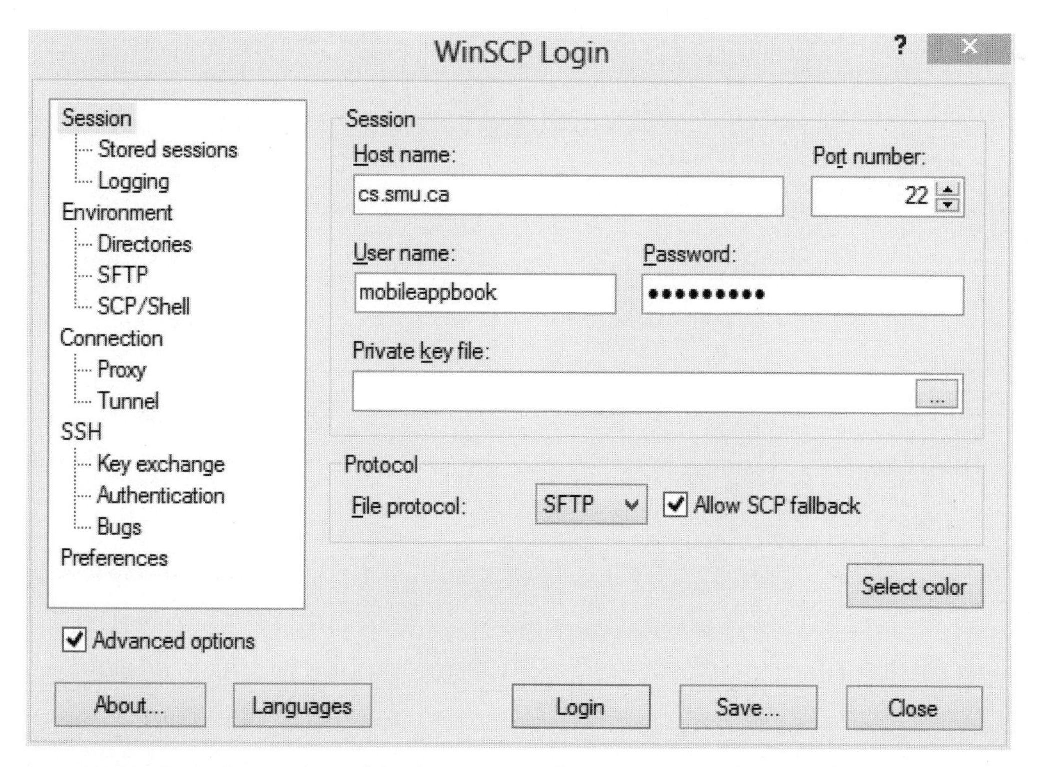

FIGURE 2.4 Screenshot of the login screen for WinSCP to upload the web pages
Follow: http://winscp.net

will prompt you for the password. Once you have logged in you will see two panels as shown in Figure 2.5. The left panel shows the folder on your personal computer, and the right one is the server that we have connected to (cs.smu.ca in our case). We can then drag and drop files from one panel to the other. We have dragged the ch03 folder under the directory public_html. That may or may not be enough for us to see the web pages through our browser. We have to make sure that the web server software has sufficient privileges to display the web pages. The details for setting the privileges depend on the operating system.

We will have a brief overview of permissions in the UNIX environment, which includes Linux and Mac OS X. The UNIX environment is a popular choice for web servers. Typically, the users are divided into three sets. These sets are not mutually exclusive; in fact, they progressively contain one another. The first set consists of a single user who is the owner of the account and all the files in the account. We will call this singleton set "owner". The second set includes the "owner" and potentially a number of other users who are collaborating on a project. This set is called "group". The set "group" contains the set owner. Finally, there is a set called "world" (also called "others") that includes every user on the system. Naturally, the set "world" includes the sets "owner" and "group".

The files from the computers called your web server are served to the world through the Internet using a program that is confusingly named as a web server as well. The web server program is considered to be a user on the computer system and usually a part of the set of users called world or others. That means in order for the web server program to serve these files, they have to be accessible to the set of users called others/world.

In UNIX systems (Linux/Mac OS), we can specify three types of permissions: read (R), write (W), and execute/run (X). Let us first look at what these permissions mean for simple files. Then we will look at their implications for directories/folders (we will use the term directory and folder, interchangeably). The meaning of R and W is rather obvious. The set of users with R permission can read the files. The set of users with W permission can write/modify the file. The X permission is relevant for the programs that can be run. The users with X permission can run the program.

Reading, writing, and executing have a little different meaning for directories. The set of users with R permission on a directory can see the names of all the files in the directory. The set of users with W permission on a directory can write to that directory, which is necessary for creating new files in that directory. The permission of X on a directory is necessary to change to the directory or any of its subdirectories. If you do not want a set of users to look at a file in a directory such as

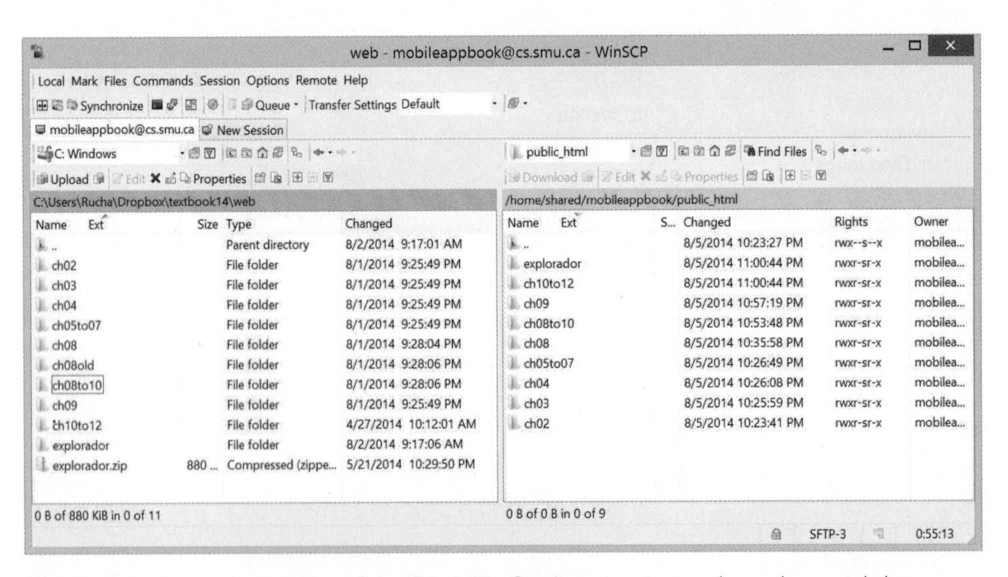

■ **FIGURE 2.5** Screenshot of the WinSCP after logging in, ready to drag and drop web pages

Follow: http://winscp.net

your home folder but want them to look at the files in the public_html directory that is underneath your home folder, you still need to provide that set of users with X permission for the home folder.

Let us summarize permission settings for permissions on our files to make them accessible to the world. We need all the files under the folder public_html and recursively all its subfolders to be readable by everyone, that is, owner, group, and others (world). We should also make all the directories executable with permission X, so the set of users called "Others" can change to these directories and through to subdirectories that are underneath. In addition, one time we need to set the home folder access for the "Others" to X.

Figure 2.6 shows the essential features of the settings for our public_html directory. Both the "Owner" and "Group" consist of a single user called mobileappbook. It has a unique numeric ID of 22609. We want all the files to be accessible to everyone on Linux that will be "World" or "Others" as shown in Figure 2.6. We need to make sure that the web server software can enter the folder; that is why we have checked the box "Add X to directories". We also check the box "Set group, owner and permissions recursively" to make sure that all the underlying folders and files are equally accessible. Readers who know more about the UNIX file system will recognize the 644 in the box labeled "Octal", for setting up these permissions. In addition, we have also checked the box to set all the "GID" (Group IDs) to mobileappbook.

The home folder permission setting can be tricky and is best done through a command line interface using the command "chmod 711". You may want to seek the help of your system administrator if you are not comfortable with the command line interface.

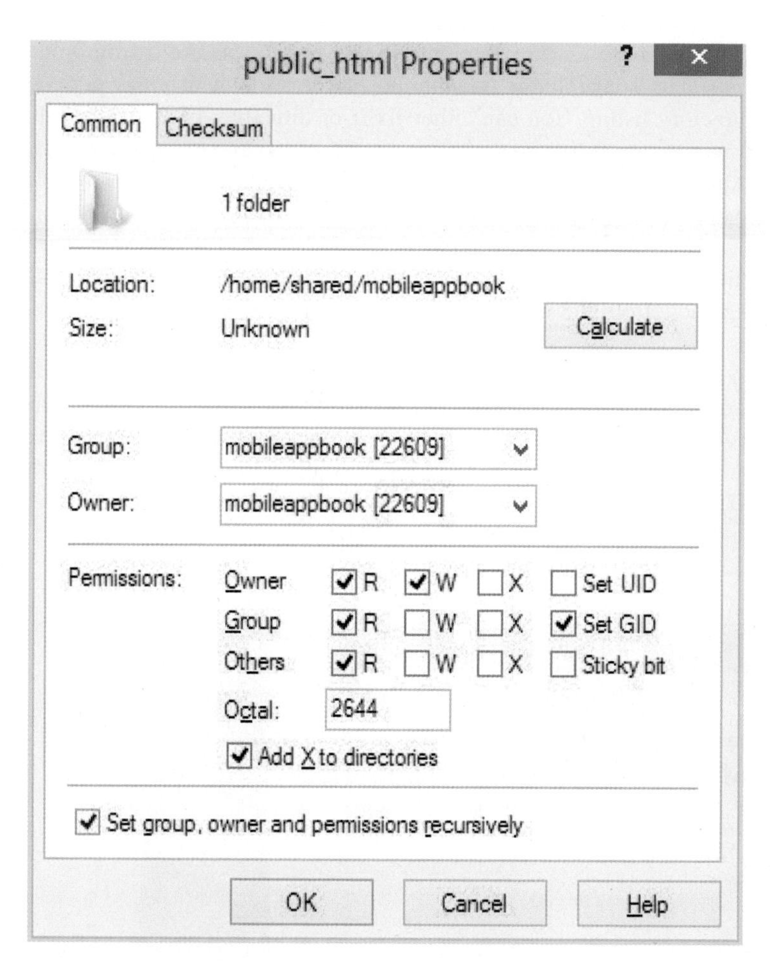

FIGURE 2.6 Screenshot of the WinSCP for setting the access privileges of files and folders

Follow: http://winscp.net

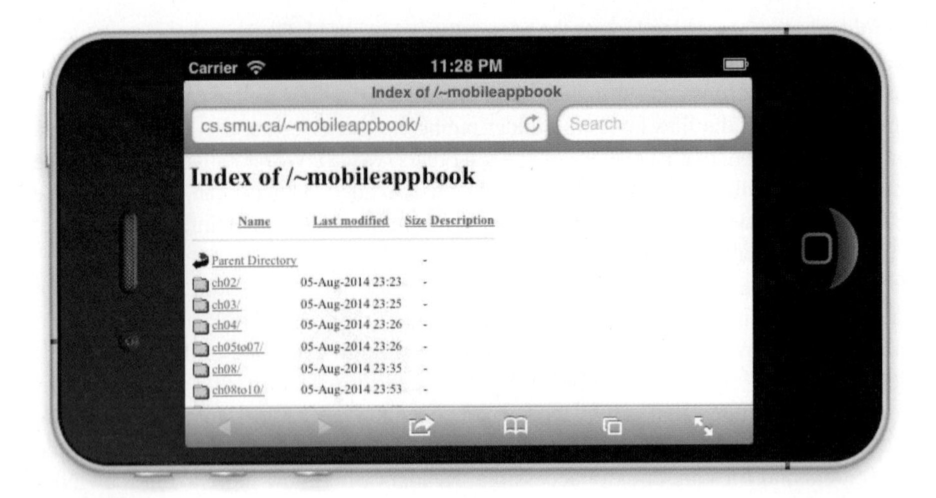

FIGURE 2.7 Screenshot of the uploaded folders in the browser

Now that we have uploaded the files to the web server and modified the access privileges, we can type in our URL, http://cs.smu.ca/~mobileappbook, in the browser's address window. Hopefully, we will see the list of folders as shown in Figure 2.7. If you do not see the copied folders in the browser and get some kind of error messages, please seek help from your system administrator. It may have something to do with the .htaccess file, which may prevent you from viewing the directory listing. You can either fix it or directly type the full URL for the web page, which in our case is `http://cs.smu.ca/~mobileappbook/ch02/physicsProjectileApp0/version1/physicsProjectileApp.html`

If you are able to see the folders, you can navigate to the correct folder by first clicking on ch02, then on physicsProjectileApp0, and then on version1. We will see the HTML5 web page: physicsProjectileApp.html as shown in Figure 2.8. Clicking on physicsProjectileApp.html will bring up the web page now from the web server as opposed to your personal computer, as shown in Figure 2.9.

Just because we can view the page properly in one browser does not mean that it will work on all the browsers on all the devices. Later on this book discusses how to view the web page on different devices/browsers and save it as a home screen icon for offline viewing. At this time we will alternate between the Safari browser on an Apple iOS device and Chrome browser on an Android OS device.

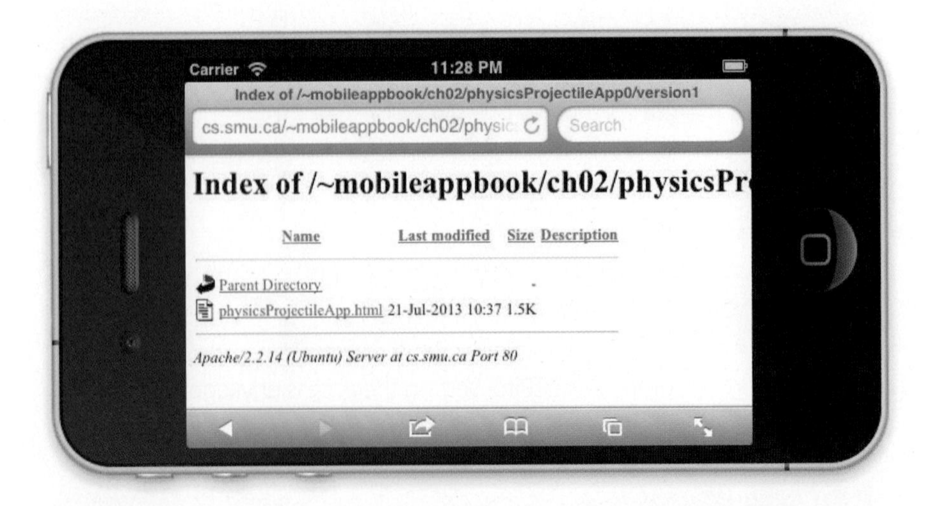

FIGURE 2.8 Screenshot after navigating to our web page in the browser

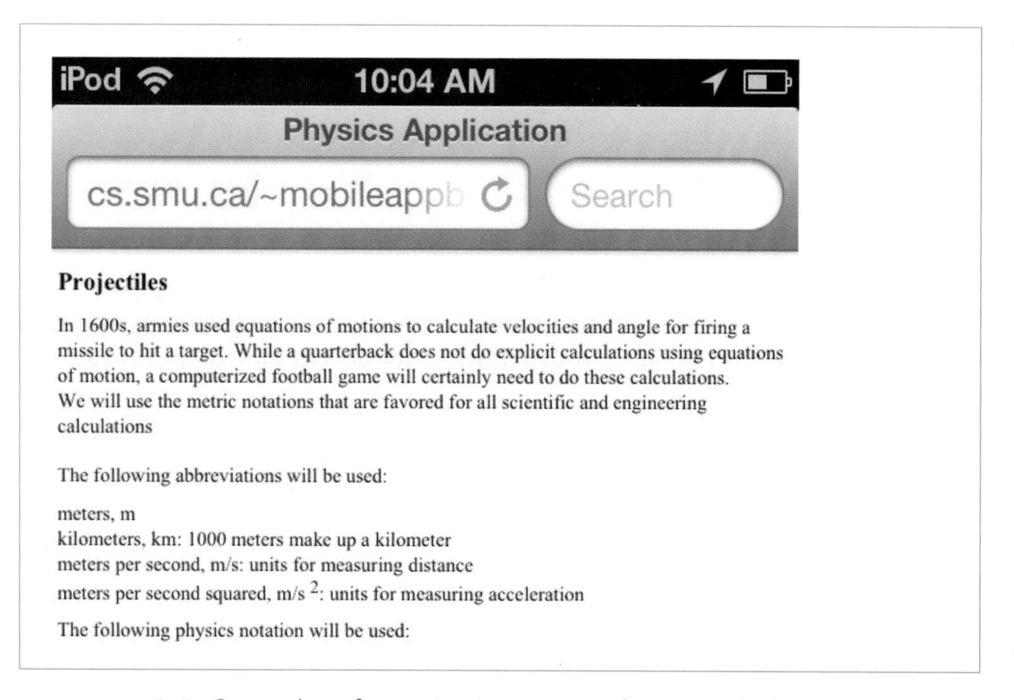

FIGURE 2.9 Screenshot after navigating to our web page in the browser
Follow: ch02/physicsProjectileApp0/version1/physicsProjectileApp.html

2.4 More HTML5 formatting

In the previous section, we created a basic web page using HTML5 and created an app on various devices that could help us see the page directly with a tap on the screen. In addition to the essential HTML5 code, we used three HTML5 tags for formatting, namely, `<p>...</p>` to enclose paragraphs, `<br>` for adding a line break, and `<sup>...</sup>` for creating superscripts. In this section, we are using the same structure of the web page and enhancing the formatting of the text. Figure 2.10 shows a high-level view of the newly formatted part of the web page. Figure 2.11 on page 27 shows the screenshot of a newer version of the web page that is in the web directory under ch02/physicsProjectileApp0/version2. The new formatting consists of a bulleted list, a table, and an enumerated list. The complete code for the web page is available in the web directory.The detailed code will be discussed on the following pages. One of the new tags that we see which we had discussed before is `<h2>...</h2>`, which allows us to have a header that is a little smaller than the `<h1>...</h1>` header we had seen in the previous chapter. We see three additional tags in Figure 2.10. The pair of tags `<ul>...</ul>` allows us to format unordered or bulleted lists, while `<ol>...</ol>` pair of tags format ordered or enumerated lists—that is, lists with numbers.

Finally, we have a pair of tags `<table>...</table>` to create a table. There is additional information associated within the tag `<table border="1">` that sets the border of the table to "1". The border is said to be an attribute of the `<table>` tag. We will look at the details of these three new tags on the next page.

2.5 Unordered and ordered lists

Figure 2.12 on page 27 shows the code for unordered list. An unordered list is defined as a list that does not specify an order with numbers or alphabets. The complete list is enclosed in a pair of tags `<ul>...</ul>`. We have added four items to the list using the pair of tags called `<li>...</li>`. Each item in a list (unordered and ordered) is enclosed in the `<li>...</li>`. The resulting list is shown in Figure 2.13 on page 28. Earlier we saw use of an attribute called border for the `<table>` tag. The `<ul>...</ul>` used to come with an attribute called type with values of disc, square, and circle. It is no longer available in HTML5. However, many browsers may still support it.

In addition to the list-related tags in our HTML5 code in Figure 2.12, we see additional formatting code. We have already discussed `<h2>...</h2>` (level 2 header) and

```
<h2>Abbreviations</h2>
      <ul>
      ......
      ......
      <!-- Code for unordered list with bullets -->
      ......
      ......


      </ul>

      <h2>Notations</h2>

      <table border="1">
      ......
      ......
      <!-- Code for table -->
      ......
      ......

      </table>

      <h2>Equations</h2>
      <ol>
      ......
      ......
      <!-- Code for ordered list with numbers -->
      ......
      ......

      </ol>
```

FIGURE 2.10 HTML5 code for a web page with bullets, numbers, and table
Follow: ch02/physicsProjectileApp0/version2/physicsProjectileApp.html

^{...} (to format superscripts to get exponents). The new pair of tags we have added is <em>...</em>. The <em> tag stands for emphasize. It usually means italicize. Therefore we can see that the text within <em>...</em> appears in italics. We can also see that the ^{...} are nested within <em>...</em>. So what appears inside ^{...} is both superscripted and italicized.

Figure 2.14 on page 28 shows the code for an ordered list, which is very similar to the unordered list. An ordered list uses numbers or alphabets for listing items. The complete list is enclosed in a pair of tags <ol>...</ol>. We have again added four items to the list using the pair of tags called <li>...</li>. The resulting list is shown in Figure 2.15 on page 28. The <ol>...</ol> also comes with a number of attributes. The reversed attribute allows us to order the list in descending order, the start attribute allows you to start at a different number than 1. The type attribute allows you to specify the values of 1 (Arabic numerals), A (uppercase alphabets), a (lowercase alphabets), I (Roman numerals in uppercase), i (Roman numerals in lowercase) to change the type of the numbers. Readers are encouraged to modify the web page to see the use of these attributes.

We have also added another pair of formatting tags in our ordered list, called <mark>...</mark>; this tag allows us to highlight the text as shown in Figure 2.15. We have also nested the ^{...} tags inside the <mark>...</mark>. Consequently, what appears inside ^{...} is both superscripted and highlighted.

Projectiles

In 1600s, armies used equations of motions to calculate velocities and angle for firing a missile to hit a target. While a quarterback does not do explicit calculations using equations of motion, a computerized football game will certainly need to do these calculations.
We will use the metric notations that are favored for all scientific and engineering calculations

Abbreviations

- meters, *m*
- kilometers, *km*: 1000 meters make up a kilometer
- meters per second, *m/s*: units for measuring distance
- meters per second squared, *m/s* ^{2}: units for measuring acceleration

Notations

Abbreviation	Meaning
u	Initial velocity
v	Final velocity
a	Acceleration
t	Time
s	Distance

Equations

1. a = (v - u) / t, which can be arranged to get the following equation
2. t = (v - u) / a, which can be further arranged as
3. v = u + a * t
4. s = u * t + 0.5 * a * t ^{2} , another useful equation of motion

FIGURE 2.11 Screenshot of a web page with bullets, numbers, and table
Follow: ch02/physicsProjectileApp0/version2/physicsProjectileApp.html

```
<h2>Abbreviations</h2>
<ul>
 <li>meters, <em>m</em></li>
 <li>kilometers, <em>km</em>: 1000 meters
   make up a kilometer</li>
 <li>meters per second, <em>m/s</em>: units
   for measuring distance</li>
 <li>meters per second squared, <em>m/s
   <sup>2</sup></em>: units for measuring
   acceleration</li>
</ul>
```

FIGURE 2.12 HTML5 detailed code for an unordered list with bullets
Follow: ch02/physicsProjectileApp0/version2/physicsProjectileApp.html

Abbreviations

- meters, *m*
- kilometers, *km*: 1000 meters make up a kilometer
- meters per second, *m/s*: units for measuring distance
- meters per second squared, m/s^2: units for measuring acceleration

FIGURE 2.13 Screenshot of an unordered list with bullets

Follow: ch02/physicsProjectileApp0/version2/physicsProjectileApp.html

```
<h2>Equations</h2>
<ol>
 <li><mark>a = (v - u) / t</mark>, which can
   be arranged to get the following
   equation</li>
 <li><mark>t = (v - u) / a</mark>, which can
   be further arranged as</li>
 <li><mark>v = u + a * t</mark></li>
 <li><mark>s = u * t + 0.5 * a * t
   <sup>2</sup></mark>, another useful
   equation of motion</li>
</ol>
```

FIGURE 2.14 HTML5 detailed code for an ordered list with numbers

Follow: ch02/physicsProjectileApp0/version2/physicsProjectileApp.html

Equations

1. a = (v - u) / t, which can be arranged to get the following equation
2. t = (v - u) / a, which can be further arranged as
3. v = u + a * t
4. $s = u * t + 0.5 * a * t^2$, another useful equation of motion

FIGURE 2.15 Screenshot of an ordered list with numbers

Follow: ch02/physicsProjectileApp0/version2/physicsProjectileApp.html

2.6 HTML5 tables

The detailed code for the table in our web page is shown in Figure 2.16 with the corresponding screenshot shown in Figure 2.17. The table is one of the more elaborate and frequently used formatting features in a web page. We are only looking at some of the most essential features of a table. We already discussed the border attribute of the <table> tag that allows us to control the size of the table border. We encourage readers to look at http://www.w3schools.com for more attributes to get the table look and feel just right. Later in this chapter, we are going to look at Cascading Style Sheets (CSS), which provide another way to change the look of a table.

A table consists of rows that are enclosed in a <tr>...</tr> pair of tags in Figure 2.16. We have five pairs of <tr>...</tr> tags for five rows in our table. Columns are enclosed in either <th>...</th> or <td>...</td> pairs of tags. We use the <th>...</th> for the first

```
<h2>Notations</h2>

     <table border="1">
      <tr>
        <th>Abbreviation</th>
        <th>Meaning</th>
      </tr>
      <tr>
        <td><em>u</em></td>
        <td>Initial velocity</td>
      </tr>
      <tr>
        <td><em>v</em></td>
        <td>Final velocity</td>
      </tr>
      <tr>
        <td><em>a</em></td>
        <td>Acceleration</td>
      </tr>
      <tr>
        <td><em>t</em></td>
        <td>Time</td>
      </tr>
      <tr>
        <td><em>s</em></td>
        <td>Distance</td>
      </tr>
     </table>
```

FIGURE 2.16 HTML5 detailed code for a table

Follow: ch02/physicsProjectileApp0/version2/physicsProjectileApp.html

row to denote that it is the header row. The <th>. . .</th> pair of tags makes the contents appear in bold format. The columns in the remaining four rows are enclosed in the <td>. . .</td> pair of tags. We have two columns in each row. Therefore we have two pairs of <th>. . .</th> or <td>. . .</td> tags in each row. As mentioned before, tables in a web page have many features, and web developers use them in imaginative ways to improve the look of a web page. We will explore some of them as we develop more web pages.

Notations

Abbreviation	Meaning
u	Initial velocity
v	Final velocity
a	Acceleration
t	Time
s	Distance

FIGURE 2.17 Screenshot of a table

Follow: ch02/physicsProjectileApp0/version2/physicsProjectileApp.html

2.7 Cascading Style Sheets (CSS)

We have looked at some of the essential features of HTML5 including how to

- create a paragraph
- introduce line breaks within a paragraph
- create unordered and ordered lists
- create a table

HTML was not designed for formatting a document. Tags that we have seen so far such as `<h1>`, `<p>`, `<ol>`, `<ul>`, `<table>` are really meant to specify the sections of a document more than format it.

Later on tags like `<font>` and color attributes were added to the HTML 3.2. Adding fonts and color information to every single page in the development of large websites was a long and expensive process. As a result, the World Wide Web Consortium (W3C) created cascading style sheets (CSS) to mitigate this problem. The use of CSS has taken the presentation of a web page to a completely different level. From HTML 4.0, all formatting can be stored in a separate CSS file. The use of CSS files makes the development of apps for a wide variety of devices of different sizes and shapes much easier. Just like HTML, CSS has gone through multiple versions. The current version of CSS is called CSS3.

In this section, we will introduce the basic and minimal CSS3 syntax. Even though the CSS3 used in mobile app development is very sophisticated, thankfully, we do not have to go through all the gory details. We will be using predefined CSS3 files that will help us display our apps appropriately on the mobile devices of various sizes, shapes, and persuasion (Apple, Google, Microsoft, . . .).

Figure 2.18 shows the relevant part of our first web page that uses a CSS file for formatting the display of our web page. The complete web page is in the web directory under ch02/physicsProjectileApp0/version3. The web page is essentially the same as the one that we saw under ch02/physicsProjectileApp0/version2. However, the display of the version2 page shown in Figure 2.11 is very different from the version3 display shown in Figure 2.19. The difference between two pages is the line that adds the link to the CSS3 page: `<link rel="stylesheet" href="physicsProjectileApp.css">` as shown in Figure 2.18.

```
<!DOCTYPE html>
<html>
  <head>
    <title>Physics Application</title>
    <link rel="stylesheet"
      href="css/physicsProjectileApp.css">
  </head>
  <body>
    <div>
     ......
     ......
     <!-- Rest of the code is essentially the same as previous version -->
     ......
     ......

    </div>
  </body>
</html>
```

FIGURE 2.18 HTML5 code to use a cascading style sheet (CSS)
Follow: ch02/physicsProjectileApp0/version3/physicsProjectileApp.html

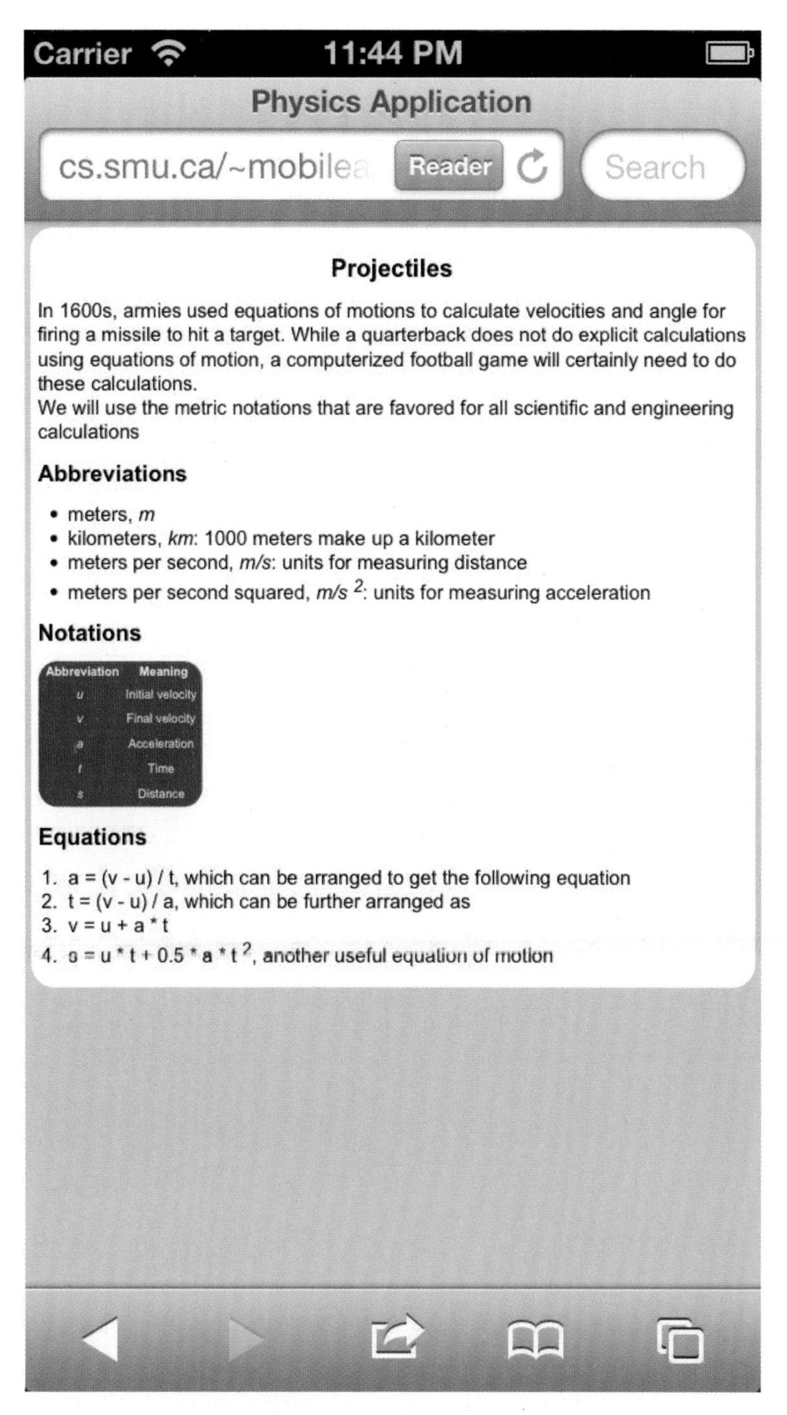

FIGURE 2.19 Screenshot of a web page that uses a cascading style sheet (CSS)
Follow: ch02/physicsProjectileApp0/version3/physicsProjectileApp.html

The `<link>` tag is used to define the relationship between an HTML5 document and an external resource. It is mostly used to link to the stylesheets. The attribute rel defines the type of relationship. We are specifying that the relationship is a "stylesheet". The other attribute we are using in the `<link>` is href, which specifies the URL of the document. We will be using the href in a few additional contexts. In this case, we are specifying the URL to be "css/physicsProjectileApp.css". Since we have not specified any protocol such as http:// as part of the URL, the server will assume that the file is in the subdirectory called "css" under the current directory. We are using a popular convention of creating a "css" directory under the directory where our web page is located, that is: ch02/physicsProjectileApp0/version3/css.

The only additional major difference worth noting is the pair of tags: `<div>...</div>`. The `<div>` tag defines a generic division or a section in an HTML document. Unlike tags such as `<h1>` and `<p>`, the `<div>` tag does not have any associated semantic. One of the most frequent uses of the `<div>` tag is for grouping a block of HTML5 elements to format them with CSS3.

Figure 2.20 shows the CSS3 file physicsProjectileApp.css from the folder ch02/physicsProjectileApp0/version3 that is used to display the version3 of our web page as shown in Figure 2.19. The format of the CSS file is rather simple. The file is showing the display style for the `<body>`, `<div>`, `<h1>`, `<table>` tags and `<td>` and `<th>` tags that are nested inside `<table>` and `<tr>`. The formatting instructions are enclosed in a pair of curly braces {...} after the name of the tag. Each formatting instruction is ended with a semicolon. The instruction consists of the attribute:value pair, and each instruction for a tag is separated by a semicolon. For example, for the body we are specifying the value #78BDFA for the attribute background color, which gives us the blue background shown in Figure 2.21. We also specify the font for the body to be Arial. The HTML5 code shown in Figure 2.18 shows that the entire

```
body
{
  background-color: #78BDFA;
  font-family: Arial;
}

div
{
  background-color: white;
  border-radius: 30px;
  padding:10px;
}

h1
{
  text-align: center;
}

table
{
  background-color:#444444;
  color: white;
  border-radius: 25px;
  text-align: center;
}

table tr td
{
  padding: 5px;
}

table tr th
{
  padding-left:8px;
}
```

FIGURE 2.20 A cascading style sheet (CSS)

Follow: File: ch02/physicsProjectileApp0/version3/css/physicsProjectileApp.css

Projectiles

■ **FIGURE 2.21** Screenshot of a web page with modified body/div section through CSS
Follow: ch02/physicsProjectileApp0/version3/physicsProjectileApp.html

contents of the web page are enclosed in the `<div>...</div>` tags. The background color for the `<div>` section is set to white, so the blue background color of the `<body>` section appears as the boundary. The padding attribute value of 10px for div specifies the size of the border for the body.

In Figure 2.20, px stands for pixel. A pixel is the smallest rectangle on the screen that can be individually changed. The rounded rectangles for the `<div>` section and the `<table>` section result from the attribute border-radius. The background of the table shown in Figure 2.22 is dark gray given by the color #444444, and the white letters come from the attribute color in the `<table>` section.

The CSS3 file in Figure 2.20 is a very simple illustration of what is possible with cascading style sheets. CSS3 plays an extremely important role in mobile app development. Fortunately, we have access to a number of open source CSS3 files, which will mean that in most cases we do not have to delve into the detailed CSS3 coding. If you need to create your own look and feel for your app, you will have to do some sophisticated CSS3 programming. However, it is beyond the scope of this book.

So far we have only worked with text within our web pages. Now it is time to add some multimedia. In the web directory under ch02/physicsProjectileApp0/version4/ we have a newer version of the web page that includes the code shown in Figure 2.23. We now have a new pair of tags in HTML5 called `<figure>...</figure>` that allows us to format a figure in our web page. This pair of tags was not available in previous versions of HTML. The `<figure>` is often used in conjunction with a nested pair of tags `<figcaption>...</figcaption>`. As the name suggests `<figcaption>` allows us to provide a caption underneath the figure as can be seen in Figure 2.24. The `<figure>` pair of tags only provides us with the framework for creating a figure. We still have to add the figure the old-fashioned way using `<img>` tag as shown in Figure 2.23.

Before we look at the `<img>` tag in greater detail, readers who are familiar with XHTML may want to notice that the format is different from the `<img/>` that was introduced in XHTML or HTML4. We are going back to the older version by dropping the trailing '/'. Note that the `<img>` does not have a closing `</img>` tag. We use attributes to specify various aspects of the `<img>` tag. The two attributes src and alt should always be provided in an `<img>` tag. The attribute src tells the web browser where to find the image. Again, we are going to follow a frequently used convention, whereby all the multimedia files are stored in a directory called

Abbreviation	Meaning
u	Initial velocity
v	Final velocity
a	Acceleration
t	Time
s	Distance

■ **FIGURE 2.22** Screenshot of a web page with modified table through CSS
Follow: ch02/physicsProjectileApp0/version3/physicsProjectileApp.html

```
<figure>
  <img src="media/projectilePic.png"
   alt="A Cannon Firing a Projectile"
   width="333" height="228">
   <figcaption><strong>Fig. 1 - A Projectile
    Fired From a Cannon</strong></figcaption>
</figure>
```

FIGURE 2.23 Code for a web page with an image and video
Follow: ch02/physicsProjectileApp0/version4/physicsProjectileApp.html

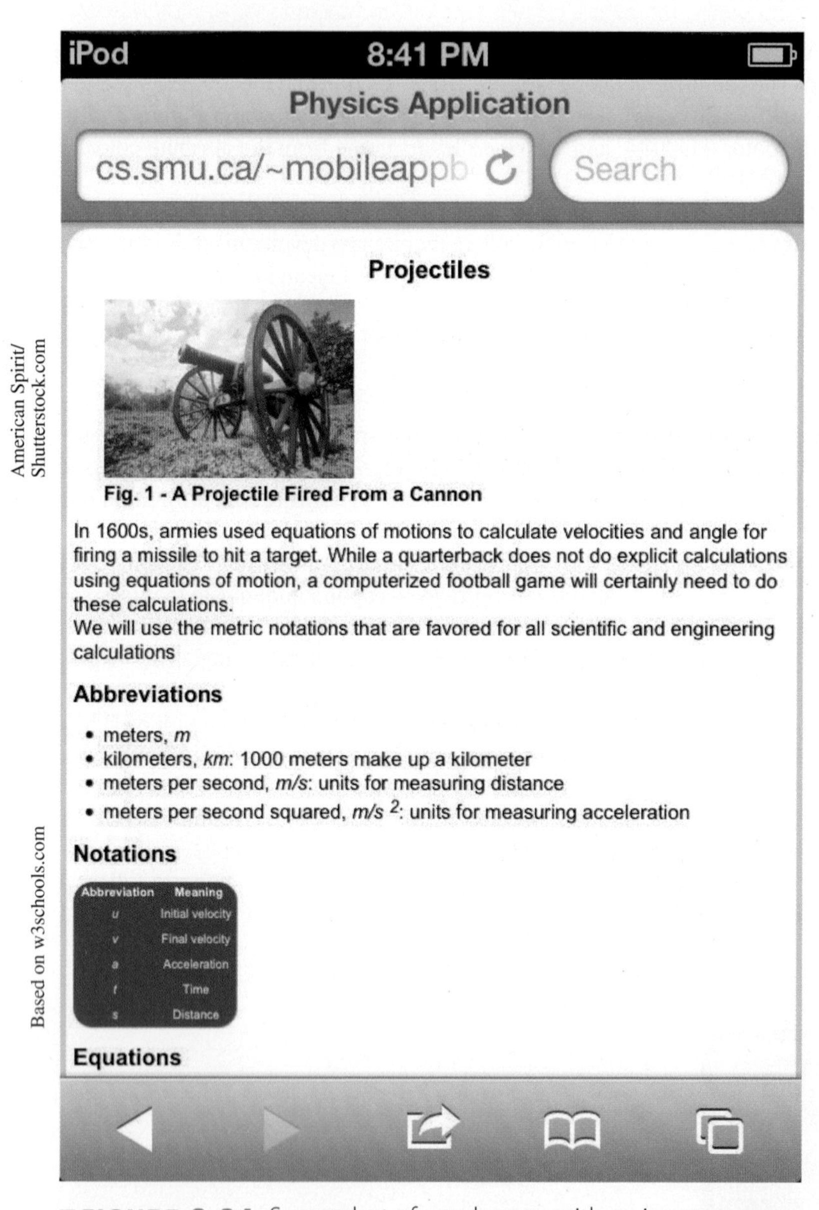

American Spirit/ Shutterstock.com

Based on w3schools.com

FIGURE 2.24 Screenshot of a web page with an image
Follow: ch02/physicsProjectileApp0/version4/physicsProjectileApp.html

```
<video controls>
  <source src="media/projectileVid.mp4"
    type="video/mp4">
  <source src="media/projectileVid.ogg"
    type="video/ogg">
</video>
```

FIGURE 2.25 Code for a web page with an image and video
Follow: ch02/physicsProjectileApp0/version5/physicsProjectileApp.html

media under the directory where our web page is located, that is ch02/physicsProjectileApp0/version4/media. The other strongly recommended attribute alt tells the browser text string to display if the image cannot be loaded for any reason. The text specified in the alt attribute will also appear in a box if you hover your mouse over the image. Two other attributes we have for the `<img>` tag are width and height. The meaning of these attributes is obvious. These attributes are optional. If we do not specify values for width and height, the browser will use the values from the image itself. The modern browsers support a number of image formats with jpg, gif, and png being some of the more popular ones.

Prior to HTML5, we did not have a standard HTML tag for including videos. That is all changed with the HTML5 pair of tags called `<video>...</video>`. We have augmented our page with a video, which can be found in the web directory under: ch02/physicsProjectileApp0/version5/.

The relevant HTML5 code is shown in Figure 2.25, and the corresponding screenshot can be found in Figure 2.26. We have kept the code to the bare minimum. The attribute controls provide the play, pause, volume, and full-screen controls as shown in Figure 2.26. The `<video>` tag must have a nested tag called `<source>` (note there is no closing counterpart `</source>`). The name of the video file is specified with the attribute src in the `<source>` tag. As mentioned before, we are using the convention of keeping all the files in a folder called media. That is why we specify media/projectileVid.mp4 as the value for src attribute. We are using the MP4 format for the first `<source>` element, which is mentioned under the attribute type as video/mp4. As readers would have noticed we provide an additional `<source>` element. The second `<source>` element provides another file called projectileVid.ogg. It is just another media file that uses the Ogg format, specified by the attribute type as video/ogg. Your browser will play the video format that it supports.

The `<video>` element supports the MP4, WebM, and Ogg video formats. Details about these formats can be found at the w3schools.com site. Unfortunately, not all formats are supported by all of the browsers. The following table adopted from w3schools.com shows which browser supports which video formats as of April 2015.

Browser	MP4	WebM	Ogg
Internet Explorer	√	X	X
Chrome	√	√	√
Firefox	√	√	√
Safari	√	X	X

Based on w3schools.com

Only Chrome and Firefox support all the formats. However, MP4 covers all the major browsers.

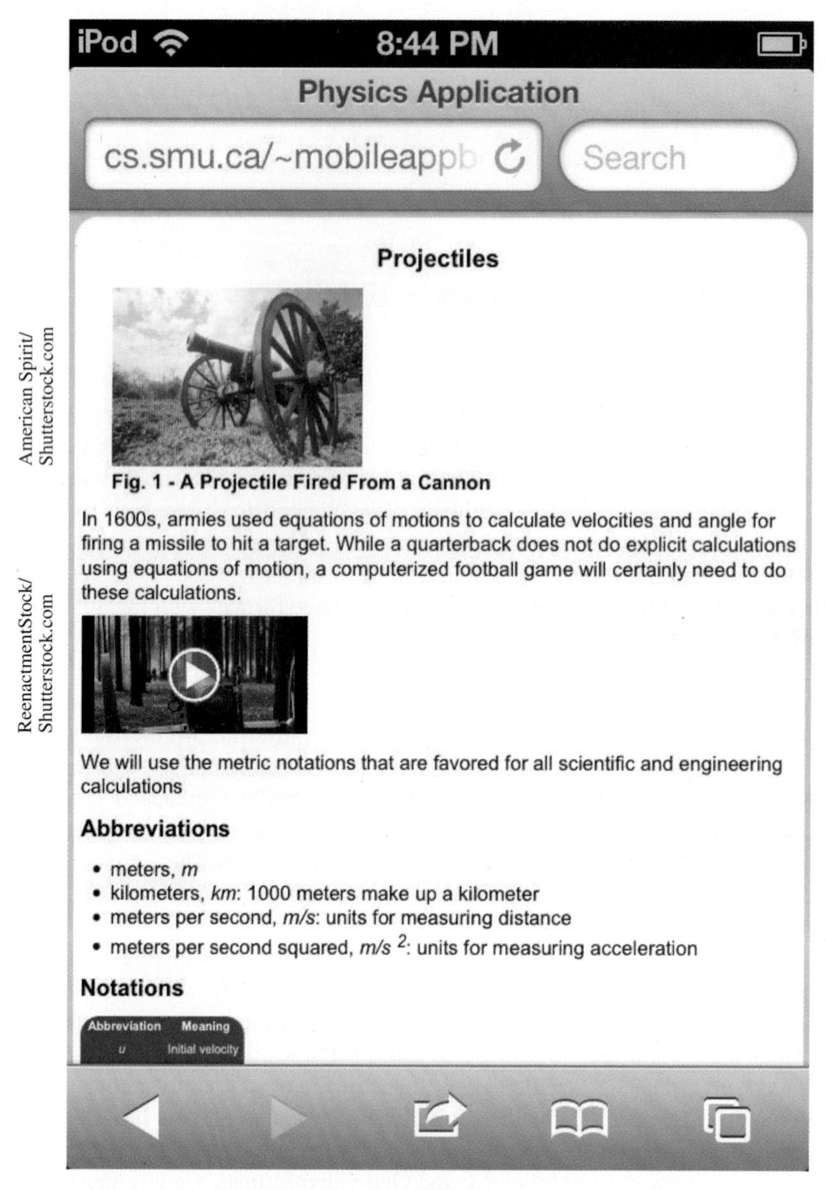

FIGURE 2.26 Screenshot of a web page with an image and video

Follow: ch02/physicsProjectileApp0/version5/physicsProjectileApp.html

Quick facts/buzzwords

HTML5: Version 5 of HyperText Markup Language.

Web server: Refers to a computer or a software that provides access to web pages over the Internet.

Shortcut on desktop/home screen: An icon on a device's desktop/home screen that can be tapped to directly access the web page.

HTML tags: Special instructions in angled brackets to the browser to display a web page.

<!DOCTYPE>: An HTML5 tag to tell the type of the document.

<html> </html>: A pair of HTML5 tags that enclose an HTML document.

<head> </head>: A pair of HTML5 tags that enclose the head section of an HTML document.

<body> </body>: A pair of HTML5 tags that enclose the body section of an HTML document.

<p> </p>: A pair of HTML5 tags that enclose a paragraph in an HTML document.

**
:** An HTML5 tag used to add a line break in an HTML document.

: A pair of HTML5 tags that are used to display a superscript in an HTML document.

http: HyperText Transfer Protocol for transmitting web pages on the Internet.

ftp: File Transfer Protocol for transmitting files on the Internet.

scp: Secure Copy Protocol, a secure alternative to ftp.

secure ftp: A secure alternative to ftp.

Self-test exercises

1. Who invented the World Wide Web?
2. What does the acronym http stand for?
3. What does the acronym ftp stand for?
4. What does the acronym scp stand for?
5. What are HTML5 tags?
6. List four HTML5 tags.
7. What is the purpose of comments in an HTML5 document? What is the syntax for a comment?
8. Which tag is used to create a table in an HTML5 document?
9. Which tag is used to create an unordered list in an HTML5 document?
10. Which tag is used to create an ordered list in an HTML5 document?
11. Which tag is used to create a list item in an HTML5 document?
12. What is the difference between an ordered and unordered list in an HTML5 document?
13. What does the acronym CSS stand for?
14. What is a stylesheet?
15. Which tag is used to include an image in an HTML5 document?
16. Which tag is used to include a video in an HTML5 document?

Programming exercises

1. Create a web page that looks like:

   # First HTML5 document

   Available on the web

   I am reading this book to learn how to develop websites that can be accessed from any device and can serve as cross-platform apps.
   The devices I will test will be running the following operating systems:

   Apple iOS
   Google Android
   Blackberry OS
   Microsoft Windows Phone OS

   Get an account from your system administrator and upload your web page to your account.
2. View your web page from as many devices and web browsers as possible.
3. Write HTML5 code that will create the following display:

   1. Apple iOS
   2. Google Android
   3. Blackberry OS
   4. Microsoft Windows Phone OS

4. Write HTML5 code that will create the following display:

    1. Apple iOS
        - iPad
        - iPhone
        - iPod touch
    2. Google Android
        - Nexus 7
        - Samsung Galaxy Notes 8
        - Samsung Galaxy Notes 4
        - HP Slate 7
    3. Blackberry OS
        - Blackberry Z10
        - Blackberry Q10
    4. Microsoft Windows Phone OS/RT
        - Nokia
        - Samsung ATIV
        - Surface

    Hint: You can nest an `<ul>...</ul>` pair of tags inside a pair of `<li>...</li>`.

5. Write HTML5 code that will create the following display:

Apple iOS	iPad	iPhone	iPod touch	
Google Android	Nexus 7	Samsung Galaxy Notes 8	Samsung Galaxy Notes 4	HP Slate 7
Blackberry OS	Blackberry Z10	Blackberry Q10		
Microsoft Windows Phone OS/RT	Nokia	Samsung ATIV	Surface	

6. Write the CSS3 and HTML5 that will create the following display:

Programming Exercises

Devices

7. Write the CSS3 for the web page from the previous exercise web page that will create the following display:

Programming Exercises

Devices

8. Write the CSS3 for your answer to PE5 that will create the following display:

Apple iOS	iPad	iPhone	iPod touch	
Google Android	Nexus 7	Samsung Galaxy Notes 8	Samsung Galaxy Notes 4	HP Slate 7
Blackberry OS	Blackberry Z10	Blackberry Q10		
Microsoft Windows Phone OS/RT	Nokia	Samsung ATIV	Surface	

Programming projects

Programming projects require a certain amount of research on your part to learn underlying theory. After you have done the research, create a web page that describes the problem you are going to solve (i.e., the mathematical solution). Upload the web page on the web, and make sure that it displays correctly on a number of devices.

1. **Compute the load distribution on a beam.** Assume that you have a one-dimensional beam resting on two rigid supports. The beam has a length described by the variable *length*. We have put a weight w at a distance d, where $d < length$, from the left end of the beam. You should find out the reaction forces on the left and right ends of the beam. You can make this project as complicated as you are capable of solving. You can have multiple weights, uniformly distributed load, and uniformly varying loads.

   Use tables, ordered and unordered lists, other relevant tags such as `<mark>`, `<em>`, `<sup>`, and CSS3 to improve the appearance of the web page.

   ● Hint: You can search for the term "load distribution on beams" on the web. You will find many relevant websites.

2. **Binary operator.** Create a web page that describes various binary operators such as "and", "or", "not", "xor".

   Use tables, ordered and unordered lists, other relevant tags such as `<mark>`, `<em>`, `<sup>`, and CSS3 to improve the appearance of the web page.

   ● Hint: You can search for the term "binary operators" on the web. You will find many relevant websites that will help you create content for your web page.

3. **Electricity calculations:** Create a web page that describes the physical relationship between current, voltage, resistance, and amount of electricity consumed.

   Use tables, ordered and unordered lists, other relevant tags such as `<mark>`, `<em>`, `<sup>`, and CSS3 to improve the appearance of the web page.

   ● Hint: You can search for the term "current voltage resistance electricity consumption" on the web. You will find many relevant websites.

4. **Amortization calculations:** Create a web page that describes how to compute the annual cost of a project that costs p and is funded by a loan at an interest i over y number of years.

   Use tables, ordered and unordered lists, other relevant tags such as `<mark>`, `<em>`, `<sup>`, and CSS3 to improve the appearance of the web page.

   ● Hint: You can search for the term "amortization calculations" on the web. You will find many relevant websites.

CHAPTER 3

Making apps more interactive through data input

WHAT WE WILL LEARN IN THIS CHAPTER

1. Introduction to JavaScript
2. Introduction to jQuery for mobile devices
3. How to embed JavaScript in a web page
4. How to use an external JavaScript file on a web page
5. How to use CSS elements designed for mobile devices
6. How to accept input through a web page
7. How to use different types of input widgets
8. How to link multiple mobile pages in an app

The end of the chapter has a Quick Facts section that lists all the technical terms discussed here.

In this chapter, we are going to have our first look at programming using JavaScript. JavaScript has emerged as a language of choice for mobile computing. We will have a brief introduction to JavaScript and follow it up with its practical usage in our app. We will look at the concept of document object model (DOM) and use it for Input/Output.

3.1 Embedding JavaScript in an HTML5 document

While HTML5 and CSS are languages that are used in computing, they do not fall in the traditional category of programming languages. HTML5 and CSS for the most part define the layout and presentation of a web page. We need a programming language that will bring these web pages alive—that is, create interactive content. A typical program accepts input from the user, manipulates the input, and returns the result to the user. Most readers of this book are likely to be familiar with a programming language. However, we will give a brief overview of the evolution of programming languages to the current state of web and mobile programming.

We normally work with languages that use English-like words, which are called high-level programming languages, and these high-level languages get translated to binary code that machines understand. Fortran was the first high-level programming language to gain popular acceptance by the programming community in the 1950s. In the 1960s the language C was popularized and has influenced the designs of many languages that followed including Java, Objective C, and C#, which are the languages of choice for native mobile app development on Android, Apple iOS, and Windows RT.

We will be using JavaScript for much of our mobile app programming. Despite the similarity between names, Java and JavaScript have very little in common except the syntactic similarities that can be traced back to the C language. JavaScript was originally developed under the name ActionScript by Netscape, which developed the first multimedia web browser. It was one of the first languages associated with web programming. Sun Microsystems (which developed Java) and Netscape mutually agreed to change the name to JavaScript. JavaScript was primarily used to write programs used by web browsers to animate the web pages. Programming on the client's device (i.e., a web browser) is called client-side programming. Client-side programming was the primary objective behind the development of JavaScript. Over time, other languages such as Perl and PHP have gained importance in the world of web programming, especially on the web server. The programming on a web server is called server-side programming.

With the advent of mobile devices, client-side computing has gained more importance, and new JavaScript libraries such as jQuery have been created. JavaScript is also gaining popularity within server-side computing with the help of the node.js platform. Most of the programming in this book will be based on JavaScript.

As mentioned before, high-level languages such as JavaScript are translated into binary code for execution. An entire program can be translated into the binary code and run many times, saving the overhead of translation every time it is run. This method is called compiling a program. JavaScript uses a simpler translation mechanism that translates the program every time it is run, one instruction at a time. This method is called interpretation. In the final chapter, we will see how the JavaScript-based apps can be compiled to avoid the overhead of translation when they are executed.

Let us begin with a simple web page that embeds JavaScript as shown in Figure 3.1. The page is in the web folder under `ch03/simpleJSapp` and is called `embeddedJS.html`. The body of the web page has a `<script>...</script>` pair of tags that encloses our first lines of JavaScript. We are specifying one attribute for the `<script>` tag called `type` that is set to `text/javascript`, indicating that the script is in standard text format written in JavaScript. JavaScript uses what is called Document Object Model (DOM). Every entity in our web page is addressable in JavaScript as an object. An object in JavaScript has methods to control the behavior of the object. The object also has properties, which are essentially data components. Both methods and properties are accessed using a period. The web page itself is referred appropriately as `"document"`. We are sending a command to the document object (i.e., the web page) using the "write" method. As a result, the string that is enclosed in quotes in the "write" method is displayed on the web page, as shown in Figure 3.2.

```
<!DOCTYPE html>
<html>
<head>
      <title>Simple Embedded JavaScript</title>
</head>
<body>
      <script type='text/javascript'>
            document.write("Hello from JavaScript");
      </script>
</body>
</html>
```

FIGURE 3.1 A simple JavaScript embedded in a web page
Follow: File: ch03/simpleJSapp/embeddedJS.html

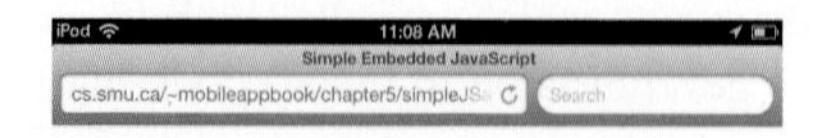

FIGURE 3.2 Screenshot of the web page after running the embedded JavaScript
Follow: File: ch03/simpleJSapp/embeddedJS.html

3.2 Using JavaScript from an external file in an HTML5 document

While we can contain the entire JavaScript program in our HTML5 web page, some programs can get very long, and it will make the web page unreadable if the JavaScript code is directly embedded. A better solution is to store our JavaScript programs (also called scripts) in another file. In fact, we will create a separate folder called scripts to store all our programs. We will look at another page called `externalJS.html` in the web folder under `ch03/simpleJSapp`. The web page is shown in Figure 3.3. The `<body>` of the web page is empty. However, in the `<head>` we have a `<script>...</script>` pair of tags to indicate the script that should be run. In addition to specifying the attribute value of `type` to `text/javascript`, we also indicate that the script is stored in the directory `scripts` and is called `simple.js` by setting the value of attribute src to `script/simple.js`. The file `simple.js` is shown in Figure 3.4, and it is more or less the same as the JavaScript we used in the file `embeddedJS.html`. We are issuing a write message to `document` to display the string that is enclosed in the quotes on the page as shown in Figure 3.3. JavaScript follows the object-oriented convention, where a period separates an object (document) and a message (write) sent to it. The corresponding screenshot is shown in Figure 3.5.

```
<!DOCTYPE html>
<html>
<head>
      <title>Simple External JavaScript</title>
      <script type='text/javascript' src='scripts/simple.js'></script>
</head>
<body >
</body>
</html>
```

FIGURE 3.3 JavaScript file included in a web page
Follow: File: ch03/simpleJSapp/externalJS.html

```
document.write("Hello from external JavaScript");
```

FIGURE 3.4 JavaScript file simple.js that was included in the web page
Follow: File: ch03/simpleJSapp/scripts/simple.js

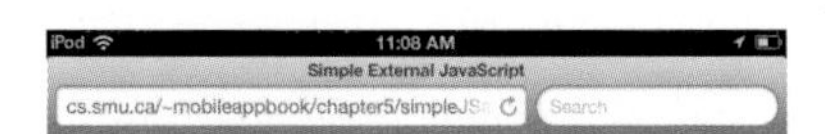

Hello from external JavaScript

FIGURE 3.5 Screenshot when JavaScript from file simple.js is run from the web page
Follow: File: ch03/simpleJSapp/externalJS.html

3.3 A JavaScript function

We continue with the same basic program in Figure 3.6, which is stored `functionJS.html` in the web folder under `ch03/simpleJSapp`. The script file that is included in the web page is called `function.js` in the folder `ch03/simpleJSapp`. Usually, we divide our programs in

```
<!DOCTYPE html>
<html>
<head>
      <title>Simple External JavaScript Function</title>
      <script type='text/javascript' src='scripts/function.js'></script>
</head>
<body onload='hello()'>
</body>
</html>
```

FIGURE 3.6 JavaScript function run at the loading included in a web page
Follow: File: ch03/simpleJSapp/functionJS.html

modules called functions. A function is a reusable block of code that has a name and performs a specific task. Our `function.js` in the scripts/directory shown in Figure 3.7 contains one function. The keyword "`function`" is followed by the name of the function "`hello`", which is further followed by open and close parentheses. The statements that will be executed in the function are enclosed in open and close curly braces. We are now familiar with the code that is in the hello function that is issuing a write message to `document` to display the string that is enclosed in the quotes. Unlike the previous scripts that were directly run as soon as we launched the web page, functions need to be explicitly called. Therefore, we add an attribute `onload` whose value is a call to the function `hello()` that leads to display of the string "`Hello from a JavaScript function`" on the web page as shown in Figure 3.8. This is a very simple introduction to JavaScript functions. We will study more refined functions later in the book.

```
function hello()
{
      document.write("Hello from a JavaScript function");
}
```

FIGURE 3.7 A JavaScript function
Follow: File: ch03/simpleJSapp/scripts/function.js

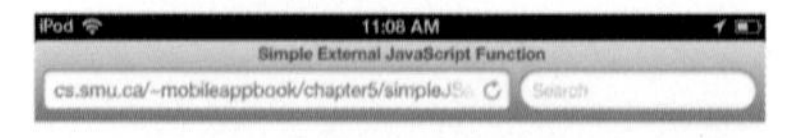

Hello from a JavaScript function

FIGURE 3.8 Screenshot of the JavaScript function run at the time of loading
Follow: File: ch03/simpleJSapp/functionJS.html

3.4 Input widget in HTML5/jQuery/jQuery Mobile

So far we were building what looked like standard web pages that could be easily displayed on any browser—on either a computer or a mobile device. There was no attempt to ensure their compatibility with the mobile devices. That all changes now. Figure 3.9 shows HTML5 code that is specifically designed to ensure that it can work on all the popular mobile devices as well as the conventional desktop/laptop web browsers. jQuery and jQuery Mobile have made this task very easy. The following quotes from the respective websites tell us what these platforms are as well as where to find them.

http://jquery.com/: "jQuery is a fast, small, and feature-rich JavaScript library. It makes things like HTML document traversal and manipulation, event handling, animation, and Ajax much simpler with an easy-to-use API that works across a multitude of browsers. With a combination of versatility and extensibility, jQuery has changed the way that millions of people write JavaScript".

```
<!DOCTYPE html>
<html>
<head>
 <title>Physics Simulator App</title>
 <link rel="stylesheet"
   href="css/jquery.mobile-1.3.1.min.css">
 <script src="scripts/jquery-1.8.3.min.js">
 </script>
 <script
  src="scripts/jquery.mobile-1.3.1.min.js">
 </script>
 <meta name="viewport"
 content="width=device-width, initial-scale=1.0">
</head>
<body>
 <div data-role="page">
  <div data-role = header>Projectile app</div>
  <div data-role="content">
   <label for="angle">Angle:</label>
   <input type='number' name='angle' id='angle'
    min='0' max='90' placeholder='In degrees'>
   <label for="velocity">Velocity:</label>
   <input type='number' name='velocity'
    id='velocity' min='0' max='299792458'
    placeholder='In metres/second'>
   <br>
   <button>Display</button>
  </div>
  <div data-role = footer>
        <small>Cross-platform mobile app demo</small>
  </div>
 </div>
</body>
</html>
```

FIGURE 3.9 HTML5 code for the accepting input in the Projectile app

Follow: File: ch03/physicsProjectileApp1/version1/physicsProjectileApp1.html

http://jquerymobile.com/: "A unified, HTML5-based user interface system for all popular mobile device platforms, built on the rock-solid jQuery and jQuery UI foundation. Its lightweight code is built with progressive enhancement, and has a flexible, easily themeable design. jQuery mobile framework takes the 'write less, do more' mantra to the next level: Instead of writing unique apps for each mobile device or OS, the jQuery mobile framework allows you to design a single highly-branded web site or application that will work on all popular smartphone, tablet, and desktop platforms".

Figure 3.9 shows that the HTML5 code that is used to create a mobile-friendly web page that is displayed in Figure 3.10 essentially has three important features: two textboxes to input the angle and velocity of the projectile respectively and a button to display these values. Eventually, we will be doing complex calculations and graphing in our app. However, in this chapter, we are going to restrict ourselves to the study of input and output. In your web folder this web page can be found at

```
ch03/physicsProjectileApp1/version1/physicsProjectileApp1.html.
```

Let us focus on the basic web page setup necessary for our web pages to be mobile-friendly. The `<head>` section of the web page includes a <link> tag that includes a stylesheet called `jquery.mobile-1.3.1.min.css` that can be downloaded from http://jquerymobile.com/ Readers are encouraged to download the latest version and store it in the `css` folder. This stylesheet is quite refined and renders the page appropriately for various devices and browsers. We also have two `<script>` tags that include JavaScript libraries

`jquery-1.8.3.min.js` (download the latest version from http://jquery.com/) and
`jquery.mobile-1.3.1.min.js` (download latest version from http://jquerymobile.com/)

These libraries are essentially a collection of JavaScript functions that simplify our programming. We will look at some of the functions as we use them in our app development. Following our convention, we have put these libraries in the `scripts` folder. We also added the following code to the `<head>` section

```
<meta name="viewport" content="width=device-width,
initial-scale=1.0">
```

This tells the mobile browsers that they only have the width of the device to render the page, and not a full-size window.

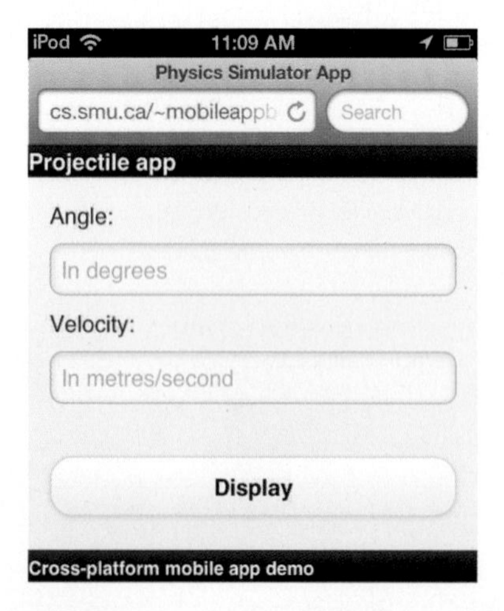

■ **FIGURE 3.10** Screenshot of the web page with input for the Projectile app
Follow: File: ch03/physicsProjectileApp1/version1/physicsProjectileApp1.html

3.5 Designing our apps

Before we delve into the specifics of the Projectile app, let us discuss how we decided on the look and feel. The most important thing to consider when designing any app is the end user and what they will want to accomplish using your app. When you have an idea for an app, consider what the main objectives of the app will be. What tasks will a user expect to complete with your app? Which will be most important to them? You should think not in terms of features for your app, but rather of tasks that your users will be able to accomplish with it. The best way to determine the answers to these questions is to talk to potential users of your application and get their feedback. Try to put yourself in the users' shoes when making decisions on the functionality and design of your app.

Once you have decided on a set of functionalities that you would like your app to help users accomplish, you should think of how you want to organize those functionalities. What pages and what sections will you divide your functionalities between? How will users navigate between these pages? When making these decisions, it is important to consider which functionality will be the most commonly used—that is, the core functionality of the app. Generally, the core functionality should be the first screen that users see when they access your app. Other functionalities should be placed in a menu of some sort to provide easy access. It is important to have a well-designed menu so that new users can see the main areas of the app that they can access, and experienced users can quickly access the functionality they desire.

Communication between your app and the user is also important to consider. There is nothing worse than an app freezing on you or doing something unexpectedly. Your app should be designed in a way that is responsive to the user's input. If possible, results or feedback should appear immediately after each user interaction. When you know something will take a long time, a loading animation or message is useful to acknowledge to users that their action has been received. When the user is about to do something nonreversible (for example, when making a purchase on a credit card), it is important to alert that user and obtain confirmation. If an error has occurred, the app should similarly alert the user with a clear description of the problem.

You also need to consider what device or devices users will be using to access your app. In this textbook, our focus is on mobile applications, and we expect that users will be using smartphones or tablets to access our apps. Consequently, we should design our apps to take advantage of the touchscreen capability of these devices. This means that buttons should be big enough to tap and that draggable sliders may be better than typing in a number. Finally, keep in mind that not all devices will render your app in exactly the same way. Certain widgets may not be supported on different platforms, and different devices have different resolutions and shapes. For this reason, it is important to test your app on as many devices as possible: iOS, Android, and other platforms, both new and old.

Our first app: physics Projectile app

For our very first app, we will be looking at the physics behind launching a projectile. In our case, the audience for our app will include users of all levels of experience, so we will be developing a simple app that is intuitive to use. Through the next few chapters, we will develop multiple versions of the app that explore different input methods (e.g., textboxes and sliders) and different output methods (e.g., tables and graphs).

The main purpose of this app is to calculate the distance of a projectile based on the angle and velocity input by the user. As a result, the calculation page will open first when you start the app. We will also have a section providing information on the theory of a projectile launch. It is expected that users will only review the theory once or twice, which is why we have decided to keep the theory page off the entry page and in a menu. This keeps the information easily accessible but does not take away the focus from the main user task.

There are many common navigation models to choose from. In our case, we chose to have a menu bar to be at the top of the page to provide navigation between the two main sections. Choosing to put the menu bar at the top is a common design that users will be familiar with from other apps. It is important to follow common design schemes that users will have seen before when possible. This results in less of a learning curve when users use the app, resulting in more usability and enjoyment. We also keep our menu bar consistent across all sections of the app so that users can easily return to other parts of the app.

Other basic principles that we will follow, and that should be kept in mind when designing any app, are that the colors and fonts you pick should be pleasing to the eye and easy to read. Headings should be in a larger font size than the body text, and all the font sizes should be large enough for all users to read.

3.6 Header, footer, and content data-role

The body of the web page is wrapped in a pair of `<div>...</div>` tags with an attribute called `data-role` whose value is set to `"page"`. The data-role attribute is an important jQuery Mobile feature. Setting it to `"page"` will display enclosed contents on a single page.

Nested inside the page, we have added three pairs of `<div>...</div>` tags with `data-role` set to `"content"`, `"header"`, and `"footer"`. This can help us distinguish content from header and footer. The effects of these three data-roles are obvious in Figure 3.11. The part of the page that was identified with `data-role` of `content` appears as the standard web page. It is not essential to put the main content of the web page in a

```
<div data-role="content">...</div>
```

pair of tags. However, doing so helps us clearly distinguish it in the code from the header and footer. Both

```
<div data-role="header">...</div> and
<div data-role="footer">...</div>
```

appear with special highlights as shown in Figure 3.11. We may use header data-role for displaying title of the app and footer for fine-print text. In order to display the text as fine print, we will need to make footer smaller by adding a pair of tags `<small>...</small>` such as the ones shown in Figure 3.12.

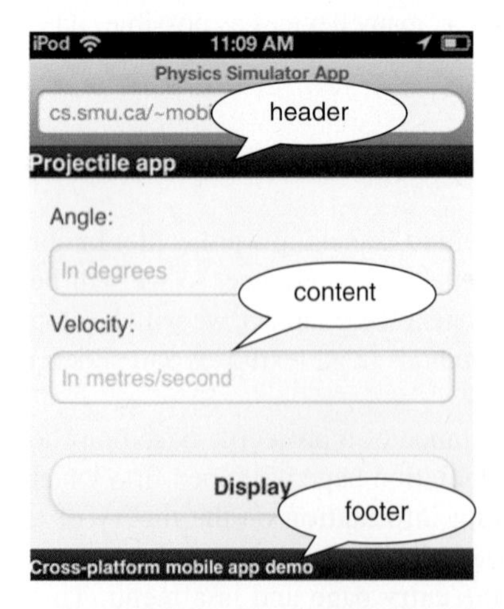

FIGURE 3.11 Screenshot of the web page with content, header, and footer

Follow: File: ch03/physicsProjectileApp1/version1/physicsProjectileApp1.html

3.7 More widgets in HTML5/jQuery/jQuery Mobile

Our app shown in Figure 3.12 in the previous section has three important features created by using what are called widgets: two textboxes to input the angle and velocity of the projectile respectively and a button to display these values. The HTML5 code for the button widget is shown in Figure 3.13. The corresponding screenshot is displayed in Figure 3.14. We are simply using a pair of tags `<button>...</button>` with the text "Display" inside the tags. We will do much more with that button in the next version of our physicsProjectileApp1.

Before the input text box we have inserted a pair of `<label>...</label>` tags that enclose the text "Angle", so the users know what is expected in the text box. We have also specified an attribute `"for"` with a value `"angle"` that associates the `<label>` with an element with an `id` of `"angle"`, which happens to be the text input box that follows.

The jQuery Mobile stylesheet and libraries allow us to create these and many other widgets in our app. A reasonably comprehensive list can be found at

```
http://api.jquerymobile.com/category/widgets/
```

```
<body>
 <div data-role="page">
  <div data-role = header>Projectile app</div>
  <div data-role="content">

  <!-- The contents of the body go here -->

  <div data-role = footer>
        <small>Cross-platform mobile app demo</small>
  </div>
 </div>
</body>
</html>
```

FIGURE 3.12 HTML5 code for different data-roles
Follow: File: ch03/physicsProjectileApp1/version1/physicsProjectileApp1.html

```
<button>Display</button>
```

FIGURE 3.13 HTML5 code for the Calculate button

Display

FIGURE 3.14 Screenshot of the Calculate button
Follow: File: ch03/physicsProjectileApp1/version1/physicsProjectileApp1.html

Let us take a closer look at the `<input>` tag that we are using for creating text input boxes in our app. Figure 3.15 shows the code for the two text input boxes, and the corresponding screenshot is shown in Figure 3.16. We have six attributes for our two `<input>` tags. We will only discuss the first of the two text boxes as the other text box has essentially the same features. Let us now look at all the six attributes that are associated with our `<input>` tag.

`type`	This describes the type of input we expect from this text box. We are using number as the type in our text input box. HTML5 provides a number of input types including password, email, telephone number, and many more. A comprehensive list can be found at the w3schools.com website. The display of some of the type values is different depending on the browser. For example, the range input type that we will be using later on in the chapter is rendered as a slider in Chrome. Some of the older browsers may still show it as a text input box. jQuery Mobile standardizes the appearance of input types such as range and search by dynamically changing their type to text.
`name`	We use this attribute to identify our element (text input box). We have assigned the value `'velocity'` to this attribute. This tag is obsolete (deprecated) and is replaced by the `id` attribute. However, you will find many web pages that use this attribute, so we are keeping it for completeness.
`id`	Similar to name, we use this attribute to identify our element (text input box). This tag is the preferred way to identify an element and must be unique in a given document. We will see how it is used in our programs later on in this chapter. We have assigned the value `'velocity'` to this attribute.
`min`	This attribute specifies the minimum possible value that can be input.
`max`	This attribute specifies the maximum possible value that can be input.
`placeholder`	The value specified for this attribute appears in the textbox when it is loaded. For angle, we display `'In degrees'` as the initial text. It will be replaced by the actual input value, when the user types it.

```
<label for="angle">Angle:</label>
   <input type='number' name='angle' id='angle'
    min='0' max='90' placeholder='In degrees'>
   <label for="velocity">Velocity:</label>
   <input type='number' name='velocity'
    id='velocity' min='0' max='299792458'
    placeholder='In metres/second'>
```

FIGURE 3.15 HTML5 code for the angle and velocity input

Follow: File: ch03/physicsProjectileApp1/version1/physicsProjectileApp1.html

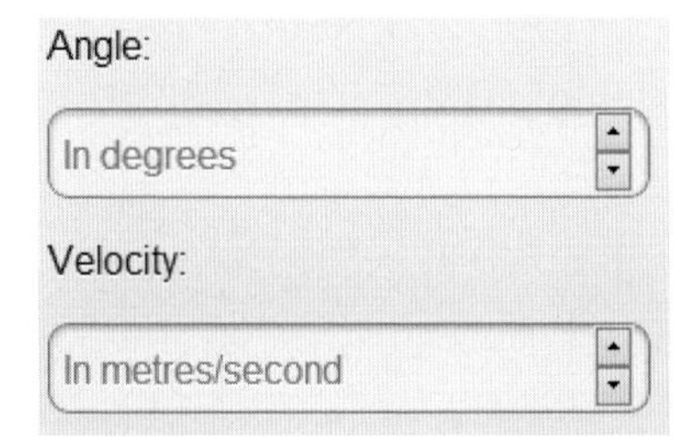

■ **FIGURE 3.16** Screenshot of the angle and velocity input
Follow: File: ch03/physicsProjectileApp1/version1/physicsProjectileApp1.html

3.8 Identifying elements from a web page

Figure 3.17 on the following page shows the HTML5 code of a web page that is in your web folder at

```
ch03/physicsProjectileApp1/version2/physicsProjectileApp1.html
```

The screenshot after launching the page is shown in Figure 3.18. Most of the code in Figure 3.17 is similar to the version1 of the app. We will focus on the new code that is highlighted.

1. The header section includes another script using the code which includes the file `physicsProjectileApp1.js` from the folder `scripts`
```
<script type='text/javascript'
src='scripts/physicsProjectileApp1.js'>
</script>
```
2. The code for the button widget now has an attribute called `onclick` that calls a JavaScript function `display` defined in `physicsProjectileApp1.js`
```
<button onclick="display();">
```
3. We also see a pair of `<table>...</table>` tags with two rows and three columns that correspond to the two lines below the Display button.
4. A feature we see throughout the HTML5 code is the `"id"` attribute. This is used by our JavaScript functions, which will be discussed on the next page.

We have reproduced the relevant parts of HTML5 code in Figure 3.19 that will be used by the JavaScript function `display()` from `physicsProjectileApp1.js`. The attribute id can be associated with almost any HTML5 element. Here we have used it to identify two `<input>`, one `<table>`, and two `<td>` elements.

The function `display()` in Figure 3.20 has two input statements and two output statements. Let us look at the first two input statements, which read angle and velocity from the `<input>` elements. Since we may use these values for a number of calculations, we should store them in memory by creating variables. In the first statement of the function `display()`, we are creating a variable called angle by saying `"var angle"`. The second input statement creates another variable called velocity to store input from the second `<input>` element. The `document.getElementByID("angle")` call returns us the `<input>` element with the `id` of `angle`.

```
<!DOCTYPE html>
<html>
<head>
 <title>Physics Simulator App</title>
 <link rel="stylesheet"
   href="css/jquery.mobile-1.3.1.min.css">
 <script type='text/javascript'
  src='scripts/physicsProjectileApp1.js'>
  </script>
 <script src="scripts/jquery-1.8.3.min.js">
 </script>
 <script
  src="scripts/jquery.mobile-1.3.1.min.js">
 </script>
 <meta name="viewport"
 content="width=device-width, initial-scale=1.0">
</head>
<body>
 <div data-role="page">
  <div data-role = header>Projectile app</div>
  <div data-role="content">
   <label for="angle">Angle:</label>
   <input type='number' name='angle' id='angle'
    min='0' max='90' placeholder='In degrees'>
   <label for="velocity">Velocity:</label>
   <input type='number' name='velocity'
    id='velocity' min='0' max='299792458'
    placeholder='In metres/second'>
   <br/>
   <button onclick="display();">
    Display
   </button>
   <br/>
   <table id="data">
    <tr>
     <td>Angle Entered:</td>
     <td id='angleDisplay'>0</td>
     <td>degrees</td>
    </tr>
    <tr>
     <td>Velocity Entered:</td>
     <td id='velocityDisplay'>0</td>
     <td>metres/second</td>
    </tr>
   </table>
  </div>
  <div data-role = footer>
        <small>Cross-platform mobile app demo</small>
  </div>
 </div>
</body>
</html>
```

FIGURE 3.17 HTML5 code for calling the JavaScript functions to read the input
Follow: File: ch03/physicsProjectileApp1/version2/physicsProjectileApp1.html

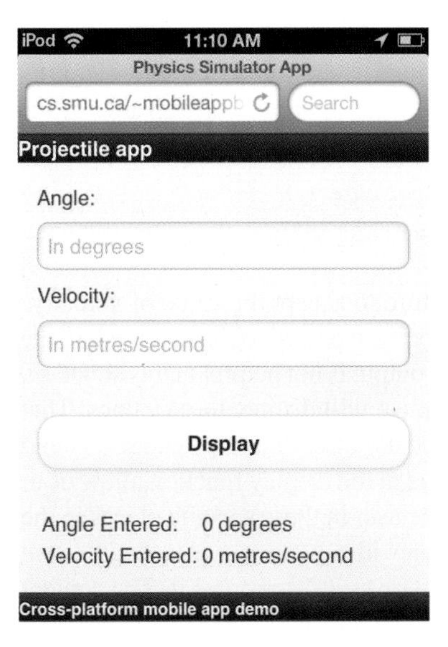

FIGURE 3.18 Screenshot of the Projectile app before the values are input

Follow: File: ch03/physicsProjectileApp1/version2/physicsProjectileApp1.html

```
<input type='number' name='angle' id='angle'
 min='0' max='90' placeholder='In degrees'>
<input type='number' name='velocity'
 id='velocity' min='0' max='299792458'
 placeholder='In metres/second'>

<!-- Some more HTML5 code -->

<table id="data">
 <tr>
  <td>Angle Entered:</td>
  <td id='angleDisplay'>0</td>
  <td>degrees</td>
 </tr>
 <tr>
  <td>Velocity Entered:</td>
  <td id='velocityDisplay'>0</td>
  <td>metres/second</td>
 </tr>
</table>
```

FIGURE 3.19 Parts of HTML5 code used by JavaScript function for I/O

Follow: File: ch03/physicsProjectileApp1/version2/physicsProjectileApp1.html

```
function display()
{
 var angle=document.getElementById("angle").value;
 var velocity=document.getElementById("velocity").value;

 document.getElementById("angleDisplay").innerHTML=angle;

 document.getElementById("velocityDisplay").innerHTML=velocity;
}
```

FIGURE 3.20 JavaScript function to read and display the input

Follow: File: ch03/physicsProjectileApp1/version2/scripts/physicsProjectileApp1.js

We use a period to access the attribute called "value" of this <input> element. That is, the call document.getElementById("angle").value brings us the value from the <input> element with the id of "angle". In the example provided in Figure 3.21, the value of the element with the id of "angle" is 40. We store it in the variable called angle using the assignment operator =, that is,

```
var angle=document.getElementById("angle").value;
```

The second input statement has essentially the same structure to accept the value of velocity.

Once we have received the values of the angle and velocity, we need to display them in the correct place. The function write that we used previously for output is not helpful in this situation, as we need to pinpoint the exact location on the app where we will display these values. That is, we want to display the angle and velocity values specifically after "Angle Entered" and "Velocity Entered" respectively. The last two statements in the display function allow us to output the values of angle and velocity that were input by the user in these specific places in the app. We know that the angle should be displayed in the column with id "angleDisplay", so we use the message document.getElementById("angleDisplay") to access it. The content of the column is accessed with an attribute called innerHTML, which is assigned the value from the variable angle with the statement: document.getElementById("angleDisplay").innerHTML=angle;

The fourth statement uses similar programming features to display the velocity.

3.9 Range slider input widget

In the previous version of the app (version2, Figures 3.17, 3.18, 3.19, 3.20, and 3.21), we used text boxes for inputting our angle and velocity using the keyboard. The HTML5 <input> element was used to provide an ability to accept the input. In a mobile device with touch-screen facility, we should provide facilities to interact without a keyboard whenever possible. Figure 3.22 shows a screenshot of a modified version of the app that uses a range slider that can be used to interact without a keyboard. The web page can be found in your web folder at

```
ch03/physicsProjectileApp1/version3/physicsProjectileApp1.html
```

The corresponding code in Figure 3.23 shows the necessary changes highlighted. The major change is that the type attribute has been changed from "number" to "range". On Chrome

FIGURE 3.21 Screenshot of the output of the velocity and angle that was input

Follow: File: ch03/physicsProjectileApp1/version2/physicsProjectileApp1.html

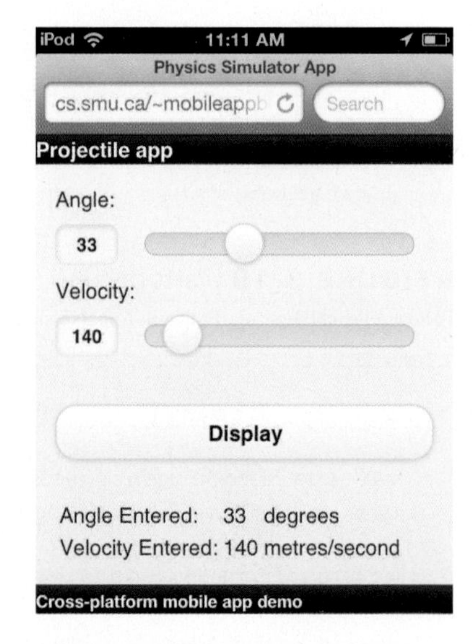

FIGURE 3.22 Screenshot of the input of the velocity and angle using slider and dialer

Follow: File: ch03/physicsProjectileApp1/version3/physicsProjectileApp1.html

```
<!DOCTYPE html>
<html>
<head>
 <title>Physics Simulator App</title>
 <link rel="stylesheet"
   href="css/jquery.mobile-1.3.1.min.css">
 <script type='text/javascript'
  src='scripts/physicsProjectileApp1.js'>
  </script>
 <script src="scripts/jquery-1.8.3.min.js">
 </script>
 <script
  src="scripts/jquery.mobile-1.3.1.min.js">
 </script>
 <meta name="viewport"
 content="width=device-width, initial-scale=1.0"></head>
<body>
 <div data-role="page">
  <div data-role = header>Projectile app</div>
  <div data-role="content">
   <label for="angle">Angle:</label>
   <input type='range' name='angle' id='angle'
    min='0' max='90'>
   <label for="velocity">Velocity:</label>
   <input type='range' name='velocity'
    id='velocity' min='0' max='999'>
   <br/>
   <button onclick="display();">
    Display
   </button>
   <br/>
   <table id="data">
    <tr>
     <td>Angle Entered:</td>
     <td id='angleDisplay'>0</td>
     <td>degrees</td>
    </tr>
    <tr>
     <td>Velocity Entered:</td>
     <td id='velocityDisplay'>0</td>
     <td>metres/second</td>
    </tr>
   </table>
  </div>
  <div data-role = footer>
       <small>Cross-platform mobile app demo</small>
  </div>
 </div>
</body>
</html>
```

FIGURE 3.23 HTML5 code for a slider to input velocity and angle

Follow: File: ch03/physicsProjectileApp1/version3/physicsProjectileApp1.html

browsers and all mobile devices, an input widget with the type range will display as a slider as shown in Figure 3.22. Another change we made was to change the maximum value for the velocity from a ridiculously large value of '299792458' (speed of light) to a more reasonable value '999'. The range slider will not have accepted such a large value. Moreover, very large values of maximum would mean that one has to move the slider very little to get even large changes in the values. The value of '999' may still be larger than what is realistic. The readers are encouraged to experiment with different values. The rest of the app works exactly as before and displays the values that were entered.

```
<!DOCTYPE html>
<html>
<head>
 <title>Physics Simulator App</title>
 <link rel="stylesheet"
   href="css/jquery.mobile-1.3.1.min.css">
 <script type='text/javascript'
  src='scripts/physicsProjectileApp1.js'>
  </script>
 <script src="scripts/jquery-1.8.3.min.js">
 </script>
 <script
  src="scripts/jquery.mobile-1.3.1.min.js">
 </script>
 <meta name="viewport"
 content="width=device-width, initial-scale=1.0">
</head>
<body>
 <div data-role="page">
  <div data-role="navbar">
   <ul>
    <li><a href="file:information.html">
     Information</a>
    </li>
    <li><a href="file:index.html"
     class="ui-btn-active ui-state-persist">
     Interface</a>
    </li>
   </ul>
  </div>
  <div data-role="content">
   <label for="angle">Angle:</label>
   <input type='range' name='angle' id='angle'
    min='0' max='90'>
   <label for="velocity">Velocity:</label>
   <input type='range' name='velocity'
    id='velocity' min='0' max='999'>
   <br>
   <!-- Rest of the code is the same as Figure 3.21 -->
```

FIGURE 3.24 HTML5 code for the navbar in the interface tab

Follow: File: ch03/physicsProjectileApp1/version4/index.html

So far we have only created a single web page. Generally, apps will have multiple pages that are connected to each other. Those who are familiar with conventional web design would expect a separate HTML5 file for each new page linked with a navigation widget. In this chapter, we will use this conventional approach based on a multi-file model. Later on in this book, we will see how to include multiple page-views in a single HTML5 file. Figure 3.24 shows the first part of the HTML5 document from the web directory

```
ch03/physicsProjectileApp1/version4/index.html
```

The corresponding screenshot is shown in Figure 3.25. We first should explain the choice of file name: index.html, which is used to indicate the main file in a folder. If we just type the URL of the folder such as

```
http://cs.smu.ca/~mobileappbook/ch03/physicsProjectileApp1/version4
```

The web server will automatically go to the index.html from the folder, if one exists. If the file index.html does not exist, we will see the listing of all the files in the folder.

Now let us look at the code in the file. Most of the code is the same as version3 of our app from Figure 3.23. We have highlighted the major change, which replaces the `<div>` section with a header data-role with a navbar data-role. This new `<div>` section contains an unordered list created using the familiar `<ul>...</ul>` pair of tags that contains two items created using the `<li>...</li>` pair of tags. Each item in the list is what is called as a hyperlink provided by a new pair of tags `<a>...</a>`. This pair of tags is used to create links to other pages. The URL for the page is specified with the attribute href. The first link in the list creates a hyperlink to the html page information.html. The second link is to the current page—that is, `index.html`. We have added another attribute to the second hyperlink. This attribute is called class and is set to the values `ui-btn-active` and `ui-state-persist`. The class attribute refers to the classes defined in the `jquery.mobile-1.3.1.min.css`. The values the class attribute is set to indicate that this link is the currently active button and `ui-state-persist` keeps it active while the page is displayed. Clicking on the first button called information will bring up the other page called information.html, which is discussed on the following page.

The information.html (Figure 3.26) page, found in the web directory in the folder

```
ch03/physicsProjectileApp1/version4,
```

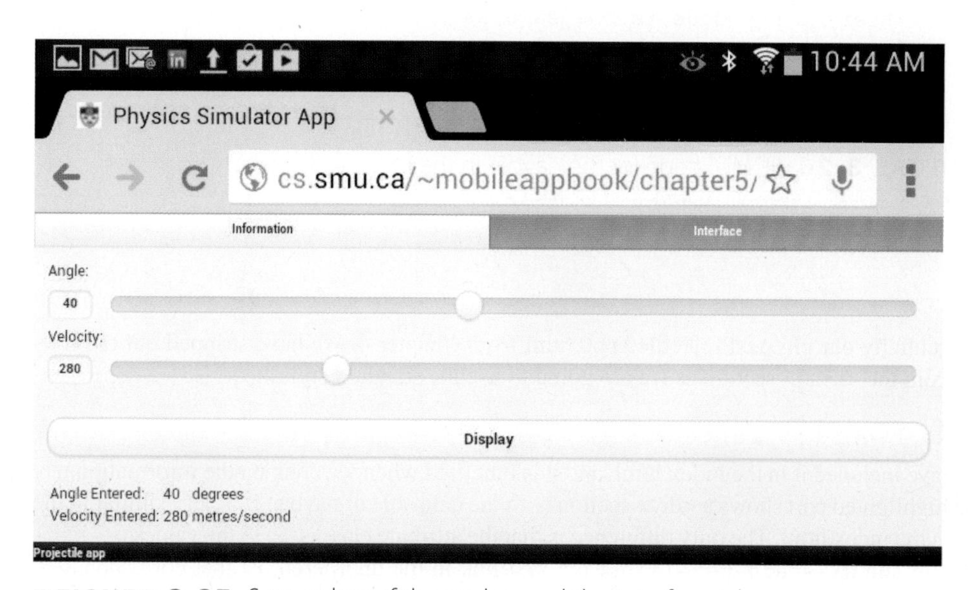

FIGURE 3.25 Screenshot of the navbar and the interface tab
Follow: File: ch03/physicsProjectileApp1/version4/index.html

```
<!DOCTYPE html>
<html>
 <head>
  <title>Physics Application</title>
 </head>
 <body>
  <div>
  <div data-role="navbar">
   <ul>
    <li><a href="file:information.html"
     class="ui-btn-active ui-state-persist">
     Information</a>
    </li>
    <li><a href="file:index.html">
     Interface</a>
    </li>
   </ul>
  </div>
  <h1>Projectiles</h1>

  <figure>
   <img src="media/projectilePic.png"
    alt="A Cannon Firing a Projectile"
    width="333" height="228">
    <figcaption><strong>Fig. 1 - A Projectile
    Fired From a Cannon</strong></figcaption>
  </figure>

  <p>In 1600s, armies used equations of motions
   to calculate velocities and angle for firing
   a missile to hit a target.

  <!--
     Rest of the code is the same as
     physicsProjectile0.html from Chapter 4
  -->
```

FIGURE 3.26 HTML5 code for the navbar in the information tab

Follow: File: ch03/physicsProjectileApp1/version4/information.html

is essentially our physicsProjectileApp0.html from Chapter 4. We have stripped out the CSS file from the <head> section as we are going to use the standard stylesheet

```
jquery.mobile-1.3.1.min.css
```

We have included it in the index.html, and it is inherited when we click on the information tab. The highlighted part shows a <div> section with the data-role of navbar. It is very similar to the one in the index.html. The only difference is that the attribute class is set to the values `ui-btn-active` and `ui-state-persist` for the first link in the unordered list that corresponds to the current file (i.e., information.html). The corresponding screenshot for the Information page is in Figure 3.27.

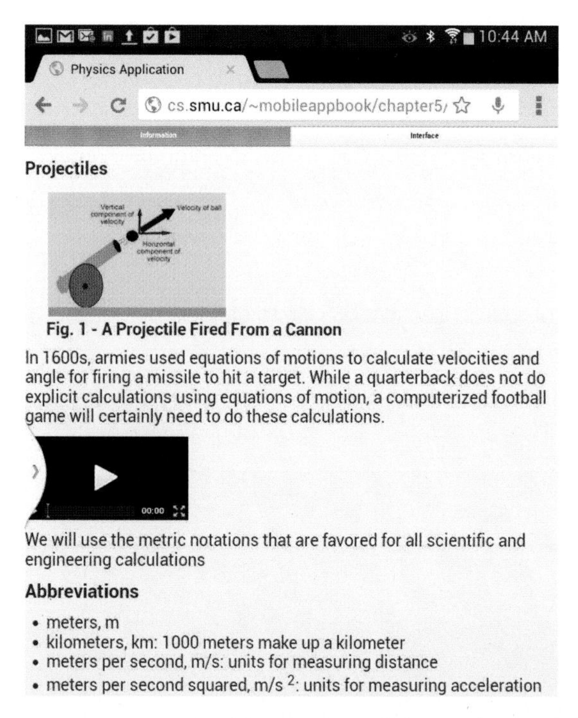

FIGURE 3.27 Screenshot of the navbar and the information tab
Follow: File: ch03/physicsProjectileApp1/version4/information.html

Quick facts/buzzwords

JavaScript: A programming language popularly used for web programming.

Client-side programming: Programs that run on the user (client) device.

Server-side programming: Programs that run on the web server.

High-level languages: Languages that use English-like words for computer programming. They need to be translated to the binary languages before they can be run on a device.

Compiled programs: Programs that are translated once and translated copy is run many times.

Interpreted programs: Programs that are translated one instruction at a time every time they are run.

Function: A program snippet that performs a particular task.

Variable: Memory location in a program to store value that is used in computations.

`<script> </script>`: A pair of HTML5 tags to include a script (program) in the web page.

jQuery Mobile: A JavaScript library that simplifies mobile app programming.

`data-role`: A jQuery-specific attribute for the `<div>` tag that allows us to specify specific roles such as `page`, `footer`, `header`, `footer`, `content`.

`<small> </small>`: A pair of HTML5 tags to reduce the font size of enclosed text.

Widget: Refers to various elements in the web page such as buttons, text boxes, or sliders.

`<input> </input>`: A pair of HTML5 tags to include widgets that facilitate input from the user.

<button> </button>: A pair of HTML5 tags to produce a button in a web page.

<label> </label>: A pair of HTML5 tags to produce a label in a web page, usually associated with a widget.

id: An attribute used in conjunction with a number of HTML5 elements to uniquely identify the element.

getElementById: A JavaScript function to access an HTML5 element using its id attribute value.

class: An attribute for an HTML5 element that is used in a CSS file to define a particular look and feel.

<a>...</a>: A pair of HTML5 tags used to include a hyperlink to other web resources.

Self-test exercises

1. What are high-level programming languages?
2. What are examples of high-level languages?
3. What are two different types of translations of high-level languages?
4. What type of translation does JavaScript use?
5. What HTML5 tag is used for including a script?
6. What are jQuery and jQuery Mobile?
7. What is client-side programming?
8. What is server-side programming?
9. What is the purpose of an id attribute?
10. What is a preferred way to add a label for an HTML5 widget?

Programming exercises

1. Write HTML5 code that will create the following display (using jQuery Mobile)

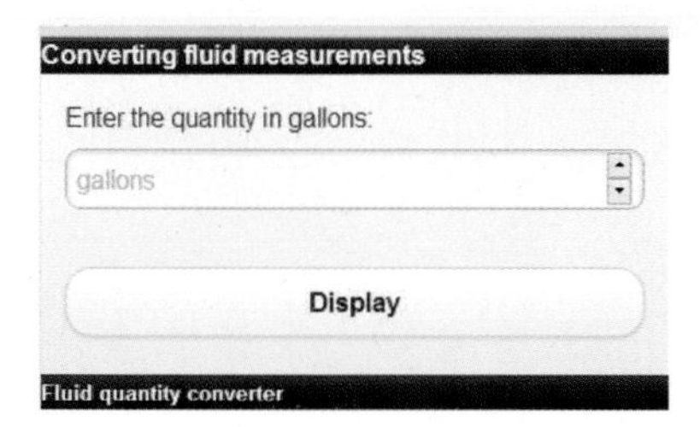

2. Take your app from PE1 and add a JavaScript function to the Display button to display the quantity entered

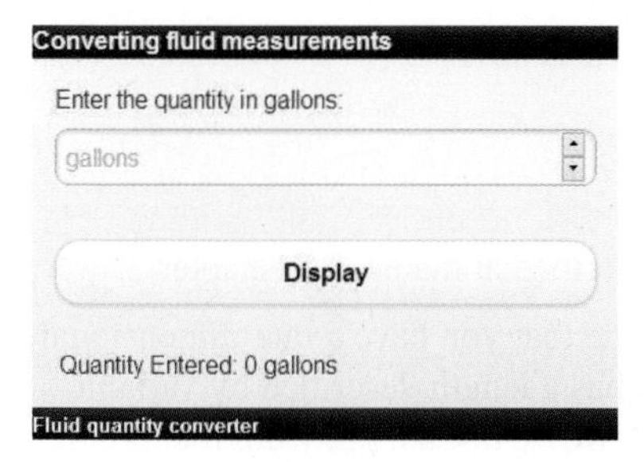

3. Modify your app from PE2 so that the quantity in gallons is input with a slider.

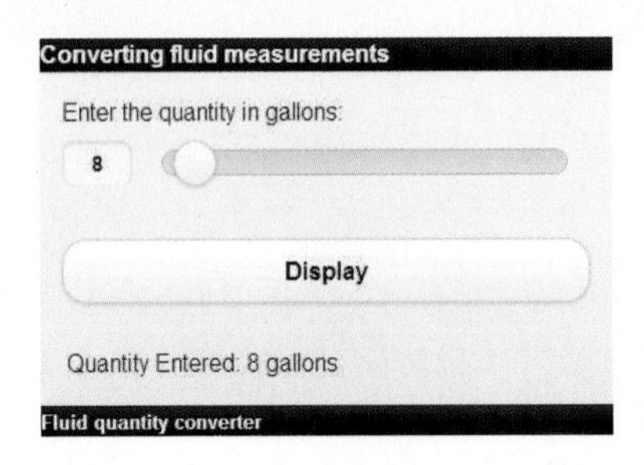

4. Write HTML5 code that will create the following linked pages. You can use the interface code from the previous exercise.

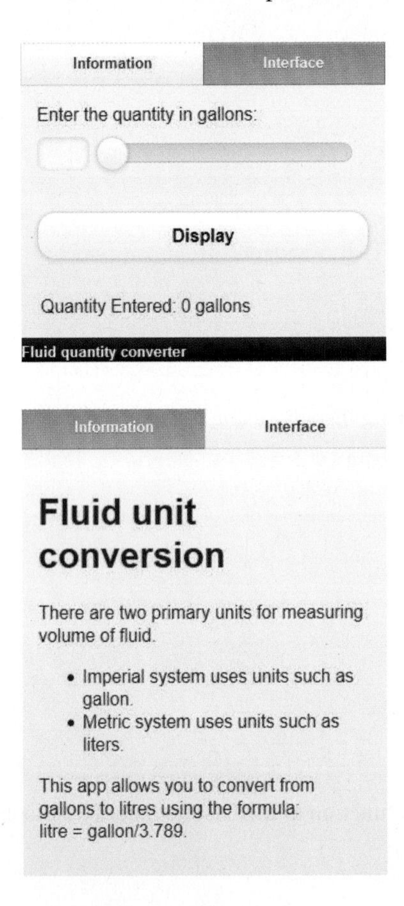

Programming projects

We build on the programming projects that were described at the end of Chapter 2.

1. **Compute load distribution on a beam**. Assume that you have a one-dimensional beam resting on two rigid supports. The beam has a length described by variable *length*. We have put a weight *w* at a distance *d*, where *d* < *length*, from the left end of the beam. You should find out the reaction forces on the left and right ends of the beam. You can make this project as complicated as you are capable of solving. You can have multiple weights, uniformly distributed load, and uniformly varying loads.

Add input widgets for all the input variables, and a button. Write a JavaScript function to display the input values when the button is clicked. Use the jQuery utility navbar to link the interface page with the information page that you have created for the project in Chapter 2. Test the app on all the mobile devices.

2. **Binary operator**. Create a web page that describes various binary operators such as "and", "or", "not", "xor".

   Add input widgets for two Boolean input variables, A and B. Add four buttons for "and", "or", "not", "xor". Write a JavaScript function to display the input values when any one of these buttons is clicked. Use the jQuery utility navbar to iink the interface page with the information page that you have created for the project in Chapter 2. Test the app on various mobile devices.

3. **Electricity calculations**: Create a web page that describes the physical relationship between, current, voltage, resistance, and amount of electricity consumed.

   Add input widgets for all the input variables and a button. Write a JavaScript function to display the input values when the button is clicked. Use the jQuery utility navbar to link the interface page with the information page that you have created for the project in Chapter 2. Test the app on various mobile devices.

4. **Amortization calculations**: Create a web page that describes how to compute the annual cost of a project that costs p and is funded by a loan at an interest i over y number of years.

   Add input widgets for all the input variables and a button. Write a JavaScript function to display the input values when the button is clicked. Use the jQuery utility navbar to link the interface page with the information page that you have created for the project in Chapter 2. Test the app on various mobile devices.

CHAPTER 4

Making apps do significant computing

WHAT WE WILL LEARN IN THIS CHAPTER

1. How to validate input
2. How to do computations based on input
3. How to do conditional computing
4. How to do iterative computing
5. How to use arrays in JavaScript

In this chapter, we are going to take our JavaScript programming to a new level. We will look at how to validate input through error checking. We will use conditional statements to validate our input. We will see how to tabulate our results through iterative computing. Finally, we will look at one of the most flexible array facility provided by JavaScript.

4.1 Temperature converter app design

The objective of our next application is to allow users to convert temperature from Fahrenheit to Celsius, and vice versa. We are designing this app to suit all types of mobile users (from inexperienced to advanced), so we must ensure that it is simple and intuitive. The app has no other functionality, so the entire app will be on one screen.

Users of our app will be able to input a temperature in one scale, either Celsius or Fahrenheit, and convert it to the other scale. Since Celsius is the most commonly used unit of temperature measurement worldwide, we will have our app set to convert temperature from Fahrenheit to Celsius by default. By setting this value as the default, we are making the most frequent task the most easily accessible within the user interface. It is important to keep in mind as a good guideline that the less work users need to do to achieve a task, the happier they will be. For example, imagine we designed the app as a series of screens, where users first have to select the temperature scale they want to convert to, then they click "next" to go to another screen where they input the temperature, then they click "next" to go to yet another screen to get the answer. This is much more tedious than having everything on one screen where users can input their value and (most likely) only have to tap "Calculate". The purpose of the app can also be grasped at a glance with this design, as the entire app can be seen on a single screen. Additionally, both the "Convert" button and the input textbox ending with the temperature unit are clearly labelled, so that no further instructions are required for the user.

Remember from when we last spoke about design that it is important to provide immediate feedback to users after any interaction. In this app, the user has the ability to toggle between the temperature scales for conversion. Whenever the user toggles between the two, there are two sets of feedback. First, the label beside the input textbox changes to the corresponding input temperature scale. Second, the dot beside the chosen temperature scale changes to show the option selected by the user. The second interaction point is the conversion button; as this calculation does not take a long time to process, there is no need for a loading indicator, and the result is displayed at the bottom of the page.

Finally, we want to keep in mind that this app is designed to be used on mobile devices. As such, our app will use big buttons that are easy to tap.

4.2 Simple JavaScript calculations for temperature conversion app

We start our exploration of JavaScript with one of the most popular scientific problems for introducing JavaScript calculations: a temperature convertor that converts temperature from Celsius to Fahrenheit and vice versa. We create a single screen that accepts the temperature and a radio button that tells us whether the temperature is in Fahrenheit or Celsius. Let us first discuss the highlighted portion of the HTML5 code from Figure 4.1. The header includes the jQuery

```
<!DOCTYPE html>
<html>
<head>
 <title>Temperature Converter</title>
 <link rel="stylesheet"
  href="css/jquery.mobile-1.3.1.min.css">
 <script src="scripts/jquery-1.8.3.min.js">
 </script>
 <script
 src="scripts/jquery.mobile-1.3.1.min.js">
 </script>
 <script type='text/javascript'
 src='scripts/temperatureConverter.js'>
 </script>
 <meta name="viewport"
 content="width=device-width, initial-scale=1.0">
</head>
<body onload="setup();">
 <div data-role="page">
 <div data-role="fieldcontain">
  <input type="number" id="temperature"
  name="temperature">
  <label id="label">&deg; F</label>
 </div>
 <fieldset data-role="controlgroup">
  <legend>Convert to:</legend>
  <input type="radio" name="units"
  id="farenheit" value="farenheit">
   <label for="farenheit">Farenheit</label>
  <input type="radio" name="units"
  id="celsius" value="celsius"
  checked="checked">
   <label for="celsius">Celsius</label>
 </fieldset>
 <input type="button" onclick="convert()"
  value="Convert">
 <p id="answer"></p>
 </div>
</body>
</html>
```

FIGURE 4.1 HTML5 code for Temperature Fahrenheit app

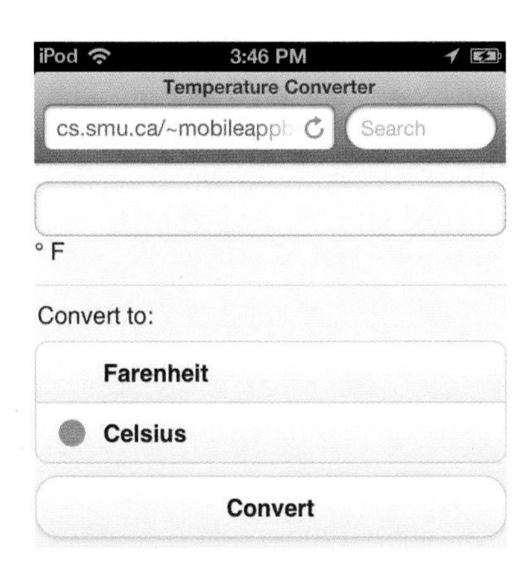

■ **FIGURE 4.2** Screenshot of Temperature Fahrenheit app

Mobile CSS file from the css subdirectory and jQuery scripts from the scripts subdirectory. In addition, we have our own JavaScript program in the file temperatureConverter js. We also added the following code to the `<head>` section

```
<meta name="viewport" content="width=device-width,
initial-scale=1.0">
```

As discussed in the previous chapter, this tells the mobile browsers that they only have the width of the device to render the page and not a full-size window. Finally, in the `<body>` tag we have an attribute called 'onload' that tells the browser to load the JavaScript function setup(), which is defined in temperatureConverter.js and will be discussed in detail later. The entire body is contained in a `<div>` section with the value of data-role attribute set to "page", which means the entire body will be a single page as we have seen earlier. We have also highlighted the closing HTML5 tags `</div>`, `</body>`, and `</html>`. Now that we have discussed the preliminary setup of the app, we will go through the body of the page and the associated JavaScript in detail.

Figure 4.3 shows the code segment for the input widget (in this case, a textbox) to accept temperature value and a corresponding descriptive label that indicates the units as Fahrenheit or Celsius. We have created a `<div>` element with the data-role defined as "fieldcontain". It is a jQuery Mobile convention to wrap a number of related form fields together with the data-role "fieldcontain". In this instance, we are grouping the textbox input element and corresponding label in such a `<div>` element. The `<input>` element has the type attribute set as "number", which indicates that it will accept numeric values. The id attribute is set to be "temperature", which will be used to refer to the element in our programs. The text of the label is set to be degree Fahrenheit (°F). Later on we will see how it is changed to Celsius (°C) using a JavaScript function. Here, we also see a special character in HTML5.

A special character in HTML5 begins with the ampersand sign (&) and ends with a semicolon (;). ° will give us the traditional symbol used to indicate temperature in degrees (°). A more detailed table of available symbols is given in Figure 4.4. The information in the table is obtained from www.w3schools.com. This website contains a complete collection of all the HTML5 special

```
<div data-role="fieldcontain">
  <input type="number" id="temperature"
    name="temperature">
  <label id="label">&deg; F</label>
</div>
```

■ **FIGURE 4.3** HTML5 code for the textbox for temperature

symbols. The table in Figure 4.4 consists of four columns. The first column shows the symbol as it will appear in the HTML5 browser. The second column shows the numeric code that can be used for the symbol. We put an ampersand followed by the hash mark (&#) before the numeric code and a semicolon after the code. It is much more readable to use a name shown in the third column as we have done for our degree symbol. The fourth column gives the definition for each of the symbols.

Character	Number	Name	Description
"	"	"	quotation mark
'	'	'	apostrophe
&	&	&	ampersand
<	<	<	less-than
>	>	>	greater-than
¢	¢	¢	cent
£	£	£	pound
¥	¥	¥	yen
©	©	©	copyright
¬	¬	¬	negation
®	®	®	registered trademark
°	°	°	degree
±	±	±	plus-or-minus
^{2}	²	²	superscript 2
^{3}	³	³	superscript 3
µ	µ	µ	micro
¼	¼	¼	fraction 1/4
½	½	½	fraction 1/2
¾	¾	¾	fraction 3/4
×	×	×	multiplication
÷	÷	÷	division

FIGURE 4.4 A table of frequently used symbols in HTML5
Based on http://www.w3schools.com

Let us now look at the HTML5 and JavaScript code behind the radio button in the app. Figure 4.5 shows the HTML5 code that is wrapped in a `<fieldset>` HTML5 element. This is an HTML5 element that is used to group related elements in a form. In our case, we are grouping two radio buttons. Immediately inside the `<fieldset>`, we have another tag `<legend>` that is used to define a caption for the `<fieldset>` element. We are using the caption "Convert to:" for our `<fieldset>`. The two radio buttons will be used to choose what is going to be the unit of the resulting converted temperature. We have two `<input>` elements in the `<fieldset>`; both of them have type attribute set to "radio", and they have the same name—that is, "units", but different values for id attribute (Fahrenheit and Celsius). The same name indicates that these two buttons form a group, and only one of them can be selected. We are also using two `<label>` elements; they specify the text that will go with these radio buttons. The "for" attribute for the label associates the label with the appropriate radio button. The value of the attribute "checked" for the Celsius button is set to "checked" by default. That means by default we will be converting Fahrenheit to Celsius. That is why, by default, the label

```
<fieldset data-role="controlgroup">
  <legend>Convert to:</legend>
  <input type="radio" name="units"
    id="farenheit" value="farenheit">
      <label for="farenheit">Farenheit</label>
  <input type="radio" name="units"
    id="celsius" value="celsius"
    checked="checked">
      <label for="celsius">Celsius</label>
</fieldset>
```

■ **FIGURE 4.5** HTML5 Code for the radio button for temperature types

for the input textbox above the radio buttons is set to °F. We can see this in the first screenshot in Figure 4.6. If we pick the Fahrenheit radio button as shown in the second screenshot in Figure 4.6, we need to change the label for the input textbox to °C. This is achieved through a couple of JavaScript functions.

We would like to draw the reader's attention to the `<body>` element of our web page shown in Figure 4.1

```
<body onload="setup();">
```

The value of the attribute "onload" tells the browser to run the JavaScript function setup(), which is shown in Figure 4.7, right away when the page is loaded. The function setup() defines what should be done when the two radio buttons are clicked. This is achieved

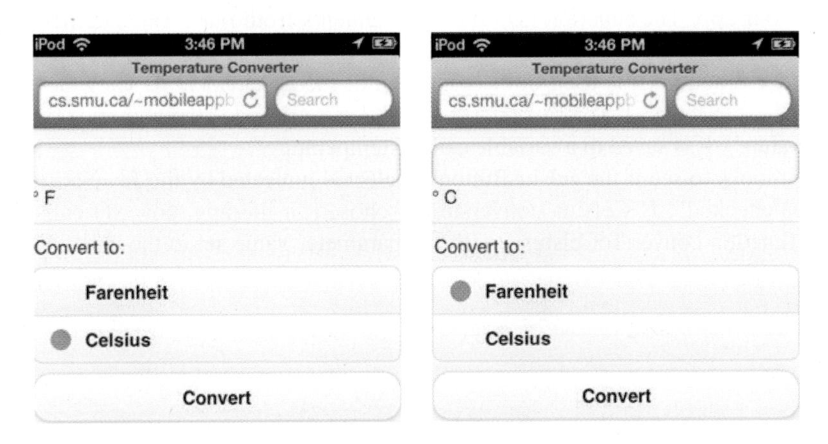

■ **FIGURE 4.6** Screenshot of switching temperature types

```
function setup()
{
  document.getElementById("farenheit").onclick=
    function(){setUnits("C");};
  document.getElementById("celsius").onclick=
    function(){setUnits("F");};
}
```

■ **FIGURE 4.7** JavaScript for the setup() function; executed when the page is loaded by first getting the element using the function `document.getElementById.` and then specifying what happens when it is clicked using the "onClick" data member. For example, `document.getElementById("farenheit").onclick` defines the action when the radio button with id "fahrenheit" is clicked. In our case, we call the function setUnits() by passing a string "C". Similarly, we indicate that the function setUnits() should be called with string "F" as a parameter, every time the user clicks on the radio button with id "Celsius".

The function setUnits() is shown in Figure 4.8. The function accepts one parameter called unit. In Figure 4.7, we saw two calls to the setUnits() function that set the unit parameter to either "C" or "F". Let us look at the body of the function setUnits(). It first retrieves the HTML5 element with the id "label", which the function stores in a variable also called label. The innerHTML component sets the text for the element with the id "label" with the unit parameter passed in the function. This makes the label for our textbox display "°F" when "F" is passed as a parameter. This result is shown by the first screenshot in Figure 4.6 when the Celsius radio button is chosen. Similarly, when the Fahrenheit radio button is selected in the second screenshot from Figure 4.6, setUnits() receives "C" as a parameter, and "°C" is displayed as the label.

Now that we have examined the JavaScript functions that take care of appropriate displays on the page, it is time to look at the elements that trigger actual computations and display results. Figure 4.9 shows the `<input>` element with attribute value of type "button", corresponding to the button at the bottom of the app. The value of the onclick attribute shows that the function convert() will be called when the button is clicked. The value attribute specifies the label that will be displayed on the button, "Convert". The paragraph element `<p>` that immediately follows the button has no text, but the id attribute is set to "answer". This id will be used by our JavaScript program to display the results in our app. Figure 4.10 shows the screenshot after entering a value with temperature of 30 with the radio button "Fahrenheit" checked, and clicking 'Convert'. The text after the convert button now shows the conversion results by displaying the text "30° Celsius converts to 86.0° Fahrenheit".

Figure 4.11 shows the JavaScript function convert() that is called when the "Convert" button is pressed in our app. The function gets the two elements from our form. The first element is the radio button with id "celsius". This element is saved in a variable called "celsiusButton" by our function. It is used in the condition of the if statement to see which type of conversion needs to be used. The second element that is retrieved is the numeric text box with id "temperature". It is saved in a variable called "temperature".

The if statement looks to see if the celsiusButton is selected indicated by the JavaScript code "celsiusButton.checked". If Celsius conversion is chosen in the app, covert() calls another JavaScript function convertToCelsius() with the parameter value set to the contents

```
function setUnits(unit)
{
  var label=document.getElementById("label");
  label.innerHTML="&deg; "+unit;
}
```

■ **FIGURE 4.8** JavaScript for the setUnits() function

```
<input type="button" onclick="convert()"
  value="Convert">
<p id="answer"></p>
```

■ **FIGURE 4.9** HTML5 Code to initiate temperature conversions and display results

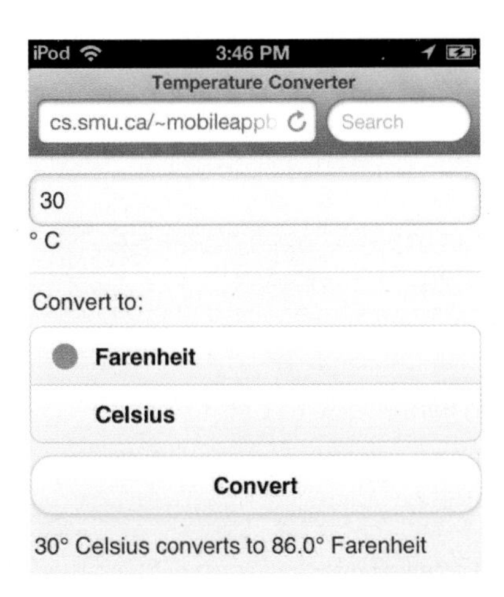

■ **FIGURE 4.10** Screenshot of the Temperature Converter app once the "Convert" button is pressed

```
function convert()
{
  var celsiusButton=document.getElementById("celsius");
  var temperature=document.getElementById("temperature");

  if(celsiusButton.checked)
  {
    convertToCelsius(temperature.value);
  }
  else
  {
    convertToFahrenheit(temperature.value);
  }
}
```

■ **FIGURE 4.11** JavaScript function that is called when the "Convert" button is pressed

of the temperature given by the JavaScript code "temperature.value". Since there are only two radio buttons—and hence only two choices—in our app, if the condition "celsiusButton.checked" is not true, it follows that Fahrenheit conversion must have been selected. Therefore, the else portion of the if statement calls the JavaScript function convertToFahrenheit() with "temperature.value" as the parameter.

Let the variable *fahrenheitTemperature* represent the temperature in Fahrenheit and the variable *celsiusTemperature* represent the temperature in Celsius. The well-known formula for the conversion from Fahrenheit to Celsius is as follows

$$celsiusTemperature = \frac{5}{9} \times (fahrenheitTemperature - 32)$$

The first line in the function convertToCelsius() implements the formula as shown in Figure 4.12a. Then the function retrieves the paragraph element with the id "answer" and sets its innerHTML attribute to the text that concatenates the two variables temperatureInFahrenheit and celsiusTemperature with the text "° Fahrenheit converts to " in between and appended by the text "° Celsius". We have already discussed that "°" gives us the degree symbol. One new feature in our text concatenation is the use of the JavaScript function toFixed(), which is used to decide how many decimal places should be used in our display. The code celsiusTemperature.toFixed(1) gives us one digit after the decimal point.

```
function convertToCelsius(temperatureInFahrenheit)
{
        var celsiusTemperature=(temperatureInFahrenheit-32)*5/9;
        document.getElementById("answer").innerHTML=
                temperatureInFahrenheit+"&deg; Fahrenheit converts to "+
                celsiusTemperature.toFixed(1)+"&deg; Celsius";
}
```

FIGURE 4.12a JavaScript function for conversion from Fahrenheit to Celsius

```
function convertToFahrenheit(temperatureInCelsius)
{
        var fahrenheitTemperature=temperatureInCelsius*9/5+32;
        document.getElementById("answer").innerHTML=
                temperatureInCelsius+"&deg; Celsius converts to "+
                fahrenheitTemperature.toFixed(1)+"&deg; Fahrenheit";
}
```

FIGURE 4.12b JavaScript function for conversion from Celsius to Fahrenheit

Conversely, the function convertToFahrenheit() shown in Figure 4.12b does the conversion from Celsius to Fahrenheit using the formula

$$fahrenheitTemperature = \frac{9}{5} \times celsiusTemperature + 32.$$

The structures of the two functions convertToCelsius() and convertToFahrenheit() other than the formulas used in the conversion are essentially the same.

4.3 Projectile app to calculate distance and height of a projectile

Now that we have seen some simple computations with JavaScript with the help of the temperature converter app, we will get back to our Physics Projectile app for more sophisticated and complex scientific computations. Figure 4.13 shows the HTML5 code for the second version of our Projectile app. It is very similar to the first version of the Projectile app, version 1. We have highlighted the important differences between the two. They are as follows

- Our JavaScript functions are in a file called physicsProjectileApp2.js
- The `<body>` tag has an attribute called "onload", which will call the JavaScript function "initialize()"
- The button in the form now is called "Calculate!" instead of "Display" and triggers a JavaScript function "update()"

The table after the button now will display the computed values of maximum height reached and total distance traveled if the projectile is launched. Therefore, the id attributes of the second column are set to "height" in the first row and "distance" in the second row. The readers may want to familiarize themselves with the overall structure of the HTML5 file before we look at the JavaScript functions associated with the app. A screenshot of when the app is first opened is shown in Figure 4.14. on page 74

```
<!DOCTYPE html>
<html>
<head>
  <title>Physics Simulator App</title>
  <script type='text/javascript'
    src='scripts/physicsProjectileApp2.js'>
    </script>
  <link rel="stylesheet"
    href="file:css/jquery.mobile-1.3.1.min.css">
  <script src="scripts/jquery-1.8.3.min.js">
  </script>
  <script
    src="scripts/jquery.mobile-1.3.1.min.js">
  </script>
  <meta name="viewport"
  content="width=device-width, initial-scale=1.0">
</head>
<body onload='initialize()'>
  <div data-role="page">
    <div data-role="content">
      <label>Angle:</label>
      <input type='number' name='angle'
        id='angle' min='0' max='90'
        placeholder='In degrees'>
      <label>Velocity:</label>
      <input type='number' name='velocity'
        id='velocity' min='0'
        max='299792458'
        placeholder='In metres/second'>
      <br>
      <button onclick='update();'>
        Calculate!
      </button>
      <br>
      <table id="data">
        <tr>
          <td>Max Height:</td>
          <td id='height'>0</td>
          <td>metres</td>
        </tr>
        <tr>
          <td>Distance Traveled:</td>
          <td id='distance'>0</td>
          <td>metres</td>
        </tr>
      </table>
    </div>
  </div>
</body>
</html>
```

■ **FIGURE 4.13** HTML5 code for the second Projectile app

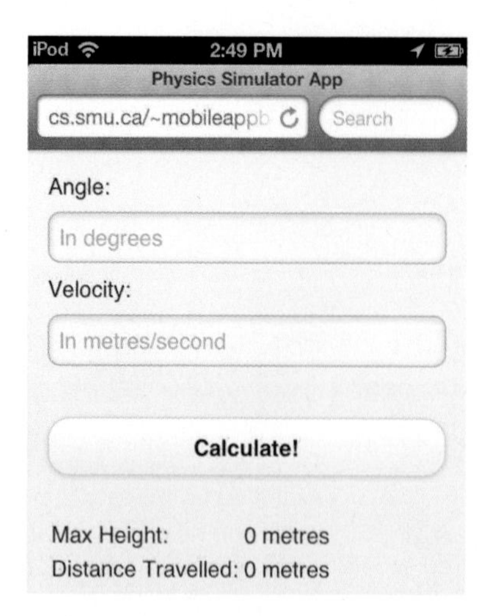

■ **FIGURE 4.14** Screenshot of the second Projectile app

4.4 Validating input values in an app using Boolean expressions in JavaScript

We have added error checking in the second Projectile app. The error checking is triggered by the JavaScript function "initialize()". The function is called when the app is loaded by the browser. The function is shown in Figure 4.15. It first gets the element with id "angle" (the numeric text box that receives the value of the angle) and saves it in a variable called angleInput. We then add an event listener for the element using the function addEventListener() as: `angleInput.addEventListener("blur", validateAngle);`

The first parameter "blur" indicates the event, which means when the element has lost focus; that is, the user enters a value in the angle textbox and then clicks away from it. The second parameter is the JavaScript function that will be called after the user has entered a value for the angle and clicks away from it. In our case, the function validateAngle() will be called. Similarly, the last two lines in the initialize() function in Figure 4.15 makes sure the function validateVelocity() will be called when the textbox with the id "velocity" is filled by the user.

Figure 4.16 shows the JavaScript function validateAngle() that is called when the user has entered the value of the angle. The first line of the function retrieves the text box element with the id "angle"; it is stored in a variable called angleInput. We then have an if statement that looks at the value of angleInput (angleInput.value). The condition of the if statement is a compound Boolean formula. It consists of two conditions

- angleInput.value < 1
- angleInput.value > 90

```
function initialize()
{
  var angleInput=document.getElementById("angle");
  angleInput.addEventListener("blur", validateAngle);

  var velocityInput=document.getElementById("velocity");
  velocityInput.addEventListener("blur", validateVelocity);
}
```

■ **FIGURE 4.15** JavaScript code for initialize() function that initiates error checking

```
function validateAngle()
{
  var angleInput=document.getElementById("angle");
  if(angleInput.value<1 || angleInput.value>90)
  {
    alert('Angle value must be between 1 and 90');
    angleInput.value="";
  }
}
```

■ **FIGURE 4.16** JavaScript code for validating angle

These expressions use the JavaScript comparison operators less than (<) and greater than (>). These operators return a true or false. A complete list of JavaScript comparison operators is given in Figure 4.17. There are a total of six comparison operators. One operator that needs particular mention is the == operator, which uses two equal signs and returns the value true when both sides of the operator are equal. It should not be confused with the operator with a single equal sign, =. The single equal sign is not used for comparison but for assigning the value on the right-hand side to the variable on the left-hand side. Readers who are familiar with C-like languages should note that JavaScript has two additional comparison operators. Three equal signs (===) means that the objects have to match in value as well as type. On the other hand, we have exclamation followed by two equal signs (!==), where either the value or type is different. This means than 0 will not match with '0', since one is a number and the other is a string.
The two conditions

- angleInput.value < 1
- angleInput.value > 90

in Figure 4.16 are joined by an operator called ||. This is an OR operator. Some of the readers may be familiar with the Boolean logic. However, we will provide a brief overview of the Boolean logic for completeness.

In Boolean logic, all values are binary, (i.e., true or false). We will use T and F for simplicity. The algebra on Boolean values typically uses three operators

1. OR, written in JavaScript as ||
2. AND, written in JavaScript as &&
3. NOT, written in JavaScript as !

Operator	Meaning
<	Less than
>	Greater than
<=	Less than or equal to
>=	Greater than or equal to
==	Equal (note two equal signs)
===	Exactly equal to (equal value and equal type) (note three equal signs)
!=	Not equal
!==	Not equal (different value or different type) (note two equal signs)

■ **FIGURE 4.17** Comparison operators in JavaScript

Let us define these three operators using two Boolean variables A and B. Just like variables in any other algebra these two variables can take any permissible value. In the case of Boolean logic, there are only two permissible values: T and F. Thus, A can take the value T or F. Similarly, B can take the value T or F. Figure 4.18 shows how the three operators work in Boolean algebra.

For two Boolean variables A and B, there are only four possible combinations as shown in Figure 4.18. Readers can take a moment to think of any other combination to convince themselves. The A||B (OR) operation is true (T) if either A or B is true (T). It is false only when both A and B are false. On the other hand, the A&&B (AND) operation is true only when both A and B are true (T). In all other cases, it is false (F). The !A (NOT) operation will take the value that is opposite of A. That is, when A is true (T) !A will be false (F) and vice versa.

Armed with this knowledge of Boolean algebra, let us revisit our condition in the function validateAngle()

```
angleInput.value < 1 || angleInput.value > 90
```

We are saying that the condition will be true if the angle entered is either less than 1 or greater than 90. If the condition is true, then the body of the if statement in curly braces will be executed. The body of the if statement consists of two statements. The first statement calls the JavaScript function alert by passing it the string 'Angle value must be between 1 and 90'. The alert function creates a pop-up window, displays the string that was passed to it, and waits for the user to press the OK button as shown in Figure 4.19. The second statement in the if condition resets the value of angle to an empty string "".

Readers may have noticed our use of both single quotes and double quotes for strings in JavaScript. Unlike some of the conventional programming languages, there is no difference between single and double quote in JavaScript. They can be used interchangeably. However, they should be used as a pair. You should not use a single quote to close a double quote and vice versa.

A	B	A || B	A && B	!A
T	T	T	T	F
T	F	T	F	F
F	T	T	F	T
F	F	F	F	T

FIGURE 4.18 Boolean operators in JavaScript

FIGURE 4.19 Screenshot with invalid angle

Figure 4.20 shows the code for the JavaScript function validateVelocity(), which ensures that the velocity is greater than 0 and less than 299,792,458 (speed of light). The structure of the function is similar to validateAngle(), which we saw in Figure 4.16. There is one important difference. We have what are called nested if statements (Figure 4.21). That is, first we check to see if the velocity is less than 0. If it is, we output an appropriate message. Otherwise, we enter the else part, where we have another (nested) if statement that checks to see if the velocity is larger than 299,792,458. If it is, we trigger an alert window with the message that the velocity is too high. This type of nesting of if statements as part of else can continue like a daisy chain if there are more conditions that need to be checked. Since we invoke the validation when the textbox has lost focus, we can see that just moving to a different element without entering a value for the velocity will give us an error message in Figure 4.21. The result of entering a very high value of velocity is shown in Figure 4.22.

```
function validateVelocity()
{
  var velocityInput=document.getElementById("velocity");
  if(velocityInput.value<1)
  {
    alert('Velocity value must be greater than 0')
    velocityInput.value="";
  }
  else if(velocityInput.value>299792458)
  {
    alert('Too fast! The velocity value cannot exceed[...]');
    velocityInput.value="";
  }
}
```

FIGURE 4.20 JavaScript code for validating angle and velocity

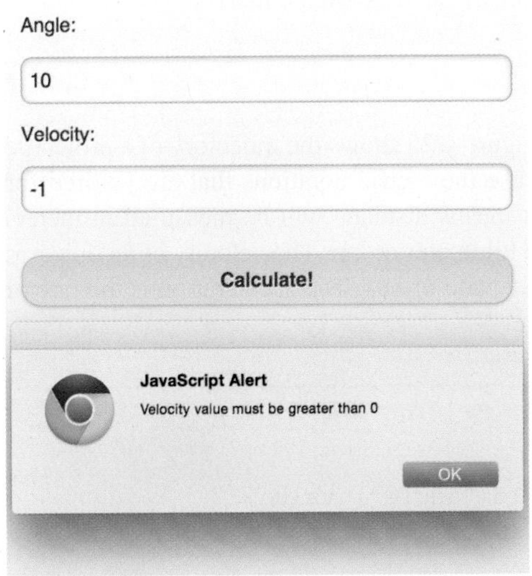

FIGURE 4.21 Screenshot when leaving velocity box without entering a value

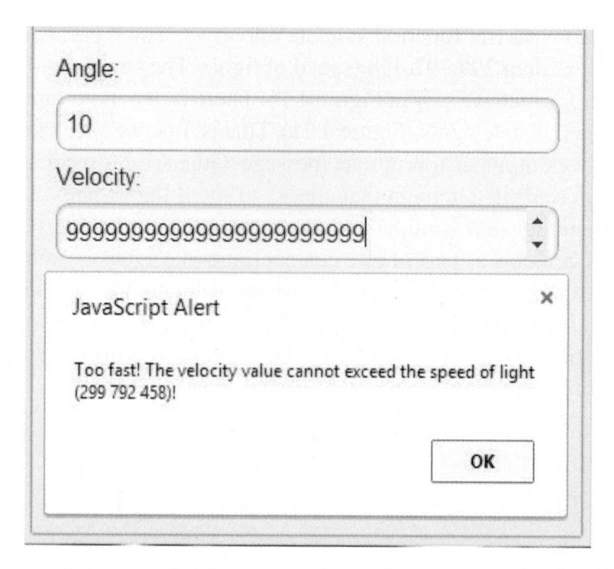

■ **FIGURE 4.22** Screenshot after entering high velocity value

4.5 Calculation of distance and height of a projectile using JavaScript Math object

Once the user has entered acceptable values for angle and velocity and pressed the "Calculate!" button, the JavaScript function update() shown in Figure 4.23 will be called. This is caused by the specification of the onclick attribute of the `<button>` element, which is set to the function update(). The update() function retrieves the value of the element with id "angle" and stores in a variable called angle. It then retrieves the value of the element with id "velocity" and stores in a variable called velocity. Another JavaScript function called calculate() is called by passing the variables angle and velocity. The function calculate(), which we will see soon, does all the calculations and displays the values of maximum height and distance as shown in Figure 4.24. The user has entered an angle of 30 and velocity of 100. The calculate() function displayed the maximum height to be 127.42 meters and distance traveled as 882.80 meters.

Physics behind projectile tracking

Let us refresh some of our physics knowledge. Figure 4.25 shows the trajectory of a projectile with velocity u fired at an angle of θ. We will use the metric notations that are favored for all scientific and engineering calculations. That means distance will be measured in meters abbreviated as m, with 1000 meters making up a kilometer or km. The velocity of an object is measured using distance in meters traveled in a second or m/s. The acceleration is the rate of

```
function update()
{
  var angle=document.getElementById("angle").value;
  var velocity=document.getElementById("velocity").value;
  calculate(angle, velocity);
}
```

■ **FIGURE 4.23** JavaScript code that is run when the calculate! button is pressed

iPod 2:49 PM
Physics Simulator App
cs.smu.ca/~mobileappt Search
Angle:
30
Velocity:
100
Calculate!
Max Height: 127.42 metres
Distance Travelled: 882.80 metres

FIGURE 4.24 Screenshot when Calculate! button is pressed

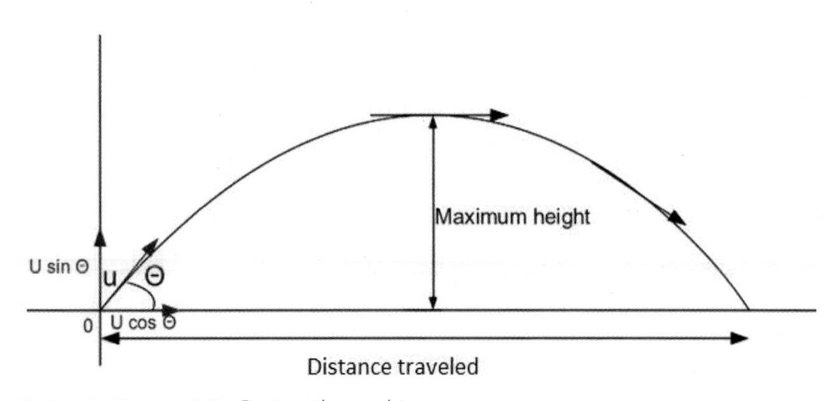

FIGURE 4.25 Projectile tracking

change of velocity measured as m/s^2. Let us use familiar physics notations: Initial velocity: u, final velocity: v, acceleration: a, time: t, distance: s
Some of the useful equations of motion for us are

1. $a = \frac{(v - u)}{t}$, which can be rearranged to get the following equation
2. $t = \frac{(v - u)}{a}$; this can be further rearranged as
3. $v = u + a \times t$.
4. $s = u \times t + \frac{1}{2} \times a \times t^2$ is another useful equation of motion. We will not derive this equation here. Readers can find a derivation of this equation in any introductory book on physics.

We will further denote the gravitational acceleration $g = 9.81\,m/s^2$ if the object is falling down. If the object is going up, the gravitational acceleration is in fact deceleration—that is, $g = -9.81\,m/s^2$.

Since the calculations are a little complicated, we will first write an algorithm for the calculations. An algorithm is written in a natural language such as English. An algorithm has two important features

1. It is a sequence of unambiguous steps.
2. When these steps are executed, the algorithm should produce correct results for all possible cases/inputs.

The algorithm shown in Figure 4.26 shows the physics calculations with a focus on calculating the maximum height and distance reached by the projectile. We assume that the projectile is shot upwards at an angle θ and initial velocity *u*.

Since the projectile is initially going upwards, gravitational acceleration is in fact deceleration given by $g = -9.81\, m/s^2$. The algorithm first computes horizontal velocity as

$$u_x = u \times cos(\theta), \text{ and vertical velocity}$$

$$u_y = u \times sin(\theta).$$

The projectile will continue until the vertical velocity $u_y = 0$.

Input: Velocity *u*, Angle θ

Known constant: Gravitational acceleration $g = -9.81\, m/s^2$

Output: *height* and *distance*

Algorithm:

1. Calculate the horizontal velocity $u_x = u \times cos(\theta)$
2. Calculate the vertical velocity $u_x = u \times sin(\theta)$
3. Calculate the time until the projectile reaches its maximum height—that is, when the final vertical velocity comes down to zero due to gravitation.

   We use the equation $t = \frac{(v - u)}{a}$, where $a = g = -9.81, v = 0, u = u_y$
4. $t_{maxHeight} = \frac{(0 - u_y)}{-9.81} = \frac{-u_y}{-9.81} = \frac{u_y}{9.81}$
5. Calculate the time until the projectile lands; it will be twice $t_{maxHeight}$ i.e., *maxHeight*
6. $t_{landing} = 2 \times t$
7. *height* is calculated using $s = u \times t + \frac{1}{2} \times a \times t^2$, where $s = height$,

   $a = g = -9.81, t = t_{maxHeight}, u = u_y$
8. $height = u_y \times t_{maxHeight} - \frac{1}{2} \times 9.81 \times t^2_{maxHeight}$
9. *distance* is calculated using $s = u \times t + \frac{1}{2} \times a \times t^2$,

   where $s = distance, a = 0, t = t_{landing}, u = u_x$

$$distance = u_x \times t_{landing} + \frac{1}{2} \times 0 \times t^2_{landing} = u_x \times t_{landing}$$

■ **FIGURE 4.26** Algorithm for calculating distance and height of a projectile

At that point we use the equation

$$t = \frac{(v - u)}{a}, \text{ where } a = g = -9.81, v = 0, u = u_y \text{ and calculate}$$

$$t_{maxHeight} = \frac{(0 - u_y)}{-9.81} = \frac{-u_y}{-9.81} = \frac{u_y}{9.81}$$

The time until the projectile lands will be twice $t_{maxHeight}$, (i.e., $t_{landing} = 2 \times t_{\max Height}$).
Now that we know the time for reaching the maximum height, *height* is calculated using

$$s = u \times t + \frac{1}{2} \times a \times t^2, \text{ where } s = height, a = g = -9.81, t = t_{maxHeight}, u = u_y$$

$$height = u_y \times t_{maxHeight} - \frac{1}{2} \times 9.81 \times t_{maxheight}^2$$

Similarly, *distance* is calculated using

$$s = u \times t + \frac{1}{2} \times a \times t^2,$$

where $s = distance$, and horizontal acceleration $a = 0, t = t_{landing}, u = u_x$.
That leads to

$$distance = u_x \times t_{landing} + \frac{1}{2} \times 0 \times t_{landing}^2 = u_x \times t_{landing}$$

Figure 4.27 shows the JavaScript calculate() function that follows the algorithm from Figure 4.26 to calculate and display maximum height reached and total distance traveled by the projectile. Let us look at each of the steps as we are using a number of new JavaScript features in this function. The first statement in the calculate() function takes the value of velocity and multiplies it by the cosine of the angle to calculate the horizontal velocity and stores it in a variable called horizontalVelocity. In order to use the trigonometric functions we use a very useful JavaScript object called Math. Let us look at the use of the Math object in the first statement in the function

```
Math.cos((anglePMath.PI)/180)
```

We are using two features of the Math object: a property and a method. The properties and methods are specified by appending a period and the name of the operator (i.e., property or method) to the Math object. Math.PI is a property or constant. It is static and will give us

```
function calculate(angle, velocity)
{
  var horizontalVelocity=velocity*Math.cos((angle*Math.PI)/180);
  var verticalVelocity=velocity*Math.sin((angle*Math.PI)/180);
  var tMaxHeight=verticalVelocity/9.81;
  var tLanding=2*tMaxHeight;

  document.getElementById('height').innerHTML=
    calcHeight(verticalVelocity, tMaxHeight);

  document.getElementById('distance').innerHTML=
    calcDistance(horizontalVelocity, tLanding);
}
```

FIGURE 4.27 JavaScript code to calculate and display height and distance of projectile

the same value in every case. Math.cos(), on the other hand, is a method or what we also refer to as a function. It returns a value that is dependent on the parameters that are passed to it. The function Math.cos() accepts angles in radians. That is why we had to convert our angle in degrees to that in radian as $\frac{angle \times \pi}{180}$ or in JavaScript as `(angle*Math.PI)/180`. Figure 4.28 gives all the constants for the JavaScript Math object from www.w3schools.com, and all the functions for the Math object are described in Figure 4.29.

Property/Constant	Description
Math.E	Euler's number
Math.LN2	natural logarithm of 2
Math.LN10	natural logarithm of 10
Math.LOG2E	base-2 logarithm of E
Math.LOG10E	base-10 logarithm of E
Math.PI	Π
Math.SQRT1_2	square root of 1/2
Math.SQRT2	square root of 2

■ **FIGURE 4.28** JavaScript Math object properties or constants
Based on http://www.w3schools.com

Method/function	Description
Math.abs(x1)	Returns the absolute value of x1
Math.acos(x1)	Returns the arccosine of x1, in radians
Math.asin(x1)	Returns the arcsine of x1, in radians
Math.atan(x1)	Returns the arctangent of x1 as a numeric value between -PI/2 and PI/2 radians
Math.atan2(x2,x1)	Returns the arctangent of the quotient of its arguments
Math.ceil(x1)	Returns x1, rounded upwards to the nearest integer
Math.cos(x1)	Returns the cosine of x1 (x1 is in radians)
Math.exp(x1)	Returns the value of E^{x1}
Math.floor(x1)	Returns x1, rounded downwards to the nearest integer
Math.log(x1)	Returns the natural logarithm (base E) of x1
Math.max(x1,x2,x3,...,xn)	Returns the number with the highest value
Math.min(x1,x2,x3,...,xn)	Returns the number with the lowest value
Math.pow(x1,x2)	Returns the value of x1 to the power of y
Math.random()	Returns a random number between 0 and 1
Math.round(x1)	Rounds x1 to the nearest integer
Math.sin(x1)	Returns the sine of x1 (x1 is in radians)
Math.sqrt(x1)	Returns the square root of x1
Math.tan(x1)	Returns the tangent of an angle

■ **FIGURE 4.29** JavaScript Math object methods or functions
Based on http://www.w3schools.com

Let us continue with our analysis of the calculate() function from Figure 4.27. The second statement uses the sine function to calculate the vertical velocity of the projectile and is stored in a variable called verticalVelocity. The third statement calculates the time required to reach the maximum height by dividing the vertical velocity by the gravitational deceleration. It is stored in a variable called tMaxHeight. We calculate the time for landing as twice the time for reaching the maximum height and store it in a variable called tLanding.

The statement

```
document.getElementById('height').innerHTML=
calcHeight(verticalVelocity, tMaxHeight);
```

gets the value of the height returned by another function called calcHeight() and assigns it to the innerHTML property of the HTML5 element with id "height" (it is a field in our app shown in Figure 4.13).

Finally, the last statement in calculate() function gets the value of the distance returned by another function called calcDistance() and assigns it to the innerHTML property of the HTML5 element with id "distance" (it is a field in our app shown in Figure 4.13).

The calculations of height and distance are relatively simple. However, we have separated them into two JavaScript functions calcHeight() (shown in Figure 4.30) and calcDistance() (shown in Figure 4.31). The calculations in these functions are based on our algorithm from Figure 4.26. These separate functions make our calculate() function a little less complex and follow good coding practices of modularity. More importantly, these are generic functions for calculation of height and distance that can be reused. We will next see how they can be recycled in the next version of the app, where we output various values of height and distance from time of launch to time of landing.

```
function calcDistance(horizontalVelocity, time)
{
  var distance=horizontalVelocity*time;
  return distance;
}
```

■ **FIGURE 4.30** JavaScript code to calculate distance of projectile

```
function calcHeight(verticalVelocity, time)
{
  var height=(verticalVelocity*time)-(0.5*9.81*time*time);
  return height;
}
```

■ **FIGURE 4.31** JavaScript code to calculate height of projectile

4.6 JavaScript arrays for storing distances and heights over time of a projectile

We are now going to modify our app one last time in this chapter to track the projectile throughout its flight. We will tabulate the height and distance traveled by the projectile. In a subsequent chapter, we will revisit this app one more time to graphically draw the projectile track. Figure 4.32 shows the HTML5 code for the app. The HTML5 code is very similar to the

```
<html>
<head>
  <title>Physics Simulator App</title>
  <script type='text/javascript' src='physicsProjectileApp2.js'>
</script>
  <link rel="stylesheet" href="file:jquery.mobile-1.3.1.css" />
  <script src="jquery-1.8.3.min.js"></script>
  <script src="jquery.mobile-1.3.1.min.js"></script>
</head>
<body onload='initialize()'>
  <div data-role="page">
    <div data-role="content">
      <p>Angle:</p>
      <input type='number' name='angle' id='angle'
        min='0' max='90' placeholder='In degrees'>
      <p>Velocity:</p>
      <input type='number' name='velocity' id='velocity'
        min='0' max='299792458'
        placeholder='In metres/second'>
      <br/>
      <button onclick='update();'>Calculate!</button>
      <br/>
      <table id="data" border="1">
      </table>
  </div>
</body>
</html>
```

FIGURE 4.32 HTML5 code for third Projectile app, which displays a table

first version of the app. The only difference is the table with id "data" (highlighted). We have deleted rows and columns in this table. The reason is that we have no a priori knowledge of the number of rows we will have in the table. It will depend on the values of inputs entered by the user. Therefore, our JavaScript function will dynamically add the rows and columns after calculation. Figure 4.33 shows a screenshot of the app after the user has entered an angle of 42 and velocity of 63 and clicked 'Calculate!'. We can see a table with columns for time, height, and distance.

Before we look at the next version of the app, we will provide an overview of arrays in JavaScript. Figure 4.34 highlights some of the aspects of the JavaScript array through a JavaScript command line session. This command line session allows us to type in JavaScript statements and see the results. Like every other variable in JavaScript, an array variable is declared as var. As soon as we declare the variable we assign it an array value. For example, in the first line in the session

```
var a = []
```

makes an array variable called a. It is assumed that readers have some exposure to arrays in another programming language. An array is a convenient data structure provided by most programming languages to store a sequence of n elements such as: $a_0, a_1, a_2, \ldots, a_{n-1}$. Note that since we started with the index of the first element to be 0, the index of the last (n-th) element is $n-1$. In this case, we do not know the value of n prior to writing the program. So we start with $n = 0$ by default by creating an empty array. We can find the length of the array by looking at the length property

```
a.length
```

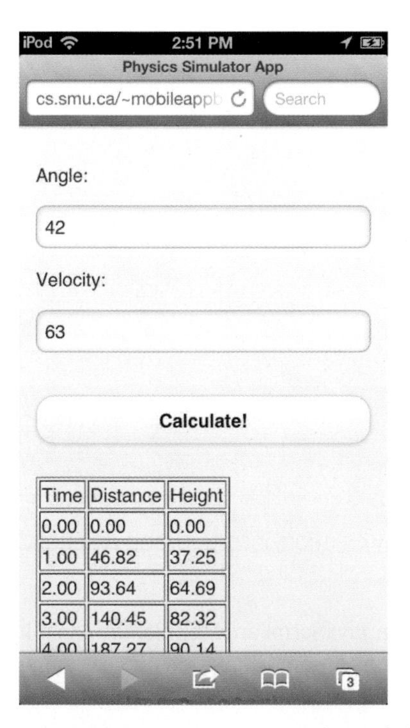

■ **FIGURE 4.33** Screenshot of third Projectile app, which displays a table

```
js> var a = [];
js> a.length
0
js> a[10] = "This is a string";
This is a string
js> a.length
11
js> a
,,,,,,,,,,This is a string
js> a[7] = 78.5
78.5
js> a[5] = 65
65
js> a
,,,,,65,,78.5,,,This is a string
js> a[3] = a[5]*a[7]
5102.5
js> a
,,,5102.5,,65,,78.5,,,This is a string
js> a.push("Appending a string");
12
js> a.length
12
js> a
,,,5102.5,,65,,78.5,,,This is a string,Appending a string
js> a.unshift(90);
13
js> a.length
13
js> a
```

```
90,,,,5102.5,,65,,78.5,,,This is a string,Appending a string
js> a.pop()
Appending a string
js> a.length
12
js> a
90,,,,5102.5,,65,,78.5,,,This is a string
js> a.shift()
90
js> a.length
11
js> a
,,,5102.5,,65,,78.5,,,This is a string
```

FIGURE 4.34 A JavaScript session to illustrate array creation, insertion, and deletion

We can see that the array length is 0 by the output. Arrays in JavaScript are flexible. We can put any data type in them. This is a major departure from the conventional compiled programming languages such as C, C++, or Java. The length of JavaScript arrays is also flexible. Let us look at the third statement in Figure 4.34

```
a[10]="This is a string"
```

Our array 'a' had a length of 0, yet we managed to insert a string at index 10. JavaScript adjusted the array to length 11. Following the arrays in C, JavaScript also begins with an index of 0. If we have elements up to and including index 10, then the array will have a length of 11. As can be seen by the following statement

```
a.length
```

which returned a length of 11. We can look at the array, which now has 10 commas for the first 10 elements indicating that these elements are empty or null. The 11th element (at index 10) is the string 'This is a string" that we assigned earlier.

We can continue to assign elements as shown in the following two statements

```
a[7]  = 78.5
a[5]  = 65
```

We now have a mix of numbers and strings in our array

```
,,,,,65,,78.5,,,This is a string
```

We can verify that a[5] and a[7] are indeed numbers and not strings of digits with some arithmetic

```
a[3]=a[5]*a[7]
```

We then verify that a[3] is indeed a result of the numeric operation—that is, 5102.5.

Most of the operations we have done so far were a little silly. It is not often that we want to mix numbers and strings in an array or randomly add an element at index 10 when originally the array was empty. We used these operations just to make readers aware of the flexibility of JavaScript arrays. Most of the time insertions and deletions will be either at the beginning or the end of the array. The statement

```
a.push("Appending a string")
```

adds a new element at the end of the array. The following two statements show that now the array length has gone up from 11 to 12, and the new element is indeed at the end of the array. We demonstrate insertion of an element at the beginning of an array with the statement

```
a.unshift(90)
```

We verify that 90 was added at the beginning of the array, and the length of the array became 13.

The deletion of an element from the end is done using

```
a.pop()
```

which deletes the last element and returns its value. We can see that now the array length is 12, and the last element is gone. For deleting the first element we use

```
a.shift()
```

which deletes the first element and returns its value. Now the array has 11 elements.

Figure 4.35 shows a few other advanced operations on our array. First, we use a.reverse() to reverse our array. If we wanted to extract part of the array we use a slice() method. For example,

```
b = a.slice(0,5)
```

```
js> a.reverse()
This is a string,,,78.5,,65,,5102.5,,,
js> a
This is a string,,,78.5,,65,,5102.5,,,
js> b = a.slice(0,5)
This is a string,,,78.5,
js> b
This is a string,,,78.5,
js> b.length
5
js> c = a.slice(5,a.length)
65,,5102.5,,,
js> c.length
6
js> d = b.concat(c)
This is a string,,,78.5,,65,,5102.5,,,
js> d
This is a string,,,78.5,,65,,5102.5,,,
js> a
This is a string,,,78.5,,65,,5102.5,,,
js> e=["c","c++","java","javascript","jquery","node.js"]
c,c++,java,javascript,jquery,node.js
js> e.push("actionscript")
7
js> e.push("c#")
8
js> e
c,c++,java,javascript,jquery,node.js,actionscript,c#
js> e.sort()
actionscript,c,c#,c++,java,javascript,jquery,node.js
js> e
actionscript,c,c#,c++,java,javascript,jquery,node.js
```

FIGURE 4.35 A JavaScript session to illustrate advanced array operations

creates a new array called b with five elements and copies elements a[0], a[1], a[2], a[3], a[4] in the array. Please note that the second argument 5 is the end point, but it is not included in the slice. Similarly, we put the second half of the array a in a new array c with the statement

```
c = a.slice(5,a.length)
```

We can look at the array c and verify that its length is 6. If we wanted to join (concatenate) array b with elements of array c, we can use the concat() method.

```
d = b.concat(c)
```

Note that concat() does not change array b. It only returns the joint array, which we assigned to d. We can verify that d is indeed the concatenated array (and the same as a).

The following statement in Figure 4.35, which creates an array called e,

```
e=["c","c++","java","javascript","jquery","node.js"]
```

shows that we can create an array and also assign values at the time of creation. It is a list of strings with names of some of the languages we have discussed in this book. You may have noticed that this array is alphabetically sorted.

The following two statements add two more strings, "actionscript" and "c#", at the end

```
e.push("actionscript")
e.push("c#")
```

If we look at the array e, it is no longer alphabetically ordered. We can reorder it as

```
e.sort()
```

Not only do we get the sorted array as a result, but e is also changed—that is, alphabetically sorted.

We have looked at examples of most of the array operations in JavaScript. A comprehensive list of array operations from www.w3schools.org is reproduced in Figure 4.36. www.w3schools.org provides further details about each of these operations.

Method	Description
concat()	Returns a copy of the joined arrays
indexOf()	Searches the array for an element and returns its position
join()	Joins all elements of an array into a string
lastIndexOf()	Searches the array for an element, starting at the end, and returns its position
pop()	Removes the last element of an array
push()	Adds new elements to the end of an array

reverse()	Reverses the order of the elements in an array
shift()	Removes the first element of an array and returns that element
slice()	Selects a part of an array and returns the new array
sort()	Sorts the elements of an array
splice()	Adds/Removes elements from an array
toString()	Converts an array to a string and returns the result
unshift()	Adds new elements to the beginning of an array and returns the new length
valueOf()	Returns the primitive value of an array

FIGURE 4.36 Operations on an array object in JavaScript
Based on http://www.w3schools.com

4.7 JavaScript for loop for repeating computations

Now armed with the knowledge of JavaScript arrays, we can study the enhanced JavaScript functions that draw a table that tracks the movement of the projectile over time. In this version of the app, we use most of the JavaScript functions from the previous version, including

- initialize()
- validateAngle()
- validateVelocity()
- update()
- calcDistance()
- calcHeight()

We will rewrite the calculate() function and add one more function called updateTable(), which will be called from the calculate() function. Figure 4.37 shows the modified version of our calculate() function. The first four statements in the function are the same as the previous calculate() function. The first statement takes the value of the velocity and multiplies it by the cosine of the angle to calculate the horizontal velocity and stores it in a variable called horizontalVelocity. The second statement uses the sine function to calculate the vertical velocity of the projectile and is stored in a variable called verticalVelocity. The third statement calculates the time required to reach the maximum height by dividing the vertical velocity by the gravitational deceleration. It is stored in a variable called tMaxHeight. We calculate the time for landing as twice the time for reaching the maximum height and store it in a variable called tLanding in the fourth statement.

The next three statements create three array variables called heightArray, distanceArray, and timeArray. All the three arrays are empty since we do not know how many elements each of these arrays will have. That decision depends on the value of the time of landing, that is, tLanding.

We are going to divide the time from 0 to tLanding into reasonable time intervals. The following set of if. . .else if. . .else statements in Figure 4.37 show that if the time for landing is less than 2, we will set the time interval to be relatively small at 0.1. Otherwise, if the time for landing is greater than or equal to 2 and less than 20, we set the time interval to be 1. Finally,

```
function calculate(angle, velocity)
{
  var horizontalVelocity=velocity*Math.cos((angle*Math.PI)/180);
  var verticalVelocity=velocity*Math.sin((angle*Math.PI)/180);
  var tMaxHeight=verticalVelocity/9.81;
  var tLanding=2*tMaxHeight;
  var heightArray=[];
  var distanceArray=[];
  var timeArray=[];

  if(tLanding<2)
  {
    var interval=0.1;
  }
  else if(tLanding<20)
  {
    var interval=1;
  }
  else
  {
    var interval=10;
  }

  for (var time=0; time<=tLanding+interval; time+=interval)
  {
    timeArray.push(time);

    var height=calcHeight(verticalVelocity, time);

    if(height<0)
    {
      height=0;
    }

    heightArray.push(height);
    var distance=calcDistance(horizontalVelocity, time)

    if(distance<0)
    {
      distance=0;
    }

    distanceArray.push(distance);
  }

  updateTable(timeArray, distanceArray,heightArray);
}
```

FIGURE 4.37 HTML5 code for third Projectile app, which displays a table

if the time for landing is greater than or equal to 20, we set the time interval to be 10. Now we are ready to push the values of time, height, and distance into the appropriate arrays. We are going to do that with the help of a loop. Again, we assume a certain familiarity with repetition/iteration using programming loops. However, we will give a brief description of one of the

JavaScript loops that is used in our calculate() function. Just like other C-based languages, JavaScript provides while, do...while, and for loops. The for loop is one of the most commonly used loops. It is used for repeating a set of statements. It has the following format

```
for(initialization; condition testing; reinitialization)
{
Body;
}
```

The curly braces are not necessary if you only have a single statement. However, we strongly recommend the use of braces, even if you only have a single statement in the body, to enhance the readability of the code. The initialization statement is executed only once at the beginning of the loop. The rest of the statements are repeated. The flow of execution is as follows as long as condition testing returns true

```
initialization;
condition testing;
body;
reinitialization;
condition testing;
body;
reinitialization;
condition testing;
body;
reinitialization;
condition testing;
```

Our for loop in Figure 4.37 has the following values

```
initialization
var time = 0;
```

We initialize a variable called time to 0. We will call this variable the loop control variable. It will keep on changing its value. Usually, the loop stops when the loop control variable reaches a particular value. We will see that value in the condition testing part.

```
condition testing
time <= tLanding+interval
```

We will continue to loop as long as the variable time is less than the time of landing. We add an extra interval since the time of landing is not evenly divided into intervals to make sure that we reach the time of landing. We will make necessary adjustments in our computations to ensure that the values are acceptable (discussed when we examine the body of the loop in detail).

```
body;
```

The JavaScript code enclosed in the curly braces following the for loop makes up the body of the loop. We will examine it later on in greater detail.

```
reinitialization
time += interval;
```

This statement increments time by the amount equal to interval. This is the first time we are seeing the use of an extended operator, that is, +=. The statement is equivalent to

time = time + interval. It is an abbreviated statement that does not require us to type the variable name time twice. It can also prevent spelling errors, since we have fewer names to type. These extended operators were first proposed in C and are available in many languages that trace their origins to C. We can use such an abbreviated extended operation with other algebraic operations such as –, *, and /.
Let us now examine the body of our for loop in more detail.

The first statement in the body of the loop pushes the current value of time variable at the end of timeArray. This value of time is then passed to our previously discussed function calcHeight() along with the vertical velocity to compute the height reached at that value of time. We make sure that this value is acceptable—that is, non-negative (greater than 0), with an if statement and then push it at the end of heightArray. We repeat the same process for calculating the distance traveled at the same value of time to the function calcDistance() by passing the value of time and the horizontal velocity. We make sure that this value is acceptable—that is, non-negative (greater than 0), with an if statement and then push it at the end of distanceArray. This process is repeated until the value of time is within one interval of the time of landing.

Once we have populated the three arrays—timeArray, distanceArray, and heightArray—we pass them to a function called updateTable() shown in Figure 4.38.

```
function updateTable(timeArray, distanceArray, heightArray)
{
  var dataTable=document.getElementById('data');

  dataTable.innerHTML='';

  //Header row
  var row=dataTable.insertRow(0);
  var timeCell=row.insertCell(0);
  var distanceCell=row.insertCell(1);
  var heightCell=row.insertCell(2);

  timeCell.innerHTML='Time';
  distanceCell.innerHTML='Distance';
  heightCell.innerHTML='Height';

  //Insert data
  for(var i=0; i<timeArray.length; i++)
  {
    var row=dataTable.insertRow(-1);
    var timeCell=row.insertCell(0);
    var distanceCell=row.insertCell(1);
    var heightCell=row.insertCell(2);

    timeCell.innerHTML=timeArray[i];
    distanceCell.innerHTML=distanceArray[i];
    heightCell.innerHTML=heightArray[i];
  }
}
```

■ **FIGURE 4.38** HTML5 code for the Projectile app, which displays a table

4.8 Dynamically adding rows and cells in a table using JavaScript

The table in this version of our Projectile app shown in Figure 4.32 did not have any rows and columns when we first defined it, because we did not know how many rows and columns our table should have. Now that we have the three arrays corresponding to the three columns in the table, we are ready to add the rows and columns dynamically using our JavaScript function updateTable() shown in Figure 4.38. We first retrieve the table element with id "data". We store it in a variable dataTable. We assign an empty string to the innerHTML property of the table; all the relevant text will be assigned to the cells in the table, not the table itself. In order to do that, we need to add rows and columns. The two JavaScript functions that will be used for this insertion are insertRow() and insertCell(). These are two methods that are available to use with table element variables, such as dataTable. These functions take one parameter that corresponds to the row or cell number, respectively, which begin with 0. The functions also have a special value of −1, which inserts a new row or cell, respectively, at the last position. Let us look at the code in function updateTable(). We add the first row for the header with the statement

```
dataTable.insertRow(0)
```

We save the row element in a variable called row and add three cells (or columns) in the row at position 0, 1, and 2 using three row.insertCell() calls. The first cell is stored in a variable called timeCell, the second in a variable called distanceCell, and third in a variable called heightCell. The innerHTML property of these three cells gets appropriate string values (i.e., 'Time', 'Distance', and 'Height') as seen in the proceeding statements.

Now we populate the table with our arrays in a for loop. The for loop has a loop control variable called i that is initialized to index 0, and goes up to but not including the length of timeArray. Note that the last element in the array will be at index timeArray.length−1, which is why we go up to but not including the timeArray.length. The reinitialization statement i++ increments our index i by 1. This is another abbreviation for i = 1 + 1. The increment operator ++ also has a dual (not used in this chapter) operator, called a decrement operator, given by —. For example, if we used i—, it will decrement the value of i by 1.

Let us now look at the body of the for loop. Just like the header part, we first insert a row using the parameter −1. This will add the row at the last position. Since we put a header in the first row, array elements with index i will be in row i+1. Therefore, we could have also replaced the parameter −1 by i+1. This new row is saved in the variable row. We then add three cells at position 0, 1, and 2 using three row.insertCell() calls. The first cell is stored in a variable called timeCell, the second in a variable called distanceCell, and the third in a variable called heightCell. The innerHTML property of these three cells gets values from index i of appropriate arrays, namely, timeArray, distanceArray, and heightArray. The loop will continue to add a row for each set of array values. The table will be displayed by the update() function discussed in Figure 4.23 when the user clicks 'Calculate' as with the previous version of the app.

In this chapter, we looked at two complete apps that perform calculations based on user input. The first app converted temperatures. The second app was a continuation of our Projectile app. We looked at two versions. The first version calculated maximum height and distance reached by the projectile. The next version of the app tracked the height and distance over time. This meant displaying an array of times, heights, and distances. We saw how JavaScript can dynamically create a table to display this array.

The following chapters are going to look at an app that manages a database of information. We will see how to create such a database and how to add/update/delete records in the database. We will first see how to store the database locally on the device. We will then see how we can also save it on a server.

Quick facts/buzzwords

data-role field: Contains jQuery convention to assign a data-role to a <div> element to wrap a number of related form fields.

<fieldset>: HTML5 tag to include related radio buttons.

<legend>: HTML5 tag to define a caption for <fieldset> elements.

onload: Property of <body> element that is used to specify the function that will be executed when the web page is loaded.

special symbols: HTML5 defines special code to include special characters such as ° used in the Temperature conversion app.

addEventListener(): A JavaScript function to specify actions when an event takes place for an HTML5 element.

comparison operators: Operators to compare the values of variables such as ==, !=, <=, >=, <, <.

Boolean algebra: An algebra that is used to operate on values that are either true or false.

Boolean operators: Operations on Boolean values such as || (or), && (and), ! (not).

Math object: An object in JavaScript that has useful properties and methods for complex mathematical computations such as Math.PI or Math.cos().

Array object: A JavaScript object to store a sequence of values.

for loop: A JavaScript function for repeating computations.

extended operators: Abbreviation of operations such as x = x + y as x + = y.

increment and decrement operators: Operators ++ and − to increment or decrement the value of a variable.

CHAPTER 4

Self-test exercises

1. What is used to include a greater than (>) sign in our HTML5 document?
2. What is used to include a quotation mark in an HTML5 document?
3. What is the purpose of `<fieldset>` tag in an HTML5 document?
4. Which tag can be used to provide a caption for a `<fieldset>` element in an HTML5 document?
5. What are the two comparison operators for not equal in JavaScript?
6. What is the difference between == and === in JavaScript?
7. Fill in the values for the last four columns in the truth table given below

A	B	!A	A&&B	!(A&&B)	!A||B
T	T				
T	F				
F	T				
F	F				

8. What is the value of Math.ceil(2.3)?
9. What is the value of Math.floor(2.99)?
10. What is the value of Math.sin(Math.PI)?
11. What is the value of Math.sin(Math.PI/2)?
12. What is the value of Math.cos(Math.PI)?
13. What is the value of Math.cos(Math.PI/2)?
14. How do you join two arrays x and y in JavaScript?
15. How do you sort an array x in JavaScript?

Programming exercises

1. Modify the fluid converter app from the previous chapter so that it gives users an option of converting fluid quantity from gallons to litres or litres to gallons. You may want to model the app after the temperature converter app, and the display could look as follows

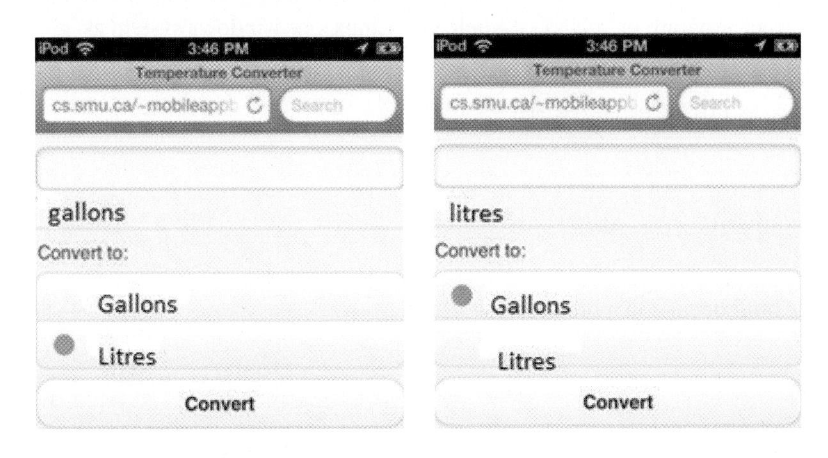

2. Write a validation function for the fluid converter app so that it only accepts values up to 1000 gallons or 4000 litres.
3. Write the JavaScript functions for the fluid converter app that will do the appropriate conversion from litres to gallons or gallons to litres based on the radio button that is checked.
4. Explain what each statement in the following piece of JavaScript code does.

```
var a = [];
var b = [];
for(i = 0; i <= 10; i++)
{
a[i] = Math.random();
b[i] = Math.random();
}
```

You can run this code using a JavaScript interpreter, such as Rhino or Node.js, available for most operating systems. You can type this program in a file called program1.js, adding the following two statements to see the values of a and b

In Rhino

```
print(a);
print(b);
```

In Node.js

```
console.log(a);
console.log(b);
```

You can run the program on the command line by typing the command

With Rhino

```
js program1.js
```

With Node.js

```
node program1.js
```

5. Modify program1.js so that instead of outputting the two arrays on a single line, it creates an html file that looks as follows in a browser. You may be able to save the output of your program on a UNIX (Apple OS, Linux), or windows system as

```
js program1.js > arraytable.html
```

or

```
node program1.js > arraytable.html
```

arraytable.html opened in a browser

Two randomly generated arrays:

i	a[i]	b[i]
0	0.03681232565193637	0.24430031627197013
1	0.6312241928723681	0.22461027067020323
2	0.45590017588702925	0.9333312050329939
3	0.22788116658351598	0.17264490536399246
4	0.5066224030843016	0.20092260032621556
5	0.8692219909741671	0.3794018308925532
6	0.8258212474924672	0.46952065824807854
7	0.580488828249338	0.7389503873679734
8	0.3174903315520794	0.3134552748606949
9	0.4598773483541462	0.3536482232117838

6. Write a JavaScript function called distance that finds the distance between the arrays a and b as

$$distance(a,b) = \frac{\sqrt{\sum_{i=0}^{n-1}(a[i] - b[i])^2}}{n},$$

where n is the length of array a. Add the function to your JavaScript file program1.js and show that it works.

Programming projects

We build on the programming projects that were described at the end of Chapter 3.

4-1. **Compute the load distribution on a beam 1:** Assume that you have a one-dimensional beam resting on two rigid supports. The beam has a length described by the variable *length*. We have put a weight *w* at a distance *d*, where $d < length$, from the left end of the beam. You should find out the reaction forces on the left and right ends of the beam.

Create an app that inputs the length of the beam and all the weights and then creates a table of distribution across the length of the beam.

4-2. **Compute the load distribution on a beam 2:** You can make the previous project as complicated as you are capable of solving. The following are three possible variations

- The user can specify the location and value of multiple weights on the beam.
- The user can specify a uniformly distributed load.
- The user can specify uniformly varying loads (i.e., value x on one end and value y on the other end). In between the load changes linearly from value x to value y.

4-3. Binary operator 1: Create an app that displays various binary operators such as "and", "or", "not", "xor" as buttons. When a user presses one of the buttons, it draws a table such as the one that follows.

An easier version could store as many tables as the number of operators and display the right table depending on the operation. For a more complex exercise, the tables should be dynamically developed using nested for loops.

A	B	A || B
1	1	1
1	0	1
0	1	1
0	0	0

4-4. Binary operator 2: An even more complicated exercise could read Boolean expressions involving 2 to 5 variables and produce a table for the Boolean expression, similar to self-test ST7.

4-5. Electricity calculations 1: Create an app that accepts voltage and resistance of an appliance as input and displays the amount of electricity consumed.

4-6. Electricity calculations 2: Create an app that accepts voltage and range of resistance as input and creates a table displaying the amount of electricity consumed.

4-7. Electricity calculations 3: Create an app that accepts the name of an appliance, voltage and resistance as input for a number of appliances, and creates a table of amount of electricity consumed. For example, users may input several models of heaters from different manufacturers. They may work on different voltages and have different resistances. The table will give a comparison of the electricity consumed by each appliance.

4-8. Amortization calculations 1: Create an app that computes the annual cost of a project that costs p and is funded by a loan at an interest i over y number of years.

4-9. Amortization calculations 2: Create an app that computes the annual cost of a project that costs p and is funded by a loan at an interest i over y number of years as before. The app should also produce a table of how much of the principal of the loan decreases with each passing year.

CHAPTER 5

A menu-driven app to monitor important indicators

WHAT WE WILL LEARN IN THIS CHAPTER

1. How to secure an app with a password
2. How to create a numeric pad for input
3. How to navigate using menus

In this chapter, we are going to develop a nontrivial app for monitoring test results of a thyroid cancer patient. The app was developed as a prototype for a project. The concepts described in this app can be used to monitor other types of quantitative test results, such as temperature and/or pressure of a boiler, wear and tear of a machine part, or the level of water in a reservoir/river. We will be discussing this app over the next three chapters. This chapter describes all the HTML5 code and demonstrates different features of the app. These features include the ability to set up password security, enter and modify basic information, enter and modify historical records of important metrics, tabulate and graph the records, and provide advice based on the history. Chapter 6 will discuss the bulk of data management scripts. We will have an in-depth discussion on managing graphics in a mobile app in Chapter 7. As part of the discussion in Chapter 7, we will illustrate the graphical analysis of the thyroid historical data.

5.1 Thyroid app design

Our next app will help medical patients with thyroid cancer track and analyze important indicators using the results of their various tests. This app is more robust than our previous apps, which only had one or two different pieces of functionality. In our design, we must consider how to appropriately divide user tasks so that the options are easy to understand and straightforward to use. Due to the highly sensitive and confidential nature of medical details, we must secure the data with password protection. In our first version of the app, the data will be stored locally on the device.

The thyroid application has many functions that, with the exception of creating a user profile, may all be accessed frequently by the user. Users of the app will be able to record important indicators from their medical test results, view the history of their records, analyze what those results mean to them, and receive basic advice on when they should seek medical help. When discussing design in earlier chapters, we said that the primary task a user will undertake should be the first thing a user sees when they open your app. However, as previously mentioned, security is very important for this app. As a result, a login screen will be the first page the user sees upon opening the app for user authentication. After the user has logged in, we will automatically check if it is a new or returning user. If it is a new user, the first thing the user must do is create a profile, so the app will take them directly to this section. Returning users will be taken to a main menu page. The menu page is the default page because this app consists of many functions that may be used equally by the user. This menu page lists all of the areas of the app in a vertical menu format that most users will be familiar with from other apps. When the user accesses one of the pages from the main menu, there is a button in the top left corner that links the user back to the main menu. This menu button is also a common design that users may have seen from other apps.

Responsiveness, as always, is a key foundation in app design, especially as the scope of your apps grows. In this app, we utilize pop-up alerts to immediately notify users of errors, such as a wrong password; to give notices, such as when they have not yet entered any records; and to provide acknowledgments, such as when we have saved the users' data. The app also incorporates data from the new records into the analysis and advice pages immediately. We must ensure that the app is providing users with accurate advice and analysis.

The same mobile design principles that we have used in the previous apps are used again in this app (e.g., large buttons that are easy to tap). For the analysis functions, the app uses graphical widgets that are optimized for the smaller screen of a mobile device. Users can view graphs that plot the test results, or look at a meter that plots their current risk level. These graphs and charts shrink and scale to fit the screen. Smaller touches, such putting the universal edit and delete icons on the record entry page, further add to usability. This is yet another way where we use a design that is already familiar to users to make our apps easier to use.

5.2 Overview of the functionality of the Thyroid app

Figure 5.1 shows the `<head>` section of the HTML5 file that contains all the screens used in our app. The `<title>` section describes the app as a "Thyroid Cancer Aide". As before, we make sure that the app will adjust the size to fit a mobile screen. We then include the jQuery Mobile cascading stylesheet as before. We are adding one more stylesheet called bootstrap.css.

```
<!DOCTYPE html>
<html lang=en>
<head>
 <title>Thyroid Cancer Aide</title>

 <!--Adjusting the page on mobile screen -->
 <meta name="viewport" content="width=device-width, initial-scale=1,
 maximum-scale=1.0, user-scalable=no">
 <meta charset="utf-8">

 <!-- CSS -->
 <link rel="stylesheet" href="css/jquery.mobile-1.3.1.min.css" />
 <link rel="stylesheet" href="css/bootstrap.css" />
 <script src="scripts/jquery-1.9.1.min.js"></script>
 <script src="scripts/jquery.mobile-1.3.1.min.js"></script>
</head>
<body>

<!--  Rest of the body goes here -->

</body>
</html>
```

FIGURE 5.1 Bird's-eye view of HTML5 code for the Thyroid app

This stylesheet was developed by Twitter developers Mark Otto and Jacob Thornton. It is an open source collection of tools containing templates for a number of user interface components including forms, buttons, navigation widgets, and aesthetically pleasing typography. Finally, we add the jQuery and jQuery Mobile scripts and set ourselves up for the design and development of our monitoring app.

As mentioned before, the app we are going to develop contains a number of features

- Password protection
- Ability to change the password
- Storing information (including password) locally on the device
- Making sure that the user has agreed to terms and conditions or disclaimers
- Acquiring basic information from the user
- Editing the basic information

Once the user has set up the app using the above-mentioned facilities, the app will allow the user to record entries of important indicators. In our case, the indicators include various test results of a thyroid cancer patient. The app will store historical values of these indicators and provide the following features for managing the history

- Creating new entries
- Editing existing entries
- Deleting individual entries
- Editing individual entries
- Clearing the entire history

Finally, the historical values of the indicators can be analyzed with the help of the following facilities

- Tabular view of the history
- Graphical view of the history
- Advice using graphical widgets

This chapter will discuss the HTML5 code that helps us create the displays of these features as well as scripts to navigate through these displays. The data management will be discussed in Chapter 6, and graphical facilities will be presented in Chapter 7.

5.3 Numeric pad for password entry

When the app is first launched, the user will be greeted with the screen shown in Figure 5.2. It consists of a prompt to enter the password. The initial password is “2345”. Figure 5.3 shows the HTML5 code for the page. We have a `<div>` section with the data-role attribute value of “page”. The page `<div>` section contains two `<div>` sections. The first has the data-role of “header”, which can be looked as the `<head>` section in a traditional HTML page with a single screen. The other `<div>` section, with the data-role of “content”, corresponds to the traditional `<body>` section. That means the contents will occupy the entire screen. This is the standard structure for jQuery Mobile pages and will be used for the majority of the pages in this app.

The header consists of a level-1 header (`<h1>`) section that has the title of the app—that is, “Thyroid Cancer Aide”. The content section has multiple widgets. We begin with the text box for entering the password with the HTML5 code

```
<input type="password" id="passcode"></input>
```

Setting the type to password means that whatever is entered by the user will not be displayed. Only dots will appear as the user types the text in this box. The id attribute of “passcode” will be used by our jQuery scripts to retrieve the password from the HTML5 page.

We then have a `<div>` section with the data-role “controlgroup” that has two buttons. Both of them are hyperlinks created as `<a> ... </a>` elements. Setting the data-role to button makes them appear as buttons (see screenshot from Figure 5.2). The first `<a>` element with id “btnEnter” has the type “submit” and will be processed by our jQuery script to move onto

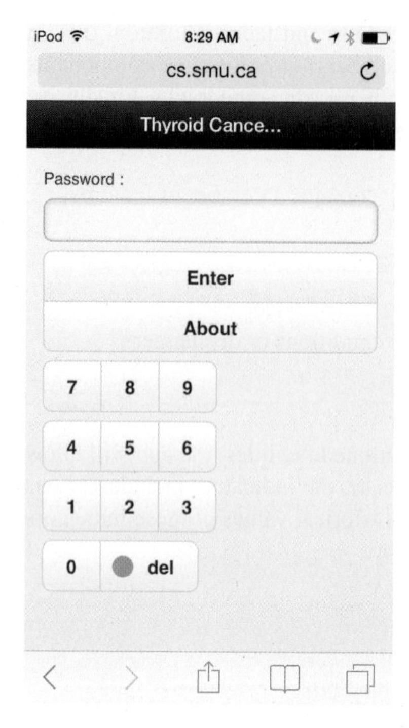

■ **FIGURE 5.2** Screenshot of the entry page in the Thyroid app

```
<!-- Start of first page -->
<div data-role="page" id="pageHome">
 <div data-role="header">
  <h1>Thyroid Cancer Aide</h1>
 </div>
 <div data-role="content">
  Password : <input type="password" id="passcode"></input>

  <div data-role="controlgroup" id="numKeyPad">
   <a data-role="button" id="btnEnter" type ="submit">Enter</a>
   <a href="#pageAbout" data-role="button">About</a>
  </div>
  <div data-role="controlgroup" data-type="horizontal">
   <a data-role="button" onclick="addValueToPassword(7)">7</a>
   <a data-role="button" onclick="addValueToPassword(8)">8</a>
   <a data-role="button" onclick="addValueToPassword(9)">9</a>
  </div>
  <div data-role="controlgroup" data-type="horizontal">
   <a data-role="button" onclick="addValueToPassword(4)">4</a>
   <a data-role="button" onclick="addValueToPassword(5)">5</a>
   <a data-role="button" onclick="addValueToPassword(6)">6</a>
  </div>
  <div data-role="controlgroup" data-type="horizontal">
   <a data-role="button" onclick="addValueToPassword(1)">1</a>
   <a data-role="button" onclick="addValueToPassword(2)">2</a>
   <a data-role="button" onclick="addValueToPassword(3)">3</a>
  </div>
```

```
    <div data-role="controlgroup" data-type="horizontal">
     <a data-role="button" onclick="addValueToPassword(0)">0</a>
     <a data-role="button" onclick="addValueToPassword('bksp')" data-
icon="delete">del</a>
    </div>
   </div>
  </div>
```

FIGURE 5.3 HTML5 code for the Thyroid app

the main screen. The second <a> element has the href attribute set to "#pageAbout", which is another screen in the same HTML5 document.

The remaining four <div> sections create a numeric pad. The data-role for these <div> sections is "controlgroup" as with the previous <div> section. We see a new value of "horizontal" for the attribute data-type. This means all the elements in this <div> section will be positioned in a single horizontal role. The first three of these sections have three buttons added with three <a> elements using the data-role "button". These buttons have labels corresponding to digits 9, 8, 7, 6, ..., 1. The onclick attribute calls a JavaScript function addValueToPassword (to be discussed later) that essentially adds an appropriate digit to the password field.

The last <div> section is slightly different from the previous three <div> sections. It only has two <a> elements with data-role "button". The onclick attribute for the button with label 0 calls the JavaScript function addValueToPassword to add the digit 0 to the password field, as with the other numbers. The onclick attribute for the second button adds the value 'bksp'. There is no such character. It is a special value that we have defined in the addValueToPassword function. Upon seeing this value, our JavaScript function will delete the last character that was entered.

The data-icon attribute for this last button is set to "delete", so it appears with a hollow dot followed by the specified label "del" as shown in Figure 5.2. However, an alternate screenshot from a Chrome browser on a desktop shows a more appropriate delete button as shown in Figure 5.4. This is an important point for us to remember. All devices will not show our elements consistently.

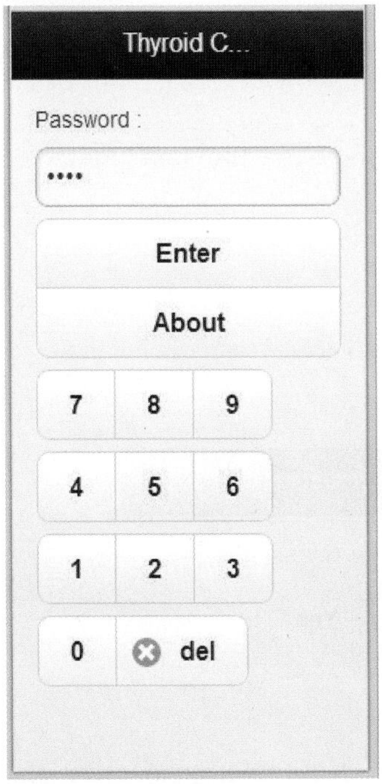

FIGURE 5.4 Alternate screenshot of the entry page in the Thyroid app

5.4 Disclaimer and help pages

When a user successfully enters the password (2345 for the first time), that user is led to the disclaimer page. The HTML5 code for the page is shown in Figure 5.5, the iOS screenshot is shown in Figure 5.6, and the Windows screenshot is shown in Figure 5.7. The `<div>` section with data-role "page" has an id of "legalNotice". There is a new attribute called data-close-btn used for this section, whose attribute value is set to "none". This value makes sure that the user is not allowed to navigate anywhere without accepting the disclaimer.

The page has two `<div>` sections: one with data-role "header" and another with data-role "content" as before. The header section has the level-1 header "Disclaimer". The content section will contain the disclaimer. We have just put a placeholder for the disclaimer. At the end of the page, we have an `<a>` element with the href attribute taking us to the "#pageUserInfo" screen upon clicking the 'Yes' button. The element has data-role "button", and data-icon is set to "forward", which just produces a hollow circle on some mobile devices as shown in Figure 5.6. However, on Windows devices we get a more meaningful icon as shown in Figure 5.7. One additional feature we have used for this button is the attribute data-mini, which is set to "false". The attribute by default is set to "true" and will make the icon smaller for smaller devices. We wanted to make is normal size for all the devices, because of the importance of the button.

Instead of entering the password, users can click on the "About" button on the entry page, which leads them to the info page. This page is also accessible from many other pages through a button at the top right. The HTML5 code for the page is shown in Figure 5.8, the iOS screenshot is shown in Figure 5.9, and the Windows screenshot is shown in Figure 5.10. The `<div>` section with data-role "page" has an id of "pageAbout". The header section has an `<a>` element with the href attribute that leads us back to the entry page. The data-role is "button", so it appears as a button. We have used "bars" as the value for the data-icon attribute. However, it

```
<!--Disclaimer Page -->
<div data-role="page" id="legalNotice" data-close-btn="none">
 <div data-role="header" >
  <h1>Disclaimer</h1>
 </div>
 <div data-role="content">
  [Insert Disclaimer Text Here]
  <br>
  <a href="#pageUserInfo" id="noticeYes" data-role="button" data-
icon ="forward" data-mini ="false">Yes</a>
 </div>
</div>
```

FIGURE 5.5 HTML5 code for the legal notice in the Thyroid app

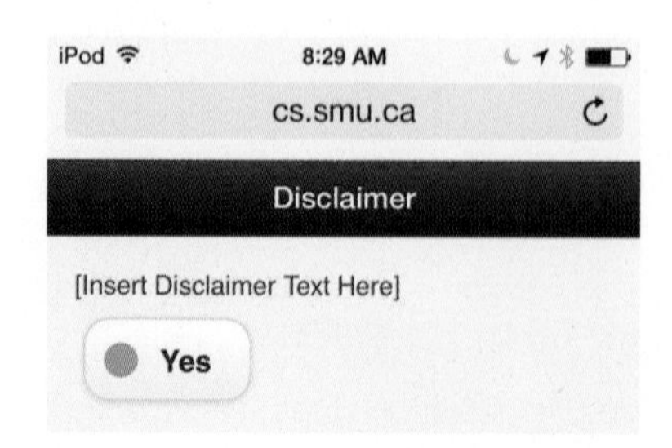

FIGURE 5.6 Screenshot of the legal notice in the Thyroid app

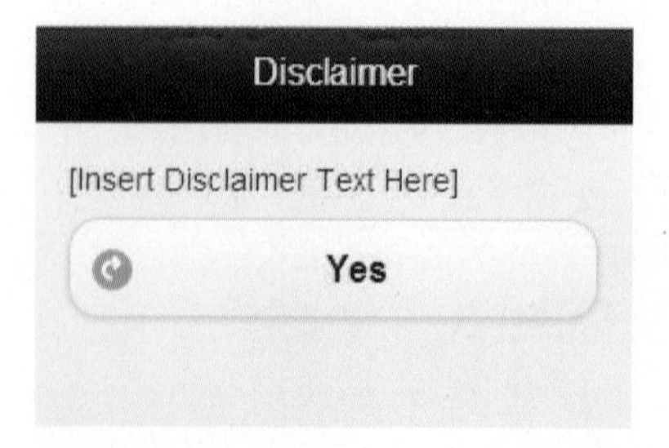

FIGURE 5.7 Alternate screenshot of the legal notice in the Thyroid app

```
<!--About Page -->
<div data-role="page" id="pageAbout">
  <div data-role="header">
    <a href="#pageHome" data-role="button" data-icon="bars" data-
iconpos="left">Home</a>
    <h1>Info</h1>
  </div>
  <div data-role="content">
    [Help and/or Website explanation/information should go here]
  </div>
</div>
```

■ **FIGURE 5.8** HTML5 code for the information page in the Thyroid app

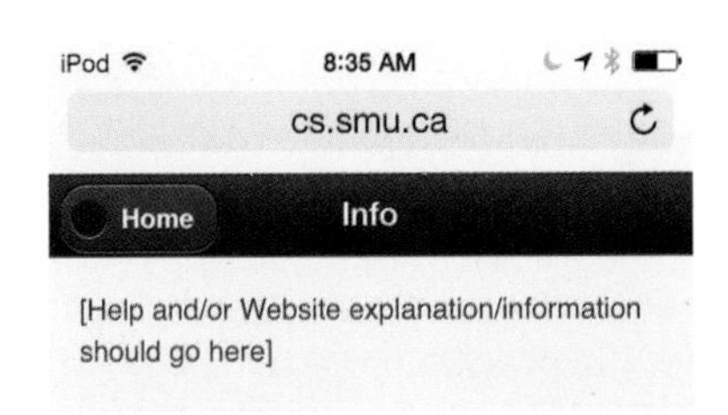

■ **FIGURE 5.9** Screenshot of the information page in the Thyroid app on an iOS device

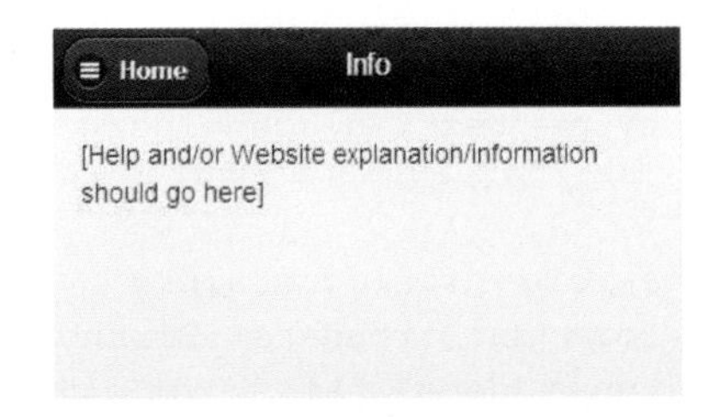

■ **FIGURE 5.10** Screenshot of the information page in the Thyroid app on a Windows device

appears as a hollow circle on iOS devices as shown in Figure 5.9, as well as Android devices. On the other hand, Windows devices have a better looking button icon. We have used the dataiconpos attribute to make sure that the button appears on the far "left". The header also has a level-1 header "Info". The content section will contain the information about the app. We have just put a placeholder for the page's content.

5.5 User information entry form

Once the user has agreed to the disclaimer using the link/button

```
<a href="#pageUserInfo" id="noticeYes" data-role="button"
data-icon="forward" data-mini ="false">Yes</a>
```

on the Disclaimer page, he or she is led to the User Information Page. We need to acquire a significant amount of information from the user. We will analyze the page in parts. Figure 5.11 shows the first half of the HTML5 code with the relevant screenshot shown in Figure 5.12, which shows fields up to the date of birth filled in. As usual, the entire page is contained in a `<div>` section with data-role "page". Its id attribute is set to "pageUserInfo". The header has two `<a>` elements with data-role "button". The first one has an href value that links to "#pageMenu", the menu page that we will examine later on in this chapter. The data-icon is set to "bars", and the button is placed to the far left by setting data-iconpos to "left". There is one new attribute called "data-inline", which is set to true. This attribute is used when you have multiple buttons that should sit side by side on the same line. Our second button on the same line will be positioned to the far "right" by setting the data-iconpos to far right. It links to the Info page with href set to "#pageAbout". The data-icon for this second button in the header is set to "info". Finally, we have the level-1 header element `<h1>` that tells the user that this is the "User Information" page.

```
<!--User Information Page/Form -->
<div data-role="page" id="pageUserInfo" >
 <div data-role="header">
  <a href="#pageMenu" data-role="button" data-icon="bars" data-
iconpos="left" data-inline="true">Menu</a>
  <a href="#pageAbout" data-role="button" data-icon="info" data-
iconpos="right" data-inline="true">Info</a>
  <h1>User Information</h1>
 <div data-role="content">
  <form id="frmUserForm" action="">
    <div data-role="fieldcontain">
     <label for="txtFirstName">First Name: </label>
     <input type="text" placeholder="First Name" name="txtFirstName" data-
mini="false" id="txtFirstName" value="" required>
    </div>
    <div data-role="fieldcontain">
     <label for="txtLastName">Last Name: </label>
     <input type="text" placeholder="Last Name" name="txtLastName" data-
mini="false" id="txtLastName" value="" required>
    </div>
    <div data-role="fieldcontain">
     <label for="datBirthdate">Birthdate: </label>
     <input type="date" name="datBirthdate" data-mini="false"
id="datBirthdate" value="" required>
    </div>
    <div data-role="fieldcontain">
     <label for="changePassword">Edit Password: </label>
     <input type="password" placeholder="New Password"
name="changePassword" data-mini="false" id="changePassword" value=""
required>
    </div>
    <div data-role="fieldcontain">
     <label for="txtHealthCardNumber">Health Card Number: </label>
     <input type="text" placeholder="Health Card Number"
name="txtHealthCardNumber" data-mini="false" id="txtHealthCardNumber"
value="" required>
    </div>
```

FIGURE 5.11 HTML5 code for the user information form in the Thyroid app—I

The content section contains only a `<form>` element with id of “frmUserForm” to get information from the user. The form consists of a total of eight input elements and a submit button. These elements are of varied types including standard text input, password, date, and dropdown boxes. We have not specified any action for the form. It will be taken care of in our JavaScript.

The part of the content section shown in Figure 5.11 has five `<div>` sections with data-role “fieldcontain”, which we have seen before; it wraps the label/form element pair. All of them have the “data-mini” attribute set to “false”, so they do not appear smaller on smaller devices. The value attribute for all of them is set to null or “”. Each `<input>` element has the attribute required. When the required attribute is set, the form cannot be submitted without entering a value for that element. Now let us look at these five elements in greater detail.

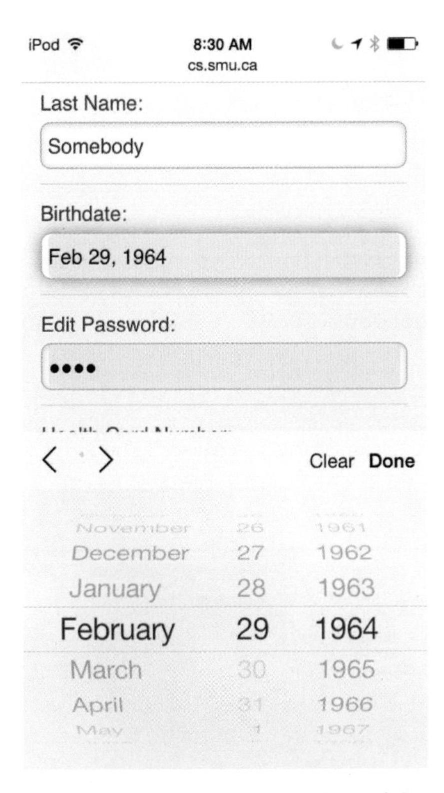

■ **FIGURE 5.12** Screenshot of the user information form in the Thyroid app—I

The first `<div>` section contains a label and a text field to accept the first name of the user. We have a placeholder attribute with a value "First Name". which will show up when the page makes its appearance. The second `<div>` section is essentially the same and accepts the last name of the user.

The third `<div>` section is meant for the birthdate. The type of the input field is set to "date". It will appear differently for different devices. Figure 5.12 shows how it will appear on an iOS device.

The fourth `<div>` element is of the type "password". We have seen this type before on the entry page. Whatever is typed in this field is not visible to the user. It appears as dots. The fifth `<div>` element is a text field just like the first two. It accepts the health insurance card number of the user.

Figure 5.13 shows the second half of the HTML5 code for the User Information Page. It consists of three dropdown lists and a submit button. Figure 5.14 shows a screenshot of one of the dropdown lists. Let us look at the code in a little more detail.

The three dropdown lists use `<div>` sections with data-role "fieldcontain" that we have seen before; it wraps the label/form element pair. All of them have the "data-mini" attribute set to "false", so they do not appear smaller on smaller devices. The three menus are created using `<select>` elements. By default, when a select button is clicked, the native select menu picker for the operating system will be used. Instead, we have set the data-native-menu attribute to "false". This setting hides the native selects. The native selects are replaced with select buttons with the look and feel of the jQuery Mobile framework. These custom menus also look better on desktop browsers where native menus tend to be small compared to the mobile menus. The `<select> ... </select>` pair contains the menu options contained in the `<option> ... </option>` pairs. We have three such dropdown menus, one for the type of the cancer, the second for the cancer stage, and the third for the reading that we wish to monitor with regular blood tests, which is called TSH. We do not need to worry about

```
        <div data-role="fieldcontain">
         <label for="slcCancerType" class="select">Cancer Type: </label>
         <select name="slcCancerType" id="slcCancerType" data-mini="false"
data-native-menu="false" required>
          <option>Select Cancer Type</option>
          <option value="Papillary">Papillary</option>
          <option value="Follicular">Follicular</option>
          <option value="Medullary">Medullary</option>
          <option value="Anaplastic">Anaplastic</option>
         </select>
        </div>
        <div data-role="fieldcontain">
         <label for="slcCancerStage" class="select">Cancer Stage: </label>
         <select name="slcCancerStage" id="slcCancerStage" data-mini="false"
data-native-menu="false" required>
          <option>Select Cancer Stage</option>
          <option value="StageOne">Stage I</option>
          <option value="StageTwo">Stage II</option>
          <option value="StageThree">Stage III</option>
          <option value="StageFour">Stage IV</option>
         </select>
        </div>
        <div data-role="fieldcontain">
         <label for="slcTSHRange" class="select">Target TSH Range: </label>
         <select name="slcTSHRange" id="slcTSHRange" data-mini="false" data-
native-menu="false" required>
          <option>Select TSH Range</option>
          <option value="StageA">A: 0.01-0.1 mIU/L</option>
          <option value="StageB">B: 0.1-0.5 mIU/L</option>
          <option value="StageC">C: 0.35-2.0 mIU/L</option>
         </select>
        </div>
       <input type="submit" id="btnUserUpdate" data-icon="check" data-
iconpos="left" value="Update" data-inline="true">
      </form>
     </div>
    </div>
```

FIGURE 5.13 HTML5 code for the user information form in the Thyroid app—II

what it means. The physician specifies a range of TSH values based on clinical examination. The three levels are

- A: 0.01 to 0.1
- B: 0.1 to 0.5
- C: 035 to 2.0

The patients will regularly enter the TSH values obtained from a blood test. The app will maintain a history of the TSH values (in addition to a couple of other values). It will plot the history of TSH values and also provide some guidance to the patient if the value falls out of the prescribed range.

Our `<form>` elements end with a submit button with id "btnUserUpdate". We have set the data-icon to be a checkmark (but not all devices display it properly as we have seen in previous chapters). The data-iconpos attribute is set to "left" along with data-inline = "true", which makes the button appear on the left. If we do not have the data-inline = "true", the button will occupy the full width of the screen.

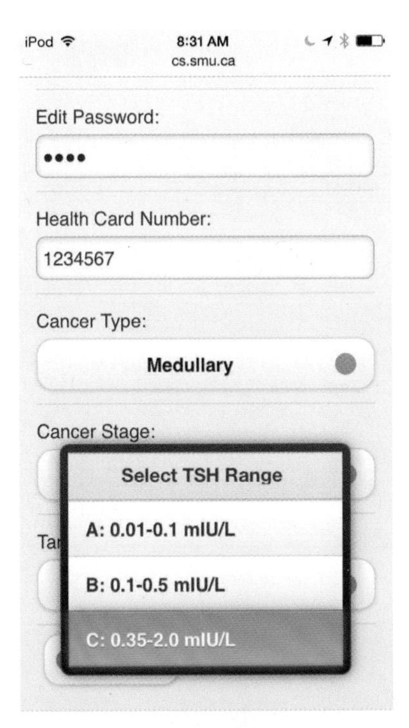

FIGURE 5.14 Screenshot of the user information form in the Thyroid app—II

5.6 Navigation with a menu

Once users have entered the basic information, they will be directed to the Menu Page. It is also accessible from most of the other pages through a button on the top right. The HTML5 code is shown in Figure 5.15, and the screenshot is shown in Figure 5.16. The id attribute for this page is set to "pageMenu". The header is similar to the one that we saw previously for the User Information Page. It has two <a> elements with data-role "button". The first one has an href value that links to "#pageMenu", the page that we are currently on. The data-icon is set to "bars", and the button is placed to the far left by setting data-iconpos to "left". The attribute data-inline is

```
<!-- Menu page -->
<div data-role="page" id="pageMenu">
  <div data-role="header">
    <a href="#pageMenu" data-role="button" data-icon="bars" data-
iconpos="left" data-inline="true">Menu</a>
    <a href="#pageAbout" data-role="button" data-icon="info"
data-iconpos="right" data-inline="true">Info</a>
    <h1>Thyroid Cancer Aide</h1>
  </div>
  <div data-role="content">
    <div data-role="controlgroup">
      <a href="#pageUserInfo" data-role="button">User Info</a>
      <a href="#pageRecords" data-role="button">Records</a>
      <a href="#pageGraph" data-role="button">Graph</a>
      <a href="#pageAdvice" data-role="button">Suggestions</a>
    </div>
  </div>
</div>
```

FIGURE 5.15 HTML5 code for the menu page in the Thyroid app

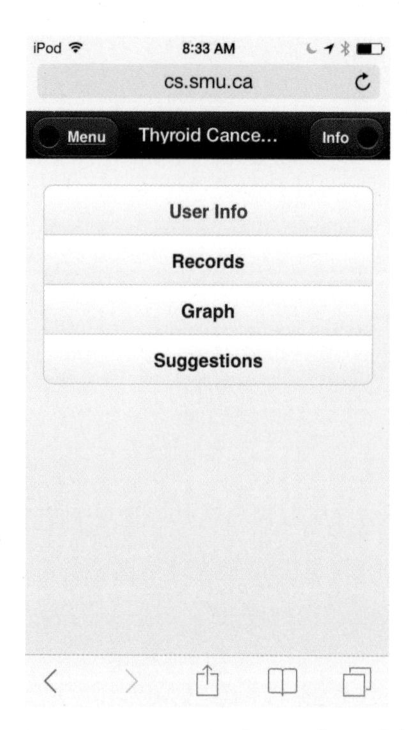

FIGURE 5.16 Screenshot of the menu page in the Thyroid app

set to true, so that multiple buttons can sit side-by-side on the same line. Our second button on the same line will be positioned to the far "right" by setting the data-iconpos to far right. It links to the Information Page with the href set to "#pageAbout". The data-icon for this second button in the header is set to "info". The `<h1>` element with "Thyroid Cancer Aide" title completes the header section.

The content `<div>` section of the menu page is relatively straightforward: there are four buttons that are wrapped in a `<div>` section with data-role "controlgroup". The buttons are provided with four `<a>` elements. The href of the button with title User Info is set to "#pageUserInfo" that will display the User Information Page. Similarly, the other three buttons will bring up the appropriate pages using the corresponding values of the href attribute.

5.7 Record display and update page

Now that we have created a user profile, we are ready to keep records of the blood test results. The HTML5 code for the records page is shown in Figure 5.17, and the screenshot appears in Figure 5.18. As usual, the entire page is contained in a `<div>` section with data-role "page". Its id attribute is set to "pageRecords". Just like the User Information Page, the header has two `<a>` elements with a data-role "button", one that takes us to the Menu Page and the other to the Information Page. We have already discussed the code for these two buttons. The `<h1>` element with "Records" title completes the header section.

The content section is divided into three parts using three `<div>` sections with data-role of "fieldcontain"

- Information about the user previously obtained from the User Information Page with id of "divUserSection", which will be used for displaying the user information through a JavaScript function.
- History of blood test results, which contains an empty table with id of "tblRecords". We have seen such an empty table in the Projectile app. It will be populated later on with a JavaScript function.

- Two <a> elements with data-role "button".
  - The first one with the label "Add New Entry" has an href value "#pageNewRecordForm" corresponding to the page that has a form for entering blood test results. We are using the data-icon "plus" for this button.
  - The second button with the label "Delete History" will trigger a JavaScript function to clear the history. It has a data-icon "delete".

```
<!-- Records page -->
<div data-role="page" id="pageRecords">
 <div data-role="header">
  <a href="#pageMenu" data-role="button" data-icon="bars" data-
iconpos="left" data-inline="true">Menu</a>
  <a href="#pageAbout" data-role="button" data-icon="info" data-
iconpos="right" data-inline="true">Info</a>
  <h1>Records</h1>
 </div>
 <div data-role="content">
  <!-- User's Information Section -->
  <div data-role="fieldcontain" id="divUserSection">
  </div>
  <h3 align="center">History</h3>
  <div data-role="fieldcontain">
   <!-- Records Table -->
   <table id="tblRecords" class="ui-responsive table-stroke">
   </table>
  </div>
  <div data-role="fieldcontain">
   <a href="#pageNewRecordForm" id="btnAddRecord" data-role="button" data-
icon="plus">Add New Entry</a>
   <a href="#" data-role="button" id="btnClearHistory" data-
icon="delete">Clear History</a>
  </div>
 </div>
</div>
```

FIGURE 5.17 HTML5 code for the records page in the Thyroid app

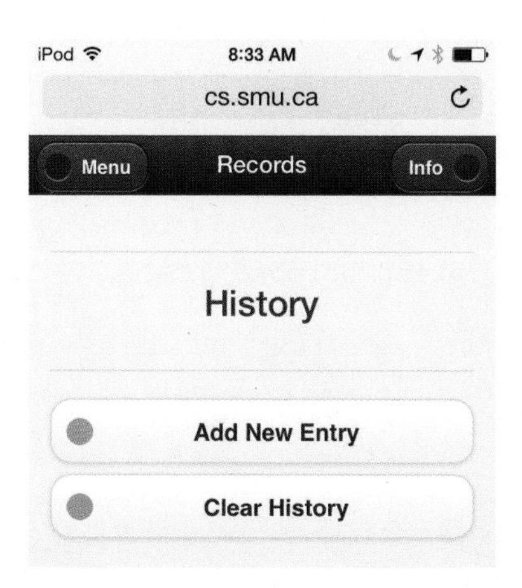

FIGURE 5.18 Screenshot of the records page in the Thyroid app

5.8 Page to add a record

From the records page, if the user presses the "Add New Entry" button, the New Record Page shown in Figures 5.19 (HTML5 code), and 5.20 (screenshot) will show up on the screen. The id attribute for this page is set to "pageNewRecordForm". Just like the User Information Page, the header has two `<a>` elements with data-role "button", one that takes us to the Menu Page and the other to the Information Page. We have already discussed the code for these two buttons. The `<h1>` element with "New Record" title completes the header section.

The content section only contains a `<form>` element with id of "frmNewRecordForm" to get information from the user. The form consists of a total of eight input elements and a submit button. These elements are of varied types including standard text input, password, date, and dropdown boxes. We have not specified any action for the form. It will be taken care of in our JavaScript.

The form has a `<div>` with data-role "fieldcontain" section to group the first four buttons (excluding the submit button). These four buttons are themselves wrapped in `<div>` sections with data-role "fieldcontain" as a label/form element pair. All of them have the "data-mini" attribute set to "false", so they do not appear smaller on smaller devices. The value attribute for all of them is set to null or "".

The first `<div>` section is the date of the blood test. The type of the input field is set to "date". As mentioned previously, it will appear differently for different devices. Figure 5.20 shows how it will appear on an iOS device.

The next three `<div>` sections contain labels and a number field to accept the values of three quantities from the blood test. We have a placeholder attribute with a value "0", which will show up when the page makes its appearance.

Our `<form>` elements ends with a submit button with id "btnSubmitRecord".

```
 <!-- New Record Form page -->
 <div data-role="page" id="pageNewRecordForm">
  <div data-role="header">
   <a href="#pageMenu" data-role="button" data-icon="bars" data-
iconpos="left" data-inline="true">Menu</a>
   <a href="#pageAbout" data-role="button" data-icon="info" data-
iconpos="right" data-inline="true">Info</a>
   <h1>New Record</h1>
  </div>
  <div data-role="content">
   <form id="frmNewRecordForm" action="">
    <div data-role="fieldcontain">
     <div data-role="fieldcontain">
      <label for="datExamDate">Date: </label>
      <input type="date" name="datExamDate" data-mini="false"
id="datExamDate" value="">
     </div>
     <div data-role="fieldcontain">
      <label for="txtTSH"><abbr title="Thyroid-stimulating hormone">TSH
(mIU/L): </abbr></label>
      <input type="number" step="0.01" placeholder="0" name="txtTSH" data-
mini="false" id="txtTSH" value="">
     </div>
     <div data-role="fieldcontain">
      <label for="txtThyroglobulin">Thyroglobulin (Âµg/L)[Optional]: </label>
      <input type="number" step="0.01" placeholder="0"
name="txtThyroglobulin" data-mini="false" id="txtThyroglobulin" value="">
```

```
      </div>
      <div data-role="fieldcontain">
       <label for="txtSynthroidDose"><abbr title="Synthroid or
Eltroxin">Synthroid* (Âµg): </abbr></label>
       <input type="number" step="0.01" placeholder="0"
name="txtSynthroidDose" data-mini="false" id="txtSynthroidDose" value="">
      </div>
     </div>
     <input type="submit" id="btnSubmitRecord" value="">
    </form>
   </div>
   <p>  100 microgram = .0001 gram</p>
  </div>
```

FIGURE 5.19 HTML5 code for the new record page in the Thyroid app

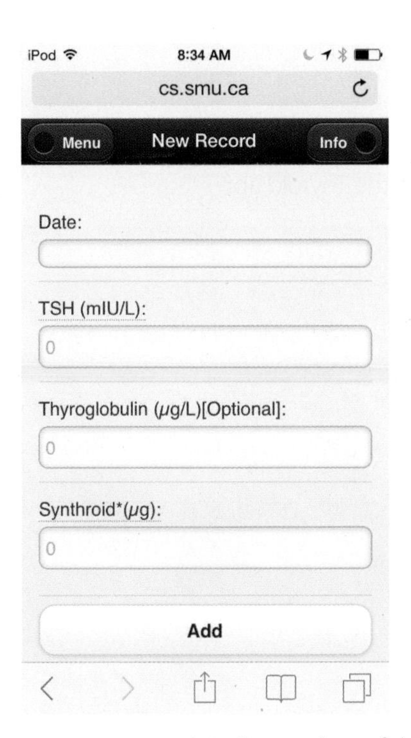

FIGURE 5.20 Screenshot of the new record page in the Thyroid app

5.9 Use of canvas and panels for graphical display

Once we have entered the blood test results, we can view them in tabular form on the Records page (Figure 5.17—HTML5 code, Figure 5.18—screenshot; note that since we do not have records, the table is not yet shown). However, sometimes a graphical representation can be more meaningful. Therefore, we have provided a Graph option in the menu that links to the Analyze page. The HTML5 code for the Analyze page is shown in Figure 5.21, and the screenshot appears in Figure 5.22. As usual, the entire page is contained in a `<div>` section with data-role "page". Its id attribute is set to "pageGraph". There are two `<div>` sections. As with the User Information Page, the header has two `<a>` elements with a data-role "button"; one takes us to the Menu Page and the other to the Information Page. We have already discussed the code for these two buttons. Finally, the header section ends with the `<h1>` element "Analyze".

```
<!--Graph Page -->
<div data-role="page" id="pageGraph" class="test">
 <div data-role="header">
  <a href="#pageMenu" data-role="button" data-icon="bars" data-iconpos="left" data-inline="true">Menu</a>
  <a href="#pageAbout" data-role="button" data-icon="info" data-iconpos="right" data-inline="true">Info</a>
  <h1>Analyze</h1>
 </div>
 <div class="panel panel-success">
  <div class="panel-heading">
 <h3 class="panel-title">TSH vs Date</h3>
   </div>
 <div class="panel-body">
 <canvas id="GraphCanvas" width="500" height="500"
  style="border:1px solid #000000;">
 </canvas>
  </div>
 </div>
</div>
```

FIGURE 5.21 HTML5 code for the graph page in the Thyroid app

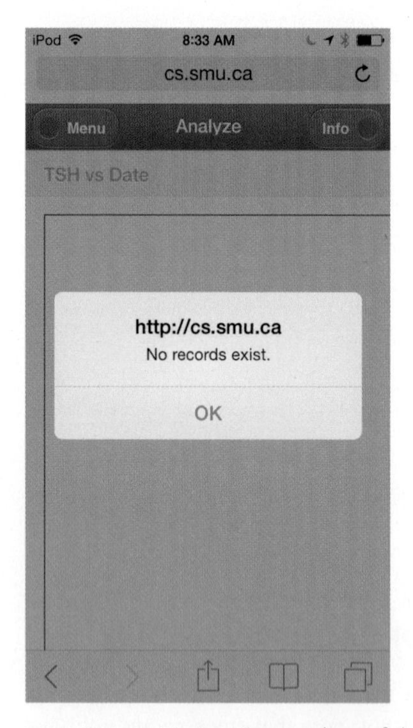

FIGURE 5.22 Screenshot of the graph page in the Thyroid app

Unlike previous pages, the <div> section that follows the header is not a <div> section with data-role "content". We are going to use a couple of new concepts in this page. The first one is the concept of a <panel> element, and the second one is <canvas> element.

The panel element is from the Bootstrap library; it provides a consistent look and feel. We want to put our content in a box with a border and some padding around the content.

Therefore, we are using the panel component. After we have discussed the Analyze page, we will take a brief detour and look at the panel component in more detail. Right now, let us focus on the panel in our Analyze page. The `<div>` section with the class "panel" encompasses our panel. We have specified the type of panel as "panel-success"; that means the panel will use the greenish theme (indicating everything is all right in our world). The panel consists of two `<div>` sections, one with class "panel-heading" and another with "panel-body". We have put the header "TSH vs Date" in the panel-heading suggesting that what appears is a graph of TSH values over time. We are using a level-3 heading with the `<h3>` element. However, we have added a twist with Bootstrap using the class "panel-title". It will use the Bootstrap look and feel to display the heading.

The `<div>` section with class "panel-body" has a canvas element that is new in HTML5. It is used to display graphics. We will provide a brief introduction to the canvas element in this chapter. We will explore it in detail in Chapter 7, when we study the JavaScript code for drawing graphics. All we see in our HTML5 is a blank canvas created as

```
<canvas id="GraphCanvas" width="500" height="500"
   style="border:1px solid #000000;">
  </canvas>
```

The id of the canvas is "GraphCanvas", with width and height set to 500 pixels and a solid black one-pixel-thick border. The number #000000 corresponds to the color black and "px" means pixel. Our JavaScript function will draw the graph. However, since we do not have any records, the function gives an error message, and the control is returned to the menu page when you click 'OK'.

As promised, we are taking a brief detour from our Thyroid app to look at the Bootstrap panel. Figure 5.23 shows the code for a page that experiments with all the five types of Bootstrap panel themes. Figure 5.24 shows the page corresponding to the code in Figure 5.23. This is a simplified version of an example from http://www.tutorialrepublic.com/. We are including the necessary Bootstrap css and JavaScript files in the header section. The body

```
<!DOCTYPE html>
<!--- Based on an example from http://www.tutorialrepublic.com/ -->
<html lang="en">
<head>
<title>Demonstrating Bootstrap Panels</title>
<link rel="stylesheet" href="css/bootstrap.min.css">
<script src="scripts/bootstrap.js"></script>
</head>
<body>
<div style="margin:42px;">
  <div class="panel panel-primary">
    <div class="panel-heading">
      <h3 class="panel-title">Primary panel</h3>
    </div>
    <div class="panel-body">
     This is how panel-primary is displayed.
      </div>
  </div>
  <div class="panel panel-info">
    <div class="panel-heading">
      <h3 class="panel-title">Information panel</h3>
    </div>
```

```
      <div class="panel-body">
       This is how panel-info is displayed.
        </div>
    </div>
    <div class="panel panel-success">
      <div class="panel-heading">
        <h3 class="panel-title">Success panel</h3>
      </div>
      <div class="panel-body">
       This is how panel-success is displayed.
    </div>
    </div>
    <div class="panel panel-warning">
      <div class="panel-heading">
        <h3 class="panel-title">Warning panel</h3>
      </div>
      <div class="panel-body">
       This is how panel-warning is displayed.
    </div>
    </div>
    <div class="panel panel-danger">
      <div class="panel-heading">
        <h3 class="panel-title">Danger panel</h3>
      </div>
      <div class="panel-body">
       This is how panel-danger is displayed.
    </div>
    </div>
  </div>
  </body>
  </html>
```

FIGURE 5.23 HTML5 code for panel example page

FIGURE 5.24 Screenshot of the panel example page

section has a <div> section that sets the margin to be 42 pixels. In this outer <div> section, we have five <div> sections each with class "panel" of five different themes

1. panel-primary: if we are not going to suggest any specific message through coloring scheme, we use this theme. It uses a dark blue coloring scheme.
2. panel-info: this theme uses light blue colors to suggest that the panel contains some information of possible use to the reader.
3. panel-success: this theme is used for indicating that the process that we are reporting is successful through a green coloring scheme.
4. panel-warning: this orange-colored theme will be used to convey warning messages to the readers.
5. panel-danger: this theme lets the user know that there is a dangerous situation using red colors.

As we saw previously, the panel consists of two <div> sections, one with class "panel-heading" and another with "panel-body". We are using a level-3 heading, employing the <h3> element and the class "panel-title" from Bootstrap in our "panel-heading" to create the content in the top section of the panel (with the colored background).

After we have entered the blood test results, our app can also provide some basic advice based on suggestions from two consulting physicians (Dr. Rajaraman and Dr. Imran from Halifax). The HTML5 code for the Advice page (which you get to by clicking 'Suggestions' in the menu) is shown in Figure 5.25, and the screenshot appears in Figure 5.26. As usual, the entire page is contained in a <div> section with data-role "page". Its id attribute is set to "pageAdvice". This page follows the standard jQuery Mobile structure with the two <div> sections: one with data-role "header" and the other with data-role "content". As with the User Information Page, the header has two <a> elements with data-role "button", one takes us to the Menu Page and the other to the Information Page. We have already discussed the code for these two buttons. The <h1> element with "Advice" title completes the header section.

The content section has a canvas element that will be explored in detail in Chapter 7. Just as with the Analyze page, all we see in our HTML5 is a blank canvas with id "GraphCanvas", width and height set to 550 pixels, and a solid black one-pixel-thick border. Our JavaScript function will create necessary graphics along with appropriate suggestions. Just as with the Analyze page, the function gives an error message saying that we do not have any records, and the control is returned to the menu page.

This chapter has essentially been about setting up a complex app that monitors important indicators. We have only looked at the HTML5 and CSS3 code of the app, which manages

```
<!--Advice Page -->
<div data-role="page" id="pageAdvice">
  <div data-role="header">
    <a href="#pageMenu" data-role="button" data-icon="bars" data-
iconpos="left" data-inline="true">Menu</a>
    <a href="#pageAbout" data-role="button" data-icon="info"
data-iconpos="right" data-inline="true">Info</a>
    <h1>Suggestions</h1>
  </div>
  <div data-role="content">
    <canvas id="AdviceCanvas" width="550" height="550"
     style="border:1px solid #000000;">
    </canvas>
  </div>
</div>
```

FIGURE 5.25 HTML5 code for the suggestions page in the Thyroid app

FIGURE 5.26 Screenshot of the suggestions page in the Thyroid app

the contents and presentation of the app. The true magic happens when JavaScript kicks in, which builds the responsive and interactive features. There are two aspects to our JavaScript functionality

1. data storage and management
2. graphics

We will study how we will store, retrieve, and manage data locally on a device in Chapter 6. The graphics will be discussed in Chapter 7. Figure 5.27 shows the HTML5 code that includes all of these JavaScript functions, which we will explore in the following two chapters.

```
<!--Custom Javascript/Jquery -->
<script src="scripts/RGraph.common.core.js"></script>
<script src="scripts/RGraph.common.effects.js"></script>
<script src="scripts/RGraph.line.js"></script>
<script src="scripts/RGraph.cornergauge.js"></script>
<script src="scripts/RGraph.hprogress.js"></script>

<script src="scripts/UserForm.js"></script>
<script src="scripts/Table.js"></script>
<script src="scripts/Navigation.js"></script>
<script src="scripts/Advice.js"></script>
<script src="scripts/GraphAnimate.js"></script>
<script src="scripts/pageLoader.js"></script>
```

FIGURE 5.27 HTML5 code for the information page in the Thyroid app

Quick facts/buzzwords

Bootstrap: This is an open source collection of tools containing templates for a number of user interface components including forms, buttons, navigation widgets, and aesthetically pleasing typography.

password: This is an input type that is used to accept text that should be secret, which will not be displayed as it is being typed. Only a dot appears when a character is entered.

data-role = button: An `<a>` element can be made to look like a button using this data-role attribute value, provided by jQuery Mobile.

data-close-btn: An attribute of `<div>` section with data-role = "page", provided by jQuery Mobile. When set to "none", the user cannot close the page. We used it so that the user cannot get out of a page without accepting the disclaimer.

data-icon: An attribute that is used to specify how a button will appear, provided by jQuery Mobile.

data-iconpos: An attribute that can be used to specify where the icon for the element will appear such as "left", "right", "top", "bottom", or "notext", provided by jQuery Mobile.

data-mini: An attribute that can be used to tell if the element should appear smaller on mobile devices, provided by jQuery Mobile.

data-inline: An attribute that is used when you want multiple buttons to appear side by side, provided by jQuery Mobile.

panel: This is a class for the `<div>` section provided by Bootstrap that is used to put contents in a box.

panel type: This attribute is used to specify how a panel will appear. The possible values include panel-primary, panel-info, panel-success, panel-warning, and panel-danger.

panel-heading: A class that is used to contain the header of a panel.

panel-body: A class that is used to contain the body of a panel.

Programming projects

1. **Boiler monitor:** Over the following three chapters, we will create an app that monitors the temperature and pressure of a boiler. You can model the app based on the Thyroid app. In this chapter, we will create the skeleton of the app, similar to the Thyroid app. Now, just create the pages and the links to navigate between them; you will implement the functionality of the pages in later chapters. The app will have

    - A password-based entry page.
    - A page to get basic information about the boiler such as the boiler ID, date of purchase, maximum allowable values of pressure and temperature, and an ability to change the password.
    - A menu page with four choices

        i. An option and corresponding page to allow you to change the basic information about the boiler.

        ii. An option and corresponding page to enter data—temperature and pressure.

        iii. An option and corresponding page to graph the data.

        iv. An option and corresponding page to make recommendations based on the values of temperature and pressure.

2. **Blood pressure monitor:** Over the following three chapters, we will create an app that monitors blood pressure. You can model the app based on the Thyroid app. In this chapter, we will create the skeleton of the app, similar to the Thyroid app. Now, just create the pages and the links to navigate between them; you will implement the functionality of the pages in later chapters. The app will have

    - A password-based entry page.
    - A page to get basic information about the person such as name, date of birth, and health insurance card number, along with an ability to change the password.
    - A menu page with four choices

        i. An option and corresponding page to allow you to change the basic information about the person.

        ii. An option and corresponding page to enter data—blood pressure.

        iii. An option and corresponding page to graph the data.

        iv. An option and corresponding page to make recommendations based on the values of blood pressure.

3. **Power consumption monitor:** Over the following three chapters, we will create an app that monitors the power consumption of a manufacturing plant. You can model the app based on the Thyroid app. In this chapter, we will create the skeleton of the app, similar to the Thyroid app. Now, just create the pages and the links to navigate between them; you will implement the functionality of the pages in later chapters. The app will have

    - A password-based entry page.
    - A page to get basic information about the plant such as the plant ID and date of installation, along with an ability to change the password.

- A menu page with four choices
    i. An option and corresponding page to allow you to change the basic information about the plant.
    ii. An option and corresponding page to enter data—power consumed.
    iii. An option and corresponding page to graph the data.
    iv. An option and corresponding page to make recommendations based on the values of power consumption.

4. **Body mass index (BMI) monitor:** Over the following three chapters, we will create an app that monitors the body mass index of a person based on height and weight. You can model the app based on the Thyroid app. In this chapter, we will create the skeleton of the app, similar to the Thyroid app. Now, just create the pages and the links to navigate between them; you will implement the functionality of the pages in later chapters. The app will have
    - A password-based entry page.
    - A page to get basic information about the person such as name, date of birth, and health insurance card number, along with an ability to change the password.
    - A menu page with four choices
        i. An option and corresponding page to allow you to change the basic information about the person.
        ii. An option and corresponding page to enter data—height (meter) and weight (kg).
        iii. An option and corresponding page to graph the data, height, and BMI.
        iv. An option and corresponding page to make recommendations based on the values of body mass index calculated using the weight and height entered.

5. **Managing a line of credit:** Over the following three chapters, we will create an app that manages our line of credit at a bank. The line of credit allows us to carry a certain maximum negative balance of p. Negative balance will be charged an interest at a specified fixed rate x. A positive balance can lead to an interest accumulation at a different fixed rate y. The app will have
    - A password-based entry page.
    - A page to get basic information about the line of credit such as the name of the bank and date of account opening, along with an ability to change the password.
    - A menu page with four choices
        i. An option and corresponding page to allow you to change the basic information about the line of credit.
        ii. An option and corresponding page to enter data—credit or debit.
        iii. An option and corresponding page to graph the balance.
        iv. An option and corresponding page to make recommendations based on the current balance.

CHAPTER 6

Storing data locally on a device for long-term use

WHAT WE WILL LEARN IN THIS CHAPTER

1. What is local storage
2. How to store and retrieve data locally on the device
3. How to validate a password
4. How to make sure that the user has accepted the disclaimer
5. How to accept and manage user profiles
6. How to store and manage an array of records

In this chapter, we continue with our exploration of a nontrivial app to monitor blood test results of a thyroid cancer patient. In the previous chapter, we looked at the presentation of the pages based on HTML5 code, using jQuery Mobile and Bootstrap. In this chapter, we will study the database management based on local storage using JavaScript/jQuery. Local storage means the data will be stored on the device. Local storage is secure and more efficient than server storage. It is also one step up from previous local storage facilities that used cookies. The data management involves the creation and destruction of the database, adding, modifying, and deleting records, and transferring information from one function to another function as well as transitioning between pages. We will look at different ways of achieving this information and process flow and also explore how some of the functions can be recycled for different but related tasks such as adding and modifying a record.

6.1 Managing numeric pad for password entry

The primary focus of this chapter is the JavaScript that is used to manage the data in our Thyroid app. In order to understand some of the references in the JavaScript, we will revisit some of the code that was previously studied in Chapter 5. We are not going to discuss the presentation aspect of the HTML5 code, which has already been studied in Chapter 5. Let us look at the HTML5 code for our entry page shown in Figure 6.1. We have an `<input>` element with the type "password" so text is not displayed, and its id is "passcode". We have a total of eleven buttons for digits ranging from 0 to 9 and one for deleting the last character.

We have not provided an id attribute for any one of these buttons, because we do not want to do anything with these buttons. Each button has the attribute onclick set to call a JavaScript function addValueToPassword(). The function receives either an appropriate digit or a string

```
<!-- Start of first page -->
<div data-role="page" id="pageHome">
  <div data-role="header">
    <h1>Thyroid Cancer Aide</h1>
  </div>
  <div data-role="content">
    Password : <input type="password" id="passcode"></input>

    <div data-role="controlgroup" id="numKeyPad">
      <a data-role="button" id="btnEnter" type ="submit">Enter</a>
      <a href="#pageAbout" data-role="button">About</a>
    </div>
    <div data-role="controlgroup" data-type="horizontal">
      <a data-role="button" onclick="addValueToPassword(7)">7</a>
      <a data-role="button" onclick="addValueToPassword(8)">8</a>
      <a data-role="button" onclick="addValueToPassword(9)">9</a>
    </div>
    <div data-role="controlgroup" data-type="horizontal">
      <a data-role="button" onclick="addValueToPassword(4)">4</a>
      <a data-role="button" onclick="addValueToPassword(5)">5</a>
      <a data-role="button" onclick="addValueToPassword(6)">6</a>
    </div>
    <div data-role="controlgroup" data-type="horizontal">
      <a data-role="button" onclick="addValueToPassword(1)">1</a>
      <a data-role="button" onclick="addValueToPassword(2)">2</a>
      <a data-role="button" onclick="addValueToPassword(3)">3</a>
    </div>
    <div data-role="controlgroup" data-type="horizontal">
      <a data-role="button" onclick="addValueToPassword(0)">0</a>
      <a data-role="button" onclick="addValueToPassword('bksp')"
data-
icon="delete">del</a>
    </div>
  </div>
</div>
```

■ **FIGURE 6.1** HTML5 code for the Thyroid app

called ‘bksp’ for the delete button. Let us look at the function shown in Figure 6.2. The function receives the value that is sent when the button is pressed, which is passed into the function through the function parameter button. The first line in the function retrieves the existing value of the `<input>` element passcode using the code $(“#passcode”).val(). This is a handy way of directly accessing an element from our HTML5 page with jQuery.

We then check whether the parameter button is equal to ‘bksp’. If it is, then we are supposed to delete the last character from our string. We achieve this deletion by extracting

```
/* Adds given text value to the password text
 * field
 */
function addValueToPassword(button)
{
  var currVal=$("#passcode").val();
```

```
    if(button=="bksp")
    {
      $("#passcode").val(currVal.substring(0,currVal.length-1));
    }
    else
    {
      $("#passcode").val(currVal.concat(button));
    }
}
```

FIGURE 6.2 JavaScript function addValueToPassword() in the Thyroid app userForm.js

the substring of the variable currVal from index 0, and one character less than the length of the string—that is, currVal.length−1. This substring is passed as a parameter to the method $(“#passcode”).val(). That logic will essentially delete the last character from the `<input>` element with id “passcode”.

If the parameter button is not ‘bksp’ then we go to the else part of the if statement. We concatenate (i.e., append or add) the value of the parameter button to the end of string currVal using the call currVal.concat(button). This new value of currVal is passed as a parameter to the method $(“#passcode”).val(). This will essentially add the digit stored in the parameter button to the `<input>` element with id “passcode”.

The button element with id “btnEnter” in Figure 6.1 will be pressed once the user enters the password. It will trigger the function `$( "#btnEnter" ).click(function(){...})` shown in Figure 6.3. This is a new way of defining and calling a function with jQuery. We are calling the click method of the element with id “btnEnter” using `$( "#btnEnter" ).click()`. The parameter to the click method is a function. Instead of writing a named function, we write the entire function as the parameter as shown in Figure 6.3. This is an anonymous function, since it does not have a name. This means that this function cannot be called from any other part of the program. We will analyze the function in great detail. The first line in the function calls another function called getPassword() and stores the value in a variable called password. The function getPassword() is shown in Figure 6.4.

```
/* On the main page, after password entry, directs
 * user to main page, legal disclaimer if they
 * have not yet agreed to it, or user entry page
 * if they have not yet completed their user info.
 */
$( "#btnEnter" ).click(function()
{
  var password=getPassword();
  if(document.getElementById("passcode").value==password)
  {
    if (localStorage.getItem("agreedToLegal")==null)
    {
      $.mobile.changePage("#legalNotice");
    }
    else if(localStorage.getItem("agreedToLegal")=="true")
    {
```

```
        if(localStorage.getItem("user")==null)
        {
          /* User has not been created, direct user
           * to User Creation page
           */
           $.mobile.changePage("#pageUserInfo");
        }
        else
        {
           $.mobile.changePage("#pageMenu");
        }
      }
    }
    else
    {
      alert("Incorrect password, please try again".);
    }
  });
```

■ FIGURE 6.3 JavaScript code when the user enters the password in the Thyroid app userForm.js

```
/*
 * Retrieves password from local storage if it
 * exists, otherwise returns the default password
 */
function getPassword()
{
  if (typeof(Storage) == "undefined")
  {
    alert("Your browser does not support HTML5 localStorage. Try
upgrading".);
  }
  else if(localStorage.getItem("user")!=null)
  {
    return JSON.parse(localStorage.getItem("user")).NewPassword;
  }
  else
  {
    /*Default password*/
    return "2345";
  }
}
```

■ FIGURE 6.4 JavaScript function getPassword()in the Thyroid app userForm.js

6.2 Local storage

Once the user has entered a password, we have to check to see if the password is correct. This means we need a way to store and retrieve the information about a user including the password. Since this information needs to be preserved permanently, we will be using a JavaScript facility called localStorage. JavaScript also provides another object called sessionStorage, which is erased when the app is closed. Storage objects are stored locally on the device. The information in the localStorage object is saved permanently until the user explicitly erases it. This information is on the device even when the app is closed and even when the device is shut down. Prior to HTML5, storage was done with the help of cookies. The localStorage object has a number of advantages

- Storage
  - The storage limit of 5MB is much larger than that provided by cookies.
- Fast
  - The data is never transferred to the web server, which reduces bandwidth requirements.
  - Since the data is not included with every server request, it does not affect the website's performance.
- Secure
  - Since the data does not flow through the network, it is less likely to be exposed to the data sniffers.
  - Each website has its own storage area, which guarantees that a website can access only data stored by itself.

Web storage is supported by all major browsers including Chrome, Safari, Opera, Firefox, and Internet Explorer 8 and above. Internet Explorer version 7 and below do not support localStorage. Before using web storage, it is always a good idea to check browser support for localStorage with the following if statement structure

```
if(typeof(Storage)=="undefined")
{
  // No Web Storage support. Cannot proceed
}
else
{
  // Code that uses localStorage
}
```

We can see this check for the availability of localStorage in the function getPassword() shown in Figure 6.4.

6.3 JSON objects for Thyroid app

The data in localStorage is stored as strings. The web browser parses these strings to determine their type (e.g., integer, Boolean, etc.) when the localStorage is used by the website or app. Another way to store data is using JSON (JavaScript Object Notation), an unordered collection of name/value pairs. JSON objects can also be stored as strings in localStorage.

For the Thyroid app, we will divide our localStorage into three parts defined by the following keys

- agreedToLegal—the value will be true once the user has agreed to the disclaimer.
- user—the value will be a JSON object with name/value pairs to store information about the user such as name, health insurance card number, and date of birth.
- tbRecords—the value will be an array of JSON objects consisting of the historical values of various test results.

This is a fairly complex storage structure. We will analyze it systematically through the rest of this chapter. Let us continue with the getPassword() function from Figure 6.4. Once we have satisfied ourselves that the browser supports localStorage, we have another *if* statement in the *else* part. It checks to see if the storage key "user" exists in localStorage using the condition:

```
localStorage.getItem("user")!=null
```

This condition uses a special value called *null*; null means there is no value. If the object exists, we will get a non-null value. If the value is non-null, then the control will go to the "else if(....)" part of the code, where we retrieve the password stored for the existing "user". Since the value of the storage key "user" is a JSON object, we need to use the JSON.parse() function to go through it. The value of "user" consists of a JSON object with the following attributes

- FirstName
- LastName
- HealthCardNumber
- NewPassword
- DOB
- CancerType
- CancerStage
- TSHRange

We get the value of NewPassword by using the code

```
JSON.parse(localStorage.getItem("user")).NewPassword
```

This value is returned to the calling function in Figure 6.3. If we do not have the JSON object associated with the storage key "user" stored on the device, we go to the else part in Figure 6.4 and return the hardcoded default password of "2345" (a little harder to guess than the ever popular "1234").

After evaluating getPassword(), we are back to Figure 6.3 and the function `$("#btnEnter").click(function()...)` that called getPassword(). The function in Figure 6.3 has stored the result of getPassword() in the variable password. The function then checks to see if the password matches the one that was entered in the textbox with the "id" passcode with the following condition:

```
document.getElementById("passcode").value==password
```

If the password is a match, we want to next make sure that the user has agreed to the disclaimer. This is done by checking whether the object with name "agreedToLegal" in our localStorage is not null. If it is null, the user will be directed to the `<div>` section with id "legalNotice", which, as discussed in Chapter 5, has a data-role "page". This is achieved using jQuery mobiles built in changePage function to navigate to the `<div>` section required. We will look at the script to handle the legalNotice soon. Let us look at the rest of the navigation in the function in Figure 6.3. If the user has agreed to the legalNotice (i.e., the JSON object "agreedToLegal" is not null), then we will have a value of "true" returned using the call

```
localStorage.getItem("agreedToLegal").
```

If it is "true", then we check if there exists a JSON object with name "user" in the localStorage with the condition

```
localStorage.getItem("user")==null
```

If the object "user" does not exist, we change the page to the `<div>` section with id "pageUserInfo". This will help us gather the user information to create the "user" object. Otherwise, we have the information stored in the "user" object, so we change to the `<div>` section with id "pageMenu". We can then proceed to use the various features in the app from the menu page. The very last "else" in Figure 6.3 corresponds to the situation where the user-entered password is not a match. In this case, we show a dialog box indicating that the password is not correct and stay with the password entry page, so the user can reenter the password.

Let us now look at the code to handle the legal notice. Figure 6.5 reminds us of the HTML5 code for the page that displays the legal notice. We have already looked at the code in detail in Chapter 5. The corresponding JavaScript code shown in Figure 6.6 consists of a single anonymous JavaScript function that will be triggered when the user clicks on the button with id "noticeYes". The function is relatively simple. It sets the localStorage storage key "agreedToLegal" to true.

```
<!--Disclaimer Page -->
<div data-role="page" id="legalNotice" data-close-btn="none">
  <div data-role="header" >
    <h1>Disclaimer</h1>
  </div>
  <div data-role="content">
    [Insert Disclaimer Text Here]
    <br>
    <a href="#pageUserInfo" id="noticeYes" data-role="button"
data-icon ="forward" data-mini ="false">Yes</a>
  </div>
</div>
```

FIGURE 6.5 HTML5 code for the legal notice in the Thyroid app userForm.js

```
/* Records that the user has agreed to the legal
 * disclaimer on this device/browser
 */
$("#noticeYes").click(function(){
  localStorage.setItem("agreedToLegal", "true");
});
```

FIGURE 6.6 JavaScript code for the legal notice in the Thyroid app userForm.js

6.4 Managing JSON objects for user information

Now we are going to look at a little more complicated data handling for managing information about the user that will be saved as an item in the localStorage with the storage key "user". For reading convenience, we have reproduced the HTML5 code for the user form page in Figures 6.7 and 6.8. We have already analyzed this code in Chapter 5. We will only use this to understand references from the corresponding JavaScript functions.

```
<!--User Information Page/Form -->
<div data-role="page" id="pageUserInfo" >
  <div data-role="header">
    <a href="#pageMenu" data-role="button" data-icon="bars" data-
iconpos="left" data-inline="true">Menu</a>
    <a href="#pageAbout" data-role="button" data-icon="info" data-
iconpos="right" data-inline="true">Info</a>
    <h1>User Information</h1>
```

```
    </div><!-- /header -->
    <div data-role="content">
      <form id="frmUserForm" action="">
          <div data-role="fieldcontain">
            <label for="txtFirstName">First Name: </label>
            <input type="text" placeholder="First Name" name="txtFirstName"
data-mini="false" id="txtFirstName" value="" required>
          </div>
          <div data-role="fieldcontain">
            <label for="txtLastName">Last Name: </label>
            <input type="text" placeholder="Last Name" name="txtLastName"
data-mini="false" id="txtLastName" value="" required>
          </div>
          <div data-role="fieldcontain">
            <label for="datBirthdate">Birthdate: </label>
            <input type="date" name="datBirthdate" data-mini="false"
id="datBirthdate" value="" required>
          </div>
          <div data-role="fieldcontain">
            <label for="changePassword">Edit Password: </label>
            <input type="password" placeholder="New Password"
name="changePassword" data-mini="false" id="changePassword" value=""
required>
          </div>
          <div data-role="fieldcontain">
            <label for="txtHealthCardNumber">Health Card Number: </label>
            <input type="text" placeholder="Health Card Number"
name="txtHealthCardNumber" data-mini="false" id="txtHealthCardNumber"
value="" required>
          </div>
```

FIGURE 6.7 HTML5 code for the User Information form in the Thyroid app—I index.html

```
          <div data-role="fieldcontain">
            <label for="slcCancerType" class="select">Cancer Type:
</label>
            <select name="slcCancerType" id="slcCancerType" data-
mini="false" data-native-menu="false" required>
              <option>Select Cancer Type</option>
              <option value="Papillary">Papillary</option>
              <option value="Follicular">Follicular</option>
              <option value="Medullary">Medullary</option>
              <option value="Anaplastic">Anaplastic</option>
            </select>
          </div>
          <div data-role="fieldcontain">
            <label for="slcCancerStage" class="select">Cancer Stage:
</label>
            <select name="slcCancerStage" id="slcCancerStage"
data-mini="false" data-native-menu="false" required>
              <option>Select Cancer Stage</option>
              <option value="StageOne">Stage I</option>
              <option value="StageTwo">Stage II</option>
              <option value="StageThree">Stage III</option>
```

```
                <option value="StageFour">Stage IV</option>
              </select>
            </div>
            <div data-role="fieldcontain">
              <label for="slcTSHRange" class="select">Target TSH Range:
</label>
              <select name="slcTSHRange" id="slcTSHRange" data-
mini="false" data-native-menu="false" required>
                <option>Select TSH Range</option>
                <option value="StageA">A: 0.01-0.1 mIU/L</option>
                <option value="StageB">B: 0.1-0.5 mIU/L</option>
                <option value="StageC">C: 0.35-2.0 mIU/L</option>
              </select>
            </div>
          <input type="submit" id="btnUserUpdate" data-icon="check" data-
iconpos="left" value="Update" data-inline="true">
        </form>
      </div>
    </div>
```

FIGURE 6.8 HTML5 code for the User Information form in the Thyroid app—II index.html

Figure 6.9 shows the JavaScript code that will be executed when our form with the id "frmUserForm" is submitted. We have an anonymous function as a parameter to `$("#frmUserForm").submit()`, which is called when the user clicks the submit button of the form (i.e., the input of type "submit"). This anonymous function only calls the function saveUserForm() and returns true. The function saveUserForm() will be analyzed a bit later. This function is quite involved with a number of new concepts. The function saveUserForm() first calls a function called checkUserForm() to make sure all the values are in order. Let us spend a little more time trying to understand the checkUserForm() function given in Figure 6.10. This function makes sure that there are values in each of the fields and the date of birth is not in the future. The most complicated part of the function is constructing the current date as a string. We start by first retrieving the current date and storing it in a variable called *d*. This is done by creating a JavaScript object of the type Date using the code

```
var d = new Date();
```

```
$("#frmUserForm").submit(function(){ //Event : submitting the
form
  saveUserForm();
  return true;
});
```

FIGURE 6.9 JavaScript code for submitting user information in the Thyroid app userForm.js

Readers who are familiar with object-oriented programming may immediately know what the above statement does. We are constructing a new object, d, with the help of the operator new. Then we are calling constructor for the Date object, that is, Date() to instantiate our new object, d, which now contains a Date object with a value corresponding to today's date. This object comes with a number of methods. We see the use of three of the important methods in the three statements that follow, getMonth(), getDate(), and getFullYear().

```
function checkUserForm()
{ //Check for empty fields in the form
  //for finding current date
  var d = new Date();
  var month = d.getMonth()+1;
  var date = d.getDate();
  var year = d.getFullYear();
  var currentDate = year + '/' +
  ((''+month).length<2 ? '0' : '') + month + '/' +
  ((''+date).length<2 ? '0' : '') + date;

  if(($("#txtFirstName").val() != "") &&
     ($("#txtLastName").val() != "") &&
     ($("#txtHealthCardNumber").val() != "") &&
     ($("#datBirthdate").val() != "") && ($("#datBirthdate").val()
<= currentDate)&&
     ($("#slcCancerType option:selected").val() != "Select Cancer
Type") &&
     ($("#slcCancerStage option:selected").val() != "Select Cancer
Stage") &&
     ($("#slcTSHRange option:selected").val() != "Select TSH
Range") )
  {
    return true;
  }
  else
  {
    return false;
  }
}
```

FIGURE 6.10 JavaScript code for checking user form in the Thyroid app userForm.js

The method d.getMonth() returns the current month to us. However, this value is between 0 to 11, where 0 stands for January and 11 for December. If we want these values to be in a more conventional range of 1 to 12, we need to add 1, that is, d.getMonth()+1, as shown in Figure 6.10. The method d.getDate() gives us current day in the conventional 1 to 31 range. d.getFullYear() gives us the four-digit value of the current year, such as 2015.

The fourth statement in checkUserForm() is a fairly complex statement that constructs a date string in the format we want and assigns it to a variable currentDate. We apologize to the readers for the complexity of the code. We do not know its original author, but it can be found often on various websites. We think it is necessary that the readers become familiar with it. If the explanation is difficult to follow, just remember that it creates a string of the form yyyy/mm/dd such as 2015/01/09 for January 9, 2015.

Let us go through it one piece at a time. These pieces are concatenated together with the help of the + operator. We will not worry about some of the trivial parts of the string such as '/', which just adds a slash between values of year, month, and date. The problem is that the length of month and date can be one character (e.g., '5') or two characters (e.g., '12'). We want to make sure that the length is always two characters by preceding the value with '0' for single character values. This is done with the help of a conditional expression. A conditional expression has three parts. The part before the question mark (?) is a condition. If the condition is true, then the expression takes the value before the colon (:). If the condition is false, the expression takes the value after the colon (:).

Let us see how ((''+month).length<2 ? '0' : '') works. We are first creating a string ''+month by concatenating month to a blank character. Next, we look at the length of our string. If the length is less than 2, the expression evaluates to '0'. If the length is not less than 2, the expression evaluates to ''.

This conditional expression can also be written with an if-else construct

```
if((''+month).length<2)
{
    currentDate = currentDate + '0'
}
else
{
    currentDate = currentDate + ''
}
```

As mentioned previously, conditional expressions are not everyone's cup of tea. All we need to know is that our currentDate now will be a string of the form yyyy/mm/dd.

After we have constructed the currentDate string in our checkUserForm() function, we check to make sure that the values of every field have been entered through a mega-condition that joins several comparisons using the && operator. We make sure that the values of the fields txtFirstName, txtLastName, txtHealthCardNumber, and datBirthDate are not equal to empty string "". We also make sure that datBirthDate is less than currentDate. For the variables with an option, such as slcCancerType, slcCancerStage, or TSHrange, we check that the values do not correspond to the first option in the list. For example, the value of slcCancerType should not be "Select Cancer Type". If all the conditions evaluate to true, the mega-condition that ties them with && operators will be true, and we return a true. Otherwise, we return false. These values are returned to the function saveUserForm() shown in Figure 6.11.

If the function checkUserForm() returns true, saveUserForm() will create the "user" JSON object and save it to the "user" storage key in localStorage. Otherwise, it will show a rather terse "Please complete the form properly" dialog box. Let us look at the construction and saving of "user" object in the device's localStorage.

```
function saveUserForm()
{
  if(checkUserForm())
  {
    var user = {
    "FirstName"  : $("#txtFirstName").val(),
    "LastName"  : $("#txtLastName").val(),
    "HealthCardNumber" : $("#txtHealthCardNumber").val(),
    "NewPassword"  : $("#changePassword").val(),
    "DOB"  : $("#datBirthdate").val(),
    "CancerType"  : $("#slcCancerType option:selected").val(),
    "CancerStage"  : $("#slcCancerStage option:selected").val(),
    "TSHRange"  : $("#slcTSHRange option:selected").val()
    };

    try
    {
      localStorage.setItem("user", JSON.stringify(user));
      alert("Saving Information");
```

```
          $.mobile.changePage("#pageMenu");
        }
        catch(e)
        {
          /* Google browsers use different error
           * constant
           */
          if (window.navigator.vendor==="Google Inc".)
          {
            if(e == DOMException.QUOTA_EXCEEDED_ERR)
            {
              alert("Error: Local Storage limit exceeds".);
            }
          }
          else if(e == QUOTA_EXCEEDED_ERR){
            alert("Error: Saving to local storage".);
          }

          console.log(e);
        }
      }
      else
      {
        alert("Please complete the form properly".);
      }

    }
```

FIGURE 6.11 JavaScript code for saving user form in the Thyroid app userForm.js

If checkUserForm() has returned a true value, saveUserForm() creates a complex variable called user, which is a JSON object. It consists of name and value pairs separated by a colon (:) as required by the JSON format. For the fields txtFirstName, txtLastName, txtHealthCardNumber, and datBirthDate, we directly use the jQuery val() method to get the value, e.g.,

```
"FirstName"           : $("#txtFirstName").val()
```

For fields with an option—that is, slcCancerType, slcCancerStage, and TSHrange—we need to use a slightly different way to access the value, for example

```
"CancerType"        : $("#slcCancerType option:selected").val()
```

6.5 Exception/error handling in JavaScript

Once we have created the "user" object using the JSON format, we are going to add it to localStorage. Working with localStorage can result in errors. For example, we may exceed available storage. Usually most input/output (I/O) operations involving secondary storage, such as disk, are fraught with potential errors. Therefore, most object-oriented programming languages, including JavaScript, which can be object oriented, mandate that I/O operations involving disk are to be done in a try and catch block. This is a special exception handling feature that takes the following form

```
try{
// Code that does the disk operations
}
catch(variable)
{
// Code if the disk operations fail for any reason
}
```

If for any reason the statements listed in the try block—that is, the code in braces {....} that follow "try"—fail, the exception (which is an error in most cases) is thrown, and the rest of the statements in the try block are skipped. The control is then transferred to the catch block—that is, the code in braces {....} that follows "catch(variable)". The exception is stored in the variable, so it can be examined in the code in the catch block. Let us see how it works in our specific case.

We first take our JSON object user that was created prior to entering the try block and convert it into a string. This is done using the JSON.stringify() function. We then associate the string returned by JSON.stringify() to the storage key that we have named "user" and put it in localStorage with the method setItem() as we had done before. Once this storage operation is performed, we update the user by displaying a dialog box with the message "Saving Information". The app then transfers the user to `<div>` section "pageMenu", which consists of various menu options.

We are using a different approach for transferring the user to a new page using the $.mobile object that is a part of the jQuery Mobile API. The object provides a number of useful methods. We use the changePage() method for changing to a page. For a complete list of functions for $.mobile object, please visit http://demos.jquerymobile.com/

Let us now look at the code in the catch block. If there was an exception/error, it will be stored in the variable *e*. We are only going to check whether we are exceeding the quota of our localStorage. Usually, most error messages have a code—a number. Instead of looking up what that number means, we are provided with meaningful names of constants. For example, we have a constant called `DOMException.QUOTA_EXCEEDED_ERR` that is set to the number 22. It means we have exceeded the localStorage quota. At the time of writing the code, Google Chrome used a different constant name than other browsers. Therefore, we check to see if `window.navigator.vendor==="Google Inc"`. If it is, we match e with `DOMException.QUOTA_EXCEEDED_ERR`. Otherwise, we match e with `QUOTA_EXCEEDED_ERR`. If e matches one of these constant, we display a message stating that the quota is exceeded. In any case, we write the error code to the console using the call `console.log(e)`. This information can be used in the debugging; however, we are not going to discuss debugging in any more detail in this book.

6.6 Displaying user information

Once the information about the user is stored, every time the app is launched and the user picks the User Info option from the menu, we want to show the information stored in our localStorage. This is achieved using the function showUserForm() in Figure 6.12. We will see where this function is called when we look at the pageLoader.js in Figure 6.13, immediately after we have studied the code for showUserForm().

The function showUserForm() starts with a try and catch block that is similar to the saveUserForm() function. Instead of using the localStorage.setItem(), we use the mirror function localStorage.getItem() for the name "user" to retrieve the contents. We use the JSON.parse() method to turn the string value of "user" into a JSON object called user. Now it is time to assign the values from the object user to appropriate fields. For the fields txtFirstName, txtLastName, txtHealthCardNumber, and datBirthDate we directly use the val() method to get the value, for example.

```
$("#txtFirstName").val(user.FirstName)
```

```
function showUserForm()
{ //Load the stored values in the form
  try
  {
    var user=JSON.parse(localStorage.getItem("user"));
  }
  catch(e)
  {
    /* Google browsers use different error
     * constant
     */
    if (window.navigator.vendor==="Google Inc".)
    {
      if(e == DOMException.QUOTA_EXCEEDED_ERR)
      {
        alert("Error: Local Storage limit exceeds".);
      }
    }
    else if(e == QUOTA_EXCEEDED_ERR){
      alert("Error: Saving to local storage".);
    }

    console.log(e);
  }

  if(user != null)
  {
    $("#txtFirstName").val(user.FirstName);
    $("#txtLastName").val(user.LastName);
    $("#txtHealthCardNumber").val(user.HealthCardNumber);
    $("#changePassword").val(user.NewPassword);
    $("#datBirthdate").val(user.DOB);
    $('#slcCancerType option[value='+user.CancerType+']').attr
('selected', 'selected');
    $("#slcCancerType option:selected").val(user.CancerType);
    $(#slcCancerType').selectmenu('refresh', true);
    $('#slcCancerStage option[value='+user.CancerStage+']').attr
('selected', 'selected');
    $("#slcCancerStage option:selected").val(user.CancerStage);
    $('#slcCancerStage').selectmenu('refresh', true);
    $('#slcTSHRange option[value='+user.TSHRange+']').
attr('selected', 'selected');
    $("#slcTSHRange option:selected").val(user.TSHRange);
    $('#slcTSHRange').selectmenu('refresh', true);
  }
}
```

FIGURE 6.12 JavaScript code for showing user form in the Thyroid app userForm.js

The fields with an option, that is, slcCancerType, slcCancerStage, and TSHrange, require a little bit more work. First we need to select the correct option. For example, for the field slcCancerType

```
$('#slcCancerType option[value='+user.CancerType+']')
.attr('selected', 'selected');
```

```
/* Runs the function to display the user information, history, graph or
 * suggestions, every time their div is shown
 */
$(document).on("pageshow", function(){
  if($('.ui-page-active').attr('id')=="pageUserInfo")
  {
    showUserForm();
  }
  else if($('.ui-page-active').attr('id')=="pageRecords")
  {
    loadUserInformation();
    listRecords();
  }
  else if($('.ui-page-active').attr('id')=="pageAdvice")
  {
    advicePage();
    resizeGraph();
  }
  else if($('.ui-page-active').attr('id')=="pageGraph")
  {
    drawGraph();
    resizeGraph();
  }
});
```

FIGURE 6.13 JavaScript code for setting up page shows in the Thyroid app pageLoader.js

Here we are first accessing the option array, `#slcCancerType option`. This option array contains all the option tags under the select tag with id "slcCancerType". We then find the specific option with the value equal to the user's previously selected cancer type with option[value=user.CancerType]. Since jQuery requires its selector argument to be in a string, the above code is concatenated as `'#slcCancerType option[value=' + user.CancerType + ']'`.

We then set the selected option to be equal to option[value=user.CancerType] by using the attr() method. The attr() method can be used to set attributes of the element. In our case, we use the attr() method to set the 'selected' attribute of `('#slcCancerType`

`option[value='+user.CancerType+']')` to 'selected'.

We have now selected the correct option to display in the app. All we need to do is refresh the field with

```
$('#slcCancerType').selectmenu('refresh', true);
```

As an aside, the attr() method can also be used to return values that have already been set on the element in addition to setting values as we saw above. attr() is an overloaded function, and the syntax to return a specific value of an attribute is $(element).attr(attribute).

Figure 6.12 describes how to show the user form every time the user chooses the User Info page from the menu. However, we did not see an actual call to that function. The call is in the pageLoader.js function and is shown in Figure 6.13. We have another anonymous function that is executed when the app is launched. This corresponds to the event `$(document).on("pageshow",...)`.

The function uses the method call `$('.ui-page-active').attr('id')` to find out which page is currently active and makes an appropriate function call according to the active page. For example, if the active page is "pageUserInfo", we call the function showUserForm() that we discussed in Figure 6.12. If the active page is "pageRecords", we have two things to display as shown in Figure 6.14. We first show the user information, and then we display the history. This means we need to call two functions as shown in Figure 6.13: loadUserInformation() to load the user information, followed by listRecords() to display the history. We will begin our exploration of the scripts involving records with loadUserInformation() from the file Table.js. We will defer the other two pages, "pageAdvice" and "pageGraph", until the following chapter.

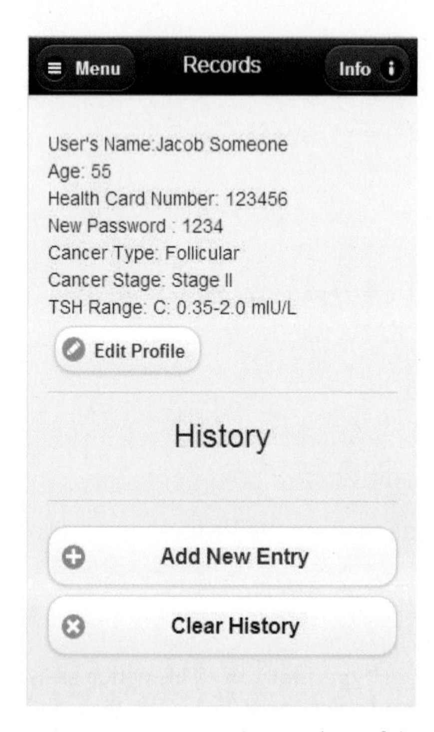

FIGURE 6.14 Screenshot of the Records page with no history in the Thyroid app

6.7 Managing the Records page

Let us refresh our memory by looking at the HTML5 code for the Records page in Figure 6.15. We have analyzed it in detail in the previous chapter. We have included the code to facilitate references to the various element ids in our JavaScript functions that follow. For displaying the user information, the page has a `<div>` section with the id "divUserSection". This is where we will display our user information using the loadUserInformation() function displayed in Figures 6.16 and 6.17. The first part of the function (Figure 6.16) shows the familiar try and

```
<!-- Records page -->
<div data-role="page" id="pageRecords">
  <div data-role="header">
    <a href="#pageMenu" data-role="button" data-icon="bars" data-
iconpos="left" data-inline="true">Menu</a>
    <a href="#pageAbout" data-role="button" data-icon="info" data-
iconpos="right" data-inline="true">Info</a>
    <h1>Records</h1>
  </div>
```

```
    <div data-role="content">
      <!-- User's Information Section -->
      <div data-role="fieldcontain" id="divUserSection">
      </div>
      <h3 align="center">History</h3>
      <div data-role="fieldcontain">
        <!-- Records Table -->
        <table id="tblRecords" class="ui-responsive
table-stroke">
        </table>
      </div>
      <div data-role="fieldcontain">
        <a href="#pageNewRecordForm" id="btnAddRecord" data-
role="button" data-icon="plus">Add New Entry</a>
        <a href="#" data-role="button" id="btnClearHistory"
data-icon="delete">Clear History</a>
      </div>
    </div>
  </div>
```

FIGURE 6.15 HTML5 code for the Records page in the Thyroid app index.html

catch block that retrieves the JSON object with the name "user" from the localStorage. We use the JSON.parse() function to convert it to a JSON object and assign it to the variable user. We use various fields from this variable in the second part of the function shown in Figure 6.17, which begins by checking to make sure that the user object is not null.

```
function loadUserInformation()
{
  try
  {
    var user=JSON.parse(localStorage.getItem("user"));
  }
  catch(e)
  {
    /* Google browsers use different error
     * constant
     */
    if (window.navigator.vendor==="Google Inc".)
    {
      if(e == DOMException.QUOTA_EXCEEDED_ERR)
      {
        alert("Error: Local Storage limit exceeds".);
      }
    }
    else if(e == QUOTA_EXCEEDED_ERR){
      alert("Error: Saving to local storage".);
    }

    console.log(e);
  }
```

FIGURE 6.16 JavaScript code for displaying user Information in the Records page in the Thyroid app—I pageLoader.js

```
    if(user != null)
    {
      $("#divUserSection").empty();
      var today=new Date();var today=new Date();
      var dob=new Date(user.DOB);
      var age=Math.floor((today-dob) / (365.25 * 24 * 60 * 60 * 1000));
      //Display appropriate Cancer Stage
      if(user.CancerStage=="StageOne")
      {
        user.CancerStage="Stage I";
      }
      else if(user.CancerStage=="StageTwo")
      {
        user.CancerStage="Stage II";
      }
      else if(user.CancerStage=="StageThree")
      {
        user.CancerStage="Stage III";
      }
      else
      {
        user.CancerStage="Stage IV";
      }

      //Display appropriate TSH Range
      if(user.TSHRange=="StageA")
      {
        user.TSHRange="A: 0.01-0.1 mIU/L";
      }
      else if(user.TSHRange=="StageB")
      {
        user.TSHRange="B: 0.1-0.5 mIU/L";
      }
      else
      {
        user.TSHRange="C: 0.35-2.0 mIU/L";
      }

      $("#divUserSection").append("User's Name:"+user.FirstName+"
  "+user.LastName+"<br>Age: "+age+"<br>Health Card Number:
  "+user.HealthCardNumber+"<br>New Password : "+user.NewPassword+"<br>Cancer
  Type: "+user.CancerType+"<br>Cancer Stage: "+user.CancerStage+"<br>TSH
  Range: "+user.TSHRange);
      $("#divUserSection").append("<br><a href='#pageUserInfo' data-
  mini='true' id='btnProfile' data-role='button' data-icon='edit' data-
  iconpos='left' data-inline='true' >Edit Profile</a>");
      $('#btnProfile').button(); // 'Refresh' the button
    }
  }
```

FIGURE 6.17 JavaScript code for displaying User Information in the Records page in the Thyroid app—II pageLoader.js

If user is not null, we have all the necessary information to display. We begin by emptying the `<div>` section with id of "divUserSection" by using the method empty() to remove all contents in the section: `$("#divUserSection").empty();`. In addition to the information stored in the user object, we also want to display the age of the user. We start by first getting the current date and storing it in the variable today. This is done by constructing a new Date object with the call: new Date(); and assigning it to the variable today. We assign the user's date of birth, given by user.DOB, to the variable dob. As the users date is stored as a string in the JSON object, we initialize a new Date object with the user.DOB attribute in order to perform our calculations. We then calculate the age as

```
Math.floor((today-dob) / (365.25 * 24 * 60 * 60 * 1000))
```

This is a standard calculation that you will find on numerous websites. We subtract the date of birth given by dob from the today variable. The subtraction gives user's age in 1000th of a second. We convert it to the number of years by dividing by 365.25 (average number of days in a year), 24 (number of hours in a day), 60 (number of minutes in a day), 60 (number of seconds in a minute), and 1000 (number of 1000th of seconds in a second). We finally round it down with the function Math.floor().

While we have the values for the fields FirstName, LastName, HealthCardNumber, age, and CancerType, we need to give a little more meaningful values to CancerStage and TSHrange. This is done through two sets of if/else if/else statements as shown in Figure 6.17.

After we obtain the data members of the object user, we add them to the `<div>` section with id "divUserSection" with the help of the append method

```
$("#divUserSection").append("User's Name:"+user.FirstName+.........
```

Once we have displayed all the information, we need to add a button that will allow users to edit their profiles. We achieve this with the help of two statements. The first statement adds the HTML5 code for the button

```
    $("#divUserSection").append("<br><a href='#pageUserInfo' ...
id='btnProfile' ... >Edit Profile</a>");
```

Then we need to style the button with the second statement

```
    $('#btnProfile').button(); // 'Refresh' the button
```

6.8 Adding a record

At the beginning of this section, we mentioned that we will store three items in our localStorage with the following storage keys: agreedToLegal, user, and tbRecords. We have already seen how to add the value "true" for the name "agreedToLegal". We have also created a complex object for value of "user", and seen how to store, retrieve, and display it. Now, we are left with the final object for the storage key "tbRecords". The value of this object will consist of an array of records.

We will spend the rest of the chapter learning how to manage this array, which includes creating the array, adding elements to this array, displaying the entire array, deleting the entire array, and displaying/editing/deleting individual elements of the array. The best place to start will be by looking at the addRecord() function, which creates the array (if it does not already exist) and then pushes a new record into the array.

First, let us see what sequence of events lead to the calling of the addRecord() function. Figure 6.15 shows the code for the Records page, which has a button element with id "btnAddRecord". When that button is clicked, the function `$("#btnAddRecord").click(function(){...}` shown in Figure 6.18 will be triggered. This is an interesting function. All it does is change the value of a button element with id "btnSubmitRecord" to 'Add' and then the button is refreshed, which transfers control to the page where the button resides, that is—the `<div>` section with data-role "page" and id "pageNewRecordForm" reproduced in Figure 6.19. This is a different way to transition to the page. It is also a clever trick. We will see later on that we can recycle the page with id "pageNewRecordForm" for editing a record by adding the value "Edit" instead of "Add". The button is refreshed only if it has the class "btn-ui-hidden", which is placed on the button after it has been viewed (and therefore initialized) previously. jQuery Mobile cannot refresh an element that has not been initialized. When the button is initialized by jQuery Mobile by navigating to the page it is automatically initialized to have the value set here. For now, let us focus on the addition of records.

```
/* The value of the Submit Record button is used
 * to determine which operation should be
 * performed
 */
$("#btnAddRecord").click(function(){
  /*.button("refresh") function forces jQuery
   * Mobile to refresh the text on the button
   */
  $("#btnSubmitRecord").val("Add");
  if($("btnSubmitRecord").hasClass("btn-ui-hidden")) {
    $("#btnSubmitRecord").button("refresh");
  }
});
```

FIGURE 6.18 JavaScript code that triggers the New Record page in the Thyroid app Table.js

```
<!-- New Record Form page -->
<div data-role="page" id="pageNewRecordForm">
  <div data-role="header">
    <a href="#pageMenu" data-role="button" data-icon="bars" data-
iconpos="left" data-inline="true">Menu</a>
    <a href="#pageAbout" data-role="button" data-icon="info" data-
iconpos="right" data-inline="true">Info</a>
    <h1>New Record</h1>
  </div>
  <div data-role="content">
    <form id="frmNewRecordForm" action="">
      <div data-role="fieldcontain">
        <div data-role="fieldcontain">
          <label for="datExamDate">Date: </label>
          <input type="date" name="datExamDate" data-mini="false"
id="datExamDate" value="">
        </div>
        <div data-role="fieldcontain">
          <label for="txtTSH"><abbr title="Thyroid-stimulating
hormone">TSH (mIU/L): </abbr></label>
          <input type="number" step="0.01" placeholder="0" name="txtTSH"
data-mini="false" id="txtTSH" value="">
```

```
            </div>
            <div data-role="fieldcontain">
              <label for="txtThyroglobulin">Thyroglobulin (Âµg/L)[Optional]:
</label>
              <input type="number" step="0.01" placeholder="0"
name="txtThyroglobulin" data-mini="false" id="txtThyroglobulin" value="">
            </div>
            <div data-role="fieldcontain">
              <label for="txtSynthroidDose"><abbr title="Synthroid or
Eltroxin">Synthroid*(Âµg): </abbr></label>
              <input type="number" step="0.01" placeholder="0"
name="txtSynthroidDose" data-mini="false" id="txtSynthroidDose" value="">
            </div>
          </div>
          <input type="submit" id="btnSubmitRecord" value="">
        </form>
      </div>
      <p>  100 microgram = .0001 gram</p>
    </div>
```

■ **FIGURE 6.19** HTML5 code for the New Record page in the Thyroid app

When the control is transferred to the page pageNewRecordForm, the following function is executed: `$("#pageNewRecordForm").on("pageshow",function(){...}` shown in Figure 6.20. As mentioned earlier, the form is going to be recycled for adding and editing the record. Here, we have set the value of the btnSubmitRecord to be "Add". This value will be assigned to the variable formOperation. Since formOperation is equal to "Add", we will call the function clearRecordForm(). This function will be discussed a little later on. It clears all the values in the form. The user can then enter the new values as shown in Figure 6.21.

```
$("#pageNewRecordForm").on("pageshow",function(){
  //We need to know if we are editing or adding a record everytime
we show this page
  //If we are adding a record we show the form with blank inputs
  var formOperation=$("#btnSubmitRecord").val();

  if(formOperation=="Add")
  {
    clearRecordForm();
  }
  else if(formOperation=="Edit")
  {
    //If we are editing a record we load the stored data in the form
    showRecordForm($("#btnSubmitRecord").attr("indexToEdit"));
  }
});
```

■ **FIGURE 6.20** JavaScript code that initiates addition of Record in the Thyroid app Table.js

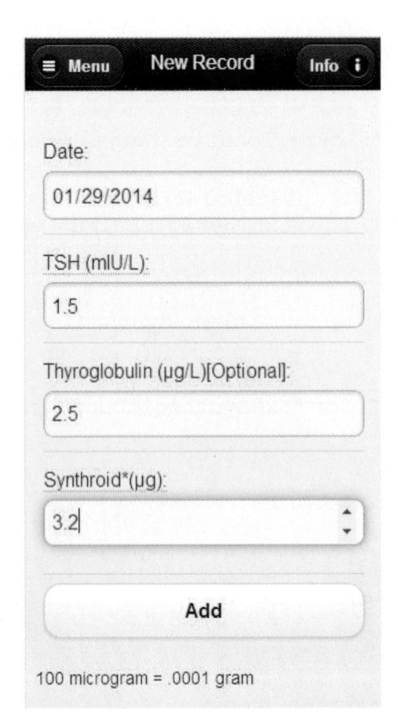

FIGURE 6.21 Screenshot of user adding a record of blood test

Once the user has entered the values as shown in Figure 6.21 and pressed the Add button, we are going to trigger the submit event for the form with id "frmNewRecordForm" shown in Figure 6.19. This leads to the method call `$("#frmNewRecordForm").submit(function(){...});` shown in Figure 6.22. As mentioned earlier, the form is going to be recycled for adding and editing the record. In this case, we have set the value of the btnSubmitRecord to be "Add". This value will be assigned to the variable formOperation as shown in Figure 6.22.

```
$("#frmNewRecordForm").submit(function(){
  var formOperation=$("#btnSubmitRecord").val();

  if(formOperation=="Add")
  {
    addRecord();
    $.mobile.changePage("#pageRecords");
  }
  else if(formOperation=="Edit")
  {
    editRecord($("#btnSubmitRecord").attr("indexToEdit"));
    $.mobile.changePage("#pageRecords");
    $("#btnSubmitRecord").removeAttr("indexToEdit");
  }

  /*Must return false, or else submitting form
   * results in reloading the page
   */
  return false;
});
```

FIGURE 6.22 JavaScript code that triggers the addition of Record in the Thyroid app Table.js

Since formOperation is equal to "Add", we will call the function addRecord(). Once the record has been added, we will change the page back to the `<div>` section with data-role "page" and id "pageRecords" (i.e., back where we originally clicked the 'Add New Entry' button) with the call to the method `$.mobile.changePage("#pageRecords")`.

We will visit Figure 6.22 again, when discussing the editing of record. Let us now turn our attention to the addRecord() function shown in Figure 6.23.

The addRecord() function begins by checking to make sure all the values in the form are appropriate using the function checkRecordForm() shown in Figure 6.24. Let us quickly go through the checkRecordForm() function before turning our attention to addRecord(). This function is more or less similar to checkUserForm() discussed in Figure 6.12. It first constructs currentDate as a string of the form yyyy/mm/dd. The code is a little complicated, and we have discussed it in detail in connection with Figure 6.12.

After we have constructed the currentDate string, we check to make sure that the values of every field have been entered through a mega-condition that joins several comparisons using the && operator. We make sure that the values of the fields txtTSH, datExamDate, and txtSynthroidDose, are not equal to empty string "". Note that there is an additional field in the form called txtThyroblobulin. Since it is optional, we do not check to see if it is not "" because

```
function addRecord()
{
  if(checkRecordForm())
  {
    var record={
      "Date"              : $('#datExamDate').val(),
      "TSH"                : $('#txtTSH').val(),
      "Tg"                 : $('#txtThyroglobulin').val(),
      "SynthroidDose"  : $('#txtSynthroidDose').val()
    };

    try
    {
      var tbRecords=JSON.parse(localStorage.getItem("tbRecords"));
      if(tbRecords == null)
      {
        tbRecords = [];
      }
      tbRecords.push(record);
      localStorage.setItem("tbRecords", JSON.stringify(tbRecords));
      alert("Saving Information");
      clearRecordForm();
      listRecords();
    }
    catch(e)
    {
      /* Google browsers use different error
       * constant
       */
      if (window.navigator.vendor==="Google Inc".)
      {
        if(e == DOMException.QUOTA_EXCEEDED_ERR)
        {
          alert("Error: Local Storage limit exceeds".);
        }
      }
```

```
        else if(e == QUOTA_EXCEEDED_ERR){
          alert("Error: Saving to local storage".);
        }

        console.log(e);
      }
    }
    else
    {
      alert("Please complete the form properly".);
    }

    return true;
  }
```

FIGURE 6.23 JavaScript code for adding a Record in the Thyroid app Table.js

it is allowed to be "". We also make sure that datExamDate is less than currentDate. In addition, the physicians asked us to validate that the value of txtSynthroidDose is less than a million. This gives us an opportunity to learn about a new method called parsefloat(). This method accepts a string and returns its value as a rational number (also called a floating point number—that is, a number with decimal point). So `parseFloat($("#txtSynthroidDose").val()` will now be a rational number. We check to make sure that it is less than 1000000. If all the conditions are met, the checkRecordForm() will return a true to the addRecord() function. Otherwise, it will return a false.

If the function checkRecordForm() returns true, addRecord() will create the "record" object and save it to localStorage. Otherwise, it will show a rather terse "Please complete the form properly" dialog box. Let us look at the construction and adding of the "record" object to the array of records and subsequently saving it in the device's localStorage.

If checkRecordForm() has returned a true value, addRecord() creates a complex variable called record, which is a JSON object. It consists of name and value pairs separated by a colon (:) as required by JSON. For example,

```
"Date"            : $('#datExamDate').val()
```

Once we have created the "record" object using the JSON format, we are going to push it to the end of the array of records. This array of records is known as tbRecords in our localStorage. In order to append the record to tbRecords, we need to first retrieve tbRecords from the localStorage. As usual, we do our localStorage operations in the familiar try and catch block to ensure we do not exceed available storage.

We first retrieve the tbRecords object, using the localStorage.getItem() method, and then use JSON.parse() to change the string into an appropriate JSON object of arrays and assign it to the variable tbRecords. If this is the first record that is being added, there is no item with the name "tbRecords" in the localStorage. In that case, the variable tbRecords will be null. If it is null, we create the array with the familiar syntax tbRecords = [].

The record is then pushed to the end of the array tbRecords using the method push(). We then stringify tbRecords again with the JSON.stringify() method and put it back in the localStorage with localStorage.setItem(). An alert() method is used to tell the user that the record was saved. A function clearRecordForm() will clean up the form, and listRecord() will be called to update the history section of the Records page. We will be looking at these methods soon. At the end, the function returns true to the calling function `$("#frmNewRecordForm").submit(function(){...});` shown in Figure 6.22. The calling function switches the focus of the app back to the `<div>` section with data-role "page" and id "pageRecords" with the call to the method `$.mobile.changePage("#pageRecords")`.

```
/* Checks that users have entered all valid info
 * and that the date they have entered is not in
 * the future
 */
function checkRecordForm()
{
  //for finding current date
  var d=new Date();
  var month=d.getMonth()+1;
  var date=d.getDate();
  var currentDate=d.getFullYear() + '/' +
  ((''+month).length<2 ? '0' : '') + month + '/' +
  ((''+date).length<2 ? '0' : '') + date;

  if(($("#txtTSH").val() != "") &&
       ($("#datExamDate").val() != "") &&
       ($("#datExamDate").val() <= currentDate)&&
      (parseFloat($("#txtSynthroidDose").val()) < 1000000) &&
      ($("#txtSynthroidDose").val() != ""))
  {
    return true;
  }
  else
  {
    return false;
  }
}
```

FIGURE 6.24 JavaScript code for verifying Record entries in the Thyroid app Table.js

6.9 Displaying all the records/history

Now that we have seen how to create the array of records, add records to it, and save it in the localStorage, it is time to look at a function that is called after every operation to populate the history section on the Records page, listRecords().

Before we look at the function, let us see the results of executing the function from a screenshot in Figure 6.25. That will help us understand the function a little better. The History section consists of a table with six columns. The header shows that the first field is the date of the blood test, the three subsequent columns have the values of three quantities, the fifth column has an edit button, and the final column has a delete button.

The function listRecords() is rather long, so we have divided it into Figures 6.26 and 6.27. The first part of the listRecords() shown in Figure 6.26 uses the familiar try and catch block to retrieve the item tbRecords from localStorage with the help of the localStorage.getItems() method. We then convert it into a JSON object using the JSON.parse() method and store it in a variable called tbRecords. If the tbRecords object is not null, we process it as shown in Figure 6.27. We want to display the records in chronological order—that is, sorted according to the date.

There is no guarantee that the user will have entered the records in chronological order. When we looked at an array of simple elements such as numbers or strings, sorting was straightforward. However, here each element has a number of fields: Date, TSH, Tg, and SynthroidDose. We want to sort using the Date field. In order to do this, we have to write a function that compares the records based on the Date field. We can see a function called

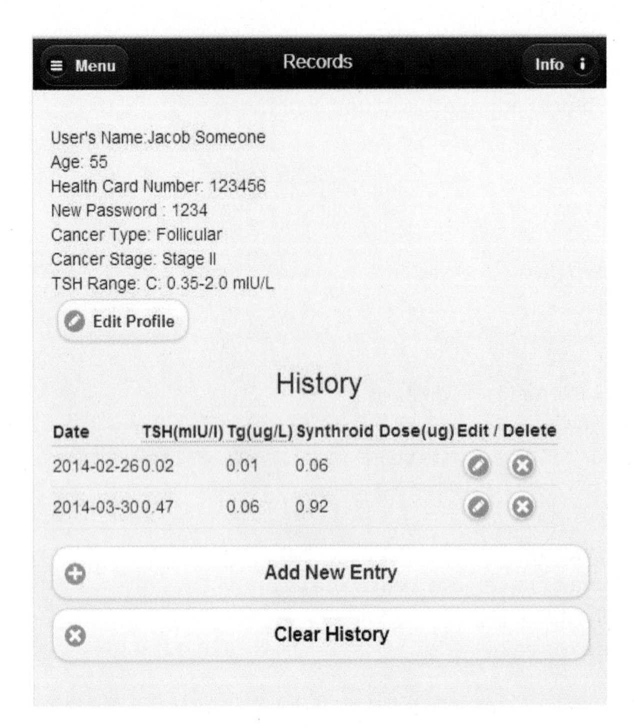

FIGURE 6.25 Screenshot of Records sections after adding records in the Thyroid app

```
function listRecords()
{
  try
  {
    var tbRecords=JSON.parse(localStorage.getItem("tbRecords"));
  }
  catch(e)
  {
    /* Google browsers use different error
     * constant
     */
    if (window.navigator.vendor==="Google Inc".)
    {
      if(e == DOMException.QUOTA_EXCEEDED_ERR)
      {
        alert("Error: Local Storage limit exceeds".);
      }
    }
    else if(e == QUOTA_EXCEEDED_ERR){
      alert("Error: Saving to local storage".);
    }

    console.log(e);
  }
```

FIGURE 6.26 JavaScript code for displaying Records in the Thyroid app—I Table.js

compareDates() in Figure 6.28. The function receives two records, a and b. It creates two variables x and y using the Date field of a and b as in

```
x = new Date(a.Date)
```

```
    //Load previous records, if they exist
    if(tbRecords != null)
    {
      //Order the records by date
      tbRecords.sort(compareDates);

      //Initializing the table
      $("#tblRecords").html(
        "<thead>"+
        "   <tr>"+
        "     <th>Date</th>"+
        "     <th><abbr title='Thyroid Stimulating Hormone'>TSH(mlU/l)</abbr></th>"+
        "     <th><abbr title='Thyroglobulin'>Tg(ug/L)</abbr></th>"+
        "     <th>Synthroid Dose(ug)</th>"+
        "     <th>Edit / Delete</th>"+
        " </tr>"+
        "</thead>"+
        "<tbody>"+
        "</tbody>"
      );

      //Loop to insert the each record in the table
      for(var i=0;i<tbRecords.length;i++)
      {
        var rec=tbRecords[i];
        $("#tblRecords tbody").append("</tr>"+
           " <td>"+rec.Date+"</td>" +
           " <td>"+rec.TSH+"</td>" +
           " <td>"+rec.Tg+"</td>" +
           " <td>"+rec.SynthroidDose+"</td>" +
           " <td><a data-inline='true' data-mini='true' data-role='button'
href='#pageNewRecordForm' onclick='callEdit("+i+")' data-icon='edit' data-
iconpos='notext'></a>"+
           " <a data-inline='true' data-mini='true' data-role='button'
href='#' onclick='callDelete("+i+")' data-icon='delete' data-
iconpos='notext'></a></td>"+
           "</tr>");
      }

      $('#tblRecords [data-role="button"]').button(); // 'Refresh' the
buttons. Without this the delete/edit buttons wont appear
    }
    else
    {
      $("#tblRecords").html("");
    }
    return true;
  }
```

FIGURE 6.27 JavaScript code for displaying Records in the Thyroid app—II Table.js

We are passing a.Date to the constructor Date(), which returns an object of the type Date. Although the a.Date attribute may already be a Date object, it may also be a string when it is initially loaded from localStorage. The function compareDates() returns 1 if x is greater than y. Otherwise, it returns −1.

```
function compareDates(a, b)
{
  var x=new Date(a.Date);
  var y=new Date(b.Date);

  if(x>y)
  {
    return 1;
  }
  else
  {
    return -1;
  }
}
```

FIGURE 6.28 JavaScript code for comparing Records using Date Field Table.js

The function compareDates() is passed as parameter to the sort method—that is, tbRecords.sort(compareDates), as shown in Figure 6.27. The sort method calls compareDates() for comparing records during the sorting process.

Once we have sorted the records, we start building the History section, which is a table with the id tblRecords. First we initialize the table by inserting code for the `<thead>` element and an empty `<tbody>` element using a call to the method `$("#tblRecords").html(...)`.

We then add the records using a for loop where our index i goes from 0 to tblRecords.length−1. We first store the element with index i in a variable named rec (a step that is not strictly necessary; we can easily work with tbRecords[i] instead of rec)—that is, `var rec=tbRecords[i];`. We then append values of different fields of rec to each row in the `<tbody>` element of the table with the method call `$("#tblRecords tbody").append()`. This is a rather involved code. We will study it in parts. The first part shown below gets the row started with the `<tr>` tag

```
$("#tblRecords tbody").append("<tr>"+
```

We then add the date, and three blood-test values in different columns as

```
" <td>"+rec.Date+"</td>" +

" <td>"+rec.TSH+"</td>" +
" <td>"+rec.Tg+"</td>" +
" <td>"+rec.SynthroidDose+"</td>" +
```

We are now going to add two buttons; the first one is used to edit the record number i:

```
" <td><a data-inline='true' data-mini='true' data-role='button'
href='#pageNewRecordForm' onclick='callEdit("+i+")' data-
icon='edit' data-iconpos='notext'></a>"
```

It is an `<a>` element with href attribute set to the page with id pageNewRecord. When clicked, it will call the function callEdit and send it the index of the record i. We will discuss this function later on.

The second button has similar code; it calls the function callDelete and passes the index of the record i as

```
" <a data-inline='true' data-mini='true' data-role='button'
href='#' onclick='callDelete("+i+")' data-icon='delete'
data-iconpos='notext'></a></td>"
```

Finally, we append the tag `</tr>` to complete the code for the row.

Once we are out of the for loop and have added all the values and buttons, we refresh the buttons with the code

```
    $('#tblRecords [data-role="button"]').button(); // 'Refresh'
the buttons. Without this the delete/edit buttons won't appear
```

The final else statement in the function corresponds to the situation where tbRecords is empty. In that case, we set the html code for the table with id tblRecords to an empty string "".

Since we have just looked at the buttons to edit and delete a record with index i, let us look at the corresponding code.

6.10 Editing a record

When we were adding a record, we had mentioned that we will recycle some of the code for editing a record. Now the time has come. For editing a record we start with the callEdit() shown in Figure 6.29. The function creates a new attribute called "indexToEdit" for the btnSubmitRecord element from the page pageNewRecordForm. The value of this new attribute is set to the index value that was passed to callEdit(). The function then sets the value of btnSubmitRecord to "Edit" and refreshes the button. It automatically takes us to the `<div>` section with data-role "page" and id pageNewRecordForm, where the button btnSubmitRecord resides.

When the control is transferred to page pageNewRecordForm, we will be executing the function

```
$("#pageNewRecordForm").on("pageshow",function(){...})
```

shown in Figure 6.30. We have seen this function while looking at the addition of record. Now we are looking at using it for editing records. As mentioned earlier, the form that is used for adding records will be recycled for editing the records. Here, we have set the value of the btnSubmitRecord to be "Edit". This value will be assigned to the variable formOperation. Since

```
function callEdit(index)
{
  $("#btnSubmitRecord").attr("indexToEdit", index);
  /*.button("refresh") function forces jQuery
   * Mobile to refresh the text on the button
   */
  $("#btnSubmitRecord").val("Edit");
  if($("btnSubmitRecord").hasClass("btn-ui-hidden")) {
    $("#btnSubmitRecord").button("refresh");
  }
}
```

FIGURE 6.29 JavaScript function called from the Edit Button in the History section in the Thyroid app Table.js

```
$("#pageNewRecordForm").on("pageshow",function(){
  //We need to know if we are editing or adding a record everytime we show
this page
  //If we are adding a record we show the form with blank inputs
  var formOperation=$("#btnSubmitRecord").val();

  if(formOperation=="Add")
  {
    clearRecordForm();
  }
  else if(formOperation=="Edit")
  {
    //If we are editing a record we load the stored data in the form
    showRecordForm($("#btnSubmitRecord").attr("indexToEdit"));
  }
});
```

FIGURE 6.30 JavaScript code that initiates Editing of Record in the Thyroid app Table.js

the formOperation is equal to "Edit", we will call the function showRecordForm(). This function, which will be discussed a little later on, opens a form containing all of the attributes from a given record and allows a user to edit and save those values.

```
$("#frmNewRecordForm").submit(function(){...}
```

We have seen this function while looking at the addition of a record in Figure 6.22. Now we are looking at it in Figure 6.31 from the editing record point of view. The value of btnSubmitRecord is stored in the variable formOperation. Since its value is now "Edit", we

```
$("#frmNewRecordForm").submit(function(){
  var formOperation=$("#btnSubmitRecord").val();

  if(formOperation=="Add")
  {
    addRecord();
    $.mobile.changePage("#pageRecords");
  }
  else if(formOperation=="Edit")
  {
    editRecord($("#btnSubmitRecord").attr("indexToEdit"));
    $.mobile.changePage("#pageRecords");
    $("#btnSubmitRecord").removeAttr("indexToEdit");
  }

  /*Must return false, or else submitting form
   * results in reloading the page
   */
  return false;
});
```

FIGURE 6.31 JavaScript code that triggers Editing of Record in the Thyroid app Table.js

will be looking at the editing part of the if statement that is highlighted. We will be calling the function editRecord() by passing the value of the attribute called "indexToEdit" for the btnSubmitRecord with

```
editRecord($("#btnSubmitRecord").attr("indexToEdit"));
```

Let us explore the rest of the code before turning our attention to the editRecord() function. Once the record with edited values has been replaced, we will change the page back to the `<div>` section with data-role "page" and id "pageRecords" with the call to the method `$.mobile.changePage("#pageRecords")`.

We will do a little bit of cleanup, namely remove the indexToEdit attribute. At the end the function returns false instead of true. If we return true, we end up staying on the same page.

The editRecord() function, shown in Figure 6.32, begins by checking to make sure all the values in the form are appropriate using the function checkRecordForm() shown in Figure 6.24. We have already discussed checkRecordForm() when we were adding a record. If checkRecordForm() has returned a true value, we use our localStorage operations in the familiar try and catch block. First the tbRecords item is retrieved from the localStorage using the localStorage.getItem() method. It is then converted to an array of records with the method JSON.parse(). The record associated with the index is then substituted with a new JSON object defined by form field values. It consists of name and value pairs separated by a colon (:) as required by JSON. For example,

```
"Date"            : $('#datExamDate').val()
```

We then stringify tbRecords, which now contains the updated record at the index, with the JSON.stringify() method and put it back in the localStorage with localStorage.setItem(). An alert() method is used to tell the user that the record was saved. A function clearRecordForm() will then clean up the form, and listRecord() will be called to update the History section of the Records page.

Let us look at two of the functions that we have used in the process of adding and editing a record. These are the two functions that are called in `$("#pageNewRecordForm").on("pageshow",function(){...})`, depending on whether we are adding or editing a record. The first function, clearRecordForm(), is used in Figure 6.32 after we save the edited record. clearRecordForm(), shown in Figure 6.33, was also used during the addition of a new record in Figure 6.20 in `$("#pageNewRecordForm").on("pageshow",function(){...})`, and Figure 6.23 in addFunction(). The function is relatively straightforward. It sets values of all the fields to an empty string "". The second function showRecordForm() is used in Figure 6.30. showRecordForm() is shown in Figure 6.34 and receives an index of the record that we want to show in the form. In the familiar try and catch block, we retrieve the tbRecords item from the localStorage using the localStorage.getItem() method. It is then converted to an array of records with the method JSON.parse(). The record at the index is then saved in a variable called rec. The values of the form fields are updated using corresponding values from rec variable, for example,

```
$('#datExamDate').val(rec.Date);
```

```
function editRecord(index)
{
  if(checkRecordForm())
  {
    try
    {
      var tbRecords=JSON.parse(localStorage.getItem("tbRecords"));
```

```
        tbRecords[index] ={
          "Date"          : $('#datExamDate').val(),
          "TSH"            : $('#txtTSH').val(),
          "Tg"            : $('#txtThyroglobulin').val(),
          "SynthroidDose" : $('#txtSynthroidDose').val()
        };//Alter the selected item in the array
        localStorage.setItem("tbRecords", JSON.
stringify(tbRecords)); //Saving array to local storage
        alert("Saving Information");
        clearRecordForm();
        listRecords();
      }
      catch(e)
      {
        /* Google browsers use different error
         * constant
         */
        if (window.navigator.vendor==="Google Inc".)
        {
          if(e == DOMException.QUOTA_EXCEEDED_ERR)
          {
            alert("Error: Local Storage limit exceeds".);
          }
        }
        else if(e == QUOTA_EXCEEDED_ERR){
          alert("Error: Saving to local storage".);
        }

        console.log(e);
      }
    }
    else
    {
      alert("Please complete the form properly".);
    }
  }
```

■ **FIGURE 6.32** JavaScript code to replace a Record in the localStorage in the Thyroid app Table.js

```
function clearRecordForm()
{
  $('#datExamDate').val("");
  $('#txtTSH').val("");
  $('#txtThyroglobulin').val("");
  $('#txtSynthroidDose').val("");
  return true;
}
```

■ **FIGURE 6.33** JavaScript code to clear a Record form in the Thyroid app Table.js

```
function showRecordForm(index)
{
  try
  {
    var tbRecords=JSON.parse(localStorage.getItem("tbRecords"));
    var rec=tbRecords[index];
    $('#datExamDate').val(rec.Date);
    $('#txtTSH').val(rec.TSH);
    $('#txtThyroglobulin').val(rec.Tg);
    $('#txtSynthroidDose').val(rec.SynthroidDose);
  }
  catch(e)
  {
    /* Google browsers use different error
     * constant
     */
    if (window.navigator.vendor==="Google Inc".)
    {
      if(e == DOMException.QUOTA_EXCEEDED_ERR)
      {
        alert("Error: Local Storage limit exceeds".);
      }
    }
    else if(e == QUOTA_EXCEEDED_ERR){
      alert("Error: Saving to local storage".);
    }

    console.log(e);
  }
}
```

FIGURE 6.34 JavaScript code to show a Record in the form in the Thyroid app Table.js

6.11 Deleting a record

The final operation involving an individual record is the deletion of the record. This will be triggered after clicking on the delete button next to the record as shown in Figure 6.25. The button calls the function callDelete() with a parameter that corresponds to the index of the record that is to be deleted. The function call is shown in Figure 6.27. The function callDelete(), shown in Figure 6.35, first calls the deleteRecord() function with index as the parameter, and then lists the records with the listRecords() function. The function deleteRecord() is shown in Figure 6.36. As usual, we do our localStorage operations in the familiar try and catch block. First the tbRecords item is retrieved from the localStorage using the localStorage.getItem() method. It is then converted to an array of JSON objects (i.e., records) with the method JSON. parse().We then invoke the splice function to remove items from the array. In our case, the call

```
tbRecords.splice(index, 1);
```

removes one item starting from the location given by the index. We then check to see if the array is empty. If it is empty, we remove the tbRecords item from localStorage with the method localStorage.removeItem(). If the leftover array is not empty, we are in the else part, where we stringify the tbRecords and add it to the localStorage using the localStorage.setItem() method.

```
function deleteRecord(index)
{
  try
  {
    var tbRecords=JSON.parse(localStorage.getItem("tbRecords"));

    tbRecords.splice(index, 1);

    if(tbRecords.length==0)
    {
      /* No items left in records, remove entire
       * array from localStorage
       */
      localStorage.removeItem("tbRecords");
    }
    else
    {
      localStorage.setItem("tbRecords", JSON.stringify(tbRecords));
    }
  }
  catch(e)
  {
    /* Google browsers use different error
     * constant
     */
    if (window.navigator.vendor==="Google Inc".)
    {
      if(e == DOMException.QUOTA_EXCEEDED_ERR)
      {
        alert("Error: Local Storage limit exceeds".);
      }
    }
    else if(e == QUOTA_EXCEEDED_ERR){
      alert("Error: Saving to local storage".);
    }

    console.log(e);
  }
}
```

FIGURE 6.35 JavaScript code to initiate deleting a Record in the Thyroid app Table.js

```
// Delete the given index and re-display the table
function callDelete(index)
{
  deleteRecord(index);
  listRecords();
}
```

FIGURE 6.36 JavaScript code to delete a Record from the localStorage in the Thyroid app Table.js

6.12 Deleting all the records/history

The very final record management function is deleting the history using the 'Clear History' button shown in Figure 6.25. Clicking on the button will trigger the function

```
$("#btnClearHistory").click(function(){...}
```

shown in Figure 6.37. We remove the tbRecords storage key from localStorage with the method localStorage.removeItem(). Then we call the function listRecords(), which will essentially put an empty string in the table with id tblRecords and display nothing in the History section of the Records page. We then display a dialog box that tells the user that all the records have been deleted.

In this chapter, we looked at record management using localStorage. In the following chapter, we will look at some of the graphics utilities associated with our Thyroid app.

```
// Removes all record data from localStorage
$("#btnClearHistory").click(function(){
  localStorage.removeItem("tbRecords");
  listRecords();
  alert("All records have been deleted".);
});
```

FIGURE 6.37 JavaScript code to delete the array of Records in the Thyroid app Table.js

Quick facts/buzzwords

localStorage: An HTML5 facility to save up to 5MB of information on a device.

localStorage.setItem(): A JavaScript function to add or replace an item in localStorage.

localStorage.getItem(): A JavaScript function to retrieve an item in localStorage.

JSON: JavaScript Object Notation for storing JavaScript objects in a name/value pair.

JSON.parse(): A method for converting a string to JSON format.

JSON.stringify(): A method for converting a JSON object to a string.

exception handling using try/catch: An object-oriented approach for handling errors/ exceptional situations.

QUOTA_EXCEEDED_ERROR: A constant that corresponds to the error, when the quota for local storage is objected.

Date object: A JavaScript object type that stores dates.

Anonymous JavaScript functions: Functions with no name that are passed to a number of event methods such as button clicks or form submits.

click(): A Method that is called when a button is clicked.

attr(): A method to set attributes of an HTML5 element.

submit(): A method to submit a form element.

conditional expression: An expression whose value depends on whether a condition is true or false.

html(): A method to add html code to an HTML5 element such as a `<div>` section or `<table>`.

append(): A method to append html code to an HTML5 element.

splice(): A method to add or delete elements from an array.

button('refresh'): A method to make a button visible. We can also use it to change the control to the page which contains the button.

$.mobile.changePage(): A jQuery Mobile method to change the control to another page.

Self-test exercises

1. What is the difference between localStorage and sessionStorage?
2. How is localStorage better than cookies?
3. How much localStorage is available for an app by default in most browsers?
4. What are the advantages of localStorage versus server storage?
5. What method allows you to save or modify the value of an object in localStorage?
6. What method allows you to retrieve the value of an object in localStorage?

Programming projects

1. **Boiler monitor:** Please continue from the previous skeleton of an app that monitors temperature and pressure of a boiler. You can model the app based on the Thyroid app. In this chapter, we will begin building out the pages in our skeleton app.
   - Implement the password-based entry page.
   - Implement the page to get basic information about the boiler such as the boiler ID, date of purchase, maximum allowable values of pressure and temperature, and an ability to change the password.
   - We already have a menu page with four choices. We will implement only the first two choices for data entry and storage
     i. Implement the page to allow you to change the basic information about the boiler.
     ii. Implement the page to enter data f– temperature and pressure
2. **Blood pressure monitor:** Please continue from the previous skeleton of an app that monitors blood pressure. You can model the app based on the Thyroid app. In this chapter, we will begin building out the pages in our skeleton app.
   - Implement the password-based entry page.
   - Implement the page to get basic information about the person such as name, date of birth, and health insurance card number, along with an ability to change the password.
   - A menu page with four choices
     i. Implement the page to allow you to change the basic information about the person.
     ii. Implement the page to enter data—blood pressure.
3. **Power consumption monitor:** Please continue from the previous skeleton of an app that monitors power consumption of a manufacturing plant. You can model the app based on the Thyroid app. In this chapter, we will begin building out the pages in our skeleton app.
   - Implement the password-based entry page.
   - Implement the page to get basic information about the plant such as the plant ID and date of installation, along with an ability to change the password.

- We already have a menu page with four choices. We will only implement the first two choices for data entry and storage
    i. Implement the page to allow you to change the basic information about the plant.
    ii. Implement the page to enter data—power consumed

4. **Body mass index (BMI) monitor:** Please continue from the previous skeleton of an app that monitors the body mass index of a person based on height and weight. You can model the app based on the Thyroid app. In this chapter, we will begin building out the pages in our skeleton app.
    - Implement the password-based entry page.
    - Implement the page to get basic information about the person such as name, date of birth, and health insurance card number, along with an ability to change the password.
    - We already have a menu page with four choices. We will only implement the first two choices for data entry and storage
        i. Implement the page to allow you to change the basic information about the person.
        ii. Implement the page to enter data—height (meters) and weight (kg). Your history should show the value of height, weight, as well as the BMI

$$BMI = \frac{weight}{(height)^2}$$

5. **Managing a line of credit:** Please continue from the previous skeleton of an app that manages our line of credit at a bank. The line of credit allows us to carry a certain maximum negative balance of p. Negative balance will be charged an interest at a specified fixed rate x. A positive balance can lead to an interest accumulation at a different fixed rate y.
    - Implement the password-based entry page.
    - Implement the page to get basic information about the line of credit such as the name of the bank and the date of account opening, along with an ability to change the password.
    - A menu page with four choices: We already have a menu page with four choices. We will only implement the first two choices for data entry and storage
        i. An option and corresponding page to allow you to change the basic information about the line of credit.
        ii. An option and corresponding page to enter data—credit or debit. The history should show the balance after the debit or credit, which should include interest charges or accumulation.

CHAPTER 7

Graphics on HTML5 canvas

WHAT WE WILL LEARN IN THIS CHAPTER

1. How to work with canvas
2. How to draw various shapes
3. How to display some of the graphical gadgets
4. How to draw graphs
5. How to create an icon for the app on the home screen
6. How to run an app locally with no Internet connection

In this chapter, we are going to focus on graphics for web and mobile pages using the canvas element in HTML5. We will begin with a couple of simple apps that help us understand some of the basic JavaScript functions to draw shapes. Then we will see how to combine these calls to compose more complicated shapes. The world of web and mobile app development is constantly changing with new open-source packages being made available to simplify the app development process. Therefore, we will look at some of the existing packages and see how we can leverage them for impressive displays of results. We will first apply these packages to enhance the analytics for our Thyroid app. We will then show how the code can be adapted for the Projectile app. This exercise is meant to show how one can recycle code from one app to another app. Finally, we will study how to launch our browser-based apps from the home screen as well as with no Internet connection.

7.1 Introduction to canvas drawing

Let us begin our exploration of graphics by drawing a line and a circle as shown in Figure 7.1. The HTML5 code for the app is shown in Figure 7.2. First, we include the script canvasIntro.js in the `<script>. . .</script>` tag. The body contains the canvas element where we will be drawing the two shapes. The height and width of the canvas is set to 250 pixels each. We set the id attribute of the canvas element to "canvasElement". The onload attribute of the `<body>` tag tells the browser to call the draw function. The JavaScript code for the app, including the draw() function, is in canvasIntro.js and is shown in Figure 7.3.

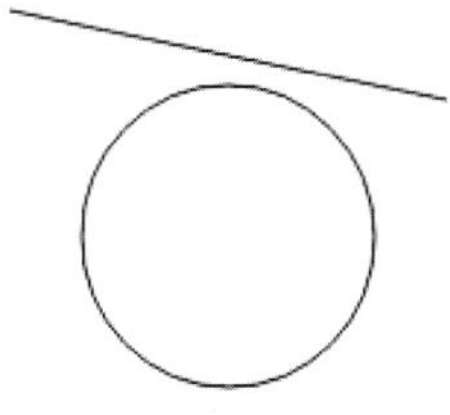

FIGURE 7.1 Screenshot of canvasIntro app

```
<!DOCTYPE html>
<html>
  <head>
    <script type="text/javascript"
      src="scripts/canvasIntro.js">
    </script>
  </head>
  <body onload="draw();">
    <canvas id="canvasElement" height="250" width="250">
    </canvas>
  </body>
</html>
```

FIGURE 7.2 HTML5 code for canvasIntro app canvasIntro.html

```
function draw()
{
  var canvas = document.getElementById("canvasElement");
  var canvasContext = canvas.getContext("2d");

  drawLine(canvasContext, 50, 50, 200, 80);
  drawCircle(canvasContext, 125, 125, 50);
}

function drawLine(canvasContext, lineStartX, lineStartY,
             lineEndX, lineEndY)
{
  canvasContext.beginPath();
  canvasContext.moveTo(lineStartX, lineStartY);
  canvasContext.lineTo(lineEndX, lineEndY);
  canvasContext.stroke();
}

function drawCircle(canvasContext, centerX, centerY, radius)
{
  var startAngleInRadians=0;
  var endAngleInRadians=2*Math.PI;

  canvasContext.beginPath();
  canvasContext.arc(centerX, centerY, radius, startAngleInRadians,
                                              endAngleInRadians);
  canvasContext.stroke();
}
```

FIGURE 7.3 JavaScript code for canvasIntro app canvasIntro.js

The first thing we do in the draw function is to retrieve the canvas using the getElementByID() method, employing the "canvasElement" as the id. The element that is returned is stored in a variable called canvas. We then call one of its methods getContext(). We always pass the "2d" as the parameter to getContext(). As new displays come along, we may pass different values to this function. It returns a built-in HTML5 object, with many properties and methods for drawing paths, boxes, circles, text, and images on the canvas. In the next section, we will discuss some of the frequently used properties and methods of the canvas element. In the draw()

function, the object returned by getContext() is stored in the variable canvasContext. We then call two functions—one to draw the line and another to draw a circle.

The drawLine() function takes the canvas element as the first parameter and the x and y coordinates of the beginning of the line as parameters two and three. The fourth and fifth parameters are the x and y coordinates of the endpoint of the line.

The drawCircle() function takes the canvas element as the first parameter and the x and y coordinates of the center of the circle as parameters two and three. The fourth parameter specifies the radius of the circle.

We will now look at the details of these two functions shown in Figure 7.3. Whenever we are going to draw any shape, we have to start with the JavaScript methods for the canvas element called beginPath() and end with the function stroke(). We see these two calls in both the drawLine() and drawCircle() functions. The drawLine() function first uses the moveTo() method of the canvas element to mark the beginning of the line by using the x and y coordinates of the starting point that were received the second and third parameters. The canvas element has a lineTo() method used to draw the line from the beginning coordinates specified by the moveTo() method to the end coordinates specified by parameters to the lineTo() method. As mentioned before, the stroke() method draws the actual path that has been established by moveTo() and lineTo().

The drawCircle() function is actually drawn by an arc function. The arc function takes five parameters: x and y coordinates of the center, radius, and the beginning and the end angles. The angles have to be specified in radians as opposed to degrees. We have set the beginning of the angle to be 0 and the end of the angle to be 360 degrees, which is $2 \times \pi$ in radians. We use the Math.PI constant from the JavaScript Math object.

7.2 Frequently used methods and properties of the canvas element

Before moving on to more refined apps involving canvas, let us look at some of the frequently used properties and methods of the HTML5 object returned by the getContext() method. The tables that list them are obtained from www.w3schools.com with minor stylistic modifications. For more information on some of the properties and methods that are not used in this book, please visit www.w3schools.com. Figure 7.4 shows a number of generic properties that are associated with the canvas element. One can either inspect the value of the property or set it in JavaScript. Figure 7.5 lists some of the methods used especially for patterns and gradients of the fill. Properties of lines that can be drawn on the canvas are shown in Figure 7.6. Methods for drawing rectangles are listed in Figure 7.7. As we saw earlier, path is the generic term for specifying lines and curves. Methods for manipulating the path are shown in Figure 7.8. You can resize and rotate a canvas using methods shown in Figure 7.9. The last method in the table also allows you to redefine where your (0,0) coordinate will be. The properties of text on a canvas are shown in Figure 7.10. You can either obtain or set the values of these properties. Methods for manipulating the text on canvas are described in Figure 7.11. There are a number of additional properties and methods. Moreover, exploiting these properties and methods to the fullest extent requires a more sophisticated understanding of their values. We encourage the readers to explore www.w3schools.com to learn more.

Property	Description
fillStyle	color, gradient, or pattern used to fill the drawing
strokeStyle	color, gradient, or pattern used for strokes
shadowColor	color to use for shadows
shadowBlur	blur level for shadows
shadowOffsetX	horizontal distance of the shadow from the shape
shadowOffsetY	vertical distance of the shadow from the shape

FIGURE 7.4 Properties related to setting colors, styles, and shadows of the canvas element
Based on http://www.w3schools.com

Method	Description
createLinearGradient()	Creates a linear gradient
createPattern()	Repeats a specified element in the specified direction
createRadialGradient()	Creates a radial/circular gradient
addColorStop()	Specifies the colors and stop positions in a gradient object

■ **FIGURE 7.5** Methods related to setting colors, styles, and shadows of a canvas element
Based on http://www.w3schools.com

Property	Description
lineCap	style of the end caps for a line
lineJoin	type of corner created when two lines meet
lineWidth	current line width
miterLimit	maximum miter length

■ **FIGURE 7.6** Properties for line styles in a canvas element
Based on http://www.w3schools.com

Method	Description
rect()	Creates a rectangle
fillRect()	Draws a filled rectangle
strokeRect()	Draws a rectangle with no fill
clearRect()	Clears the specified pixels within a given rectangle

■ **FIGURE 7.7** Methods for rectangles in a canvas element
Based on http://www.w3schools.com

Method	Description
fill()	Fills the current drawing also called path
stroke()	Draws the path that has been previously defined
beginPath()	Begins a path, or resets the current path
moveTo()	Moves the path to the specified point in the canvas, without creating a line. It is like lifting a pen and going to the point
closePath()	Ends a path, creates a path from current point to the starting point
lineTo()	Adds a new point and creates a line from that point to the last specified point in the canvas
clip()	Clips a region of any shape and size from the original canvas
quadraticCurveTo()	Creates a quadratic Bézier curve
bezierCurveTo()	Creates a cubic Bézier curve
arc()	Creates an circular arc/curve also used to create circles
arcTo()	Creates an circular arc/curve between two tangents
isPointInPath()	Returns true if the specified point is in the current path

■ **FIGURE 7.8** Methods for drawing lines and curves using path in a canvas element
Based on http://www.w3schools.com

Method	Description
scale()	Scales the current drawing bigger or smaller
rotate()	Rotates the current drawing
translate()	Remaps the (0,0) position on the canvas

■ **FIGURE 7.9** Methods for transforming drawing in a canvas element
Based on http://www.w3schools.com

Property	Description
font	current font properties for text content
textAlign	current alignment for text content
textBaseline	current text baseline used when drawing text

■ **FIGURE 7.10** Properties related to text in a canvas element
Based on http://www.w3schools.com

Method	Description
fillText()	Draws filled text on the canvas
strokeText()	Draws text on the canvas with no fill
measureText()	Returns the width of the specified text

■ **FIGURE 7.11** Methods related to text in a canvas element
Based on http://www.w3schools.com

It will not be possible to explore all the methods and properties that we listed in Figures 7.4 to 7.11. Let us experiment with some of these properties and methods to draw some refined shapes as shown in Figure 7.12. The HTML5 code for the app is shown in Figure 7.13. It is more or less the same as the canvasIntro app we saw earlier in Figure 7.2. One difference is

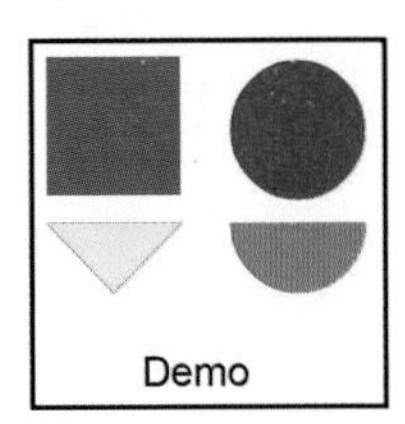

■ **FIGURE 7.12** Screenshot of the canvasShapes app

```
<html>
<head>
  <script type="text/javascript"
    src="scripts/canvasShapes.js">
  </script>
</head>
<body onload="draw()">
  <canvas id="canvasElement" width="400" height="400">
  </canvas>
</body>
</html>
```

■ **FIGURE 7.13** HTML5 code for the canvasShapes app

that we include the script canvasShapes.js. The body contains the canvas element where we will be drawing the two shapes. The height and width of the canvas are set to 400 pixels each. We set the id attribute of the canvas element to "canvasElement". The onload attribute of the the `<body>` tag tells the browser to call the draw function.

The draw function shown in Figure 7.14 proceeds in the same manner as the draw function we saw earlier. It retrieves the canvas using the getElementByID() method employing the "canvasElement" as the id. The element that is returned is stored in a variable called canvas. We then call one of its methods getContext(). It returns a built-in HTML5 object, with many properties and methods for drawing paths, boxes, circles, text, and images on the canvas. We then call six functions: one to draw the border around the canvas, four functions to draw the four shapes shown in Figure 7.12, and the sixth function to display the text. Given the length of the script, we have divided it in Figures 7.14, 7.15, and 7.16.

The drawBorder() function receives three parameters: the canvas context element and the height and width of the canvas. Before we start changing some of the properties, we want to save the current context. Otherwise, these changed properties will apply to all the subsequent drawings. We do this with the help of the method save() (note that this method was not listed in Figures 7.4–7.11). It starts with the method beginPath() as before. It then calls the rect() method with the x and y coordinates of the top left corner to be (0,0) and then specifies the height and width. We set a couple of properties for the line that will be drawn including the lineWidth of 10 and strokeStyle to be black. The stroke() method draws the actual shape. We restore the canvas context properties so that the properties lineWidth and strokeStyle will go back to their original values.

Now, it is time to draw the first shape—that is, the blue rectangle from Figure 7.12. We have passed the canvas context, the starting x and y coordinates, height, and width to the drawRectangle() function. It first sets the fillStyle to be blue, and calls the fillRect() method to draw the rectangle. This is different from the rect() method we had used previously, when we did not want our rectangle to be filled. Note that we did not use the stroke() method as it does not go with the fillRect().

```
function draw()
{
  var canvas = document.getElementById("canvasElement");
  var canvasContext = canvas.getContext("2d");

  drawBorder(canvasContext, 400, 400);
  drawRectangle(canvasContext, 20, 20, 150, 150);
  drawCircle(canvasContext, 300, 100, 75);
  drawTriangle(canvasContext, 20, 200, 150);
  drawSemicircle(canvasContext, 300, 200, 75);
  drawText(canvasContext, "Demo", 125, 375);
}

function drawBorder(canvasContext, height, width)
{
  canvasContext.save();

  canvasContext.rect(0, 0, height, width);
  canvasContext.lineWidth = 10;
  canvasContext.strokeStyle = "black";
  canvasContext.stroke();
```

```
    canvasContext.restore();
}

function drawRectangle(canvasContext, x, y, width, height)
{
    canvasContext.fillStyle = "blue";
    canvasContext.fillRect(x, y, width, height);
}
```

FIGURE 7.14 JavaScript code for the canvasShapes app—I

Drawing of the remaining three shapes is shown in Figures 7.15 and 7.16. All the shapes are going to be filled with different colors. So the first line of each of these functions starts by setting up the appropriate color for the property fillStyle. One additional common feature for all the shapes is the fact that the drawing is enclosed in the pair beginPath() and closePath(). This ensures these functions signal to the canvas that everything drawn between those two functions should be connected together. Let us now examine the details in the middle that differ for the circle, the triangle, and the semi-circle.

For the drawCircle(), we receive the x and y coordinates of the center and the radius. We set the starting angle to be 0 and the stopping angle for the arc to be $2 \times \pi$ in radians, using the Math.PI constant as before. Once we have used the arc() method with the x and y coordinates

```
function drawCircle(canvasContext, x, y, r) {
    canvasContext.fillStyle = "red";

    var start = 0;
    var stop = 2 * Math.PI;

    canvasContext.beginPath();
    canvasContext.arc(x, y, r, start, stop);
    canvasContext.stroke();
    canvasContext.fill();
    canvasContext.closePath();
}

function drawTriangle(canvasContext, startX,
    startY, length) {
    canvasContext.fillStyle = "yellow";

    canvasContext.beginPath();
    canvasContext.moveTo(startX, startY);
    canvasContext.lineTo(startX + length, startY);
    canvasContext.lineTo(startX + (length / 2),
        startY + (length / 2));
    canvasContext.lineTo(startX, startY);
    canvasContext.stroke();
    canvasContext.fill();
    canvasContext.closePath();
}
```

FIGURE 7.15 JavaScript code for the canvasShapes app—II

```
function drawSemicircle(canvasContext, x, y, r)
{
  canvasContext.fillStyle = "green";

  var start = 0;
  var stop = Math.PI;

  canvasContext.beginPath();
  canvasContext.arc(x, y, r, start, stop);
  canvasContext.stroke();
  canvasContext.fill();
  canvasContext.closePath();
}

function drawText(canvasContext, text, x, y)
{
  canvasContext.fillStyle = "black";
  canvasContext.font = "45px sans-serif";
  canvasContext.fillText(text, x, y);
}
```

FIGURE 7.16 JavaScript code for the canvasShapes app—III

of the center, the radius, and start and stop angles, we actually draw the path using the stroke() method. Since we want to fill this shape, we use the fill() method.

The drawTriangle() method receives the x and y coordinates of the starting point and length of the triangle. We use the moveTo() method to move to the starting point. The first line of the triangle is a horizontal line of the given length. Therefore, we add the length to the x coordinate of the starting point while keeping the y coordinate to be the same as the starting point resulting in the call: `canvasContext.lineTo(startX+length, startY)`. Since it is an equilateral triangle, the x coordinate of the lowest point will be half a length below and half a length to the right of the starting point. Therefore, the next line will be identified using the call `canvasContext.lineTo(startX+(length/2), startY+(length/2))`. The final line will bring us back to the starting point with the call: `canvasContext.lineTo(startX, startY)`. As before, we actually draw the path using the stroke() method and fill the resulting triangle using the fill() method.

The drawSemiCircle() function is very similar to the drawCircle. It receives the x and y coordinates of the center and the radius. The starting angle is 0. The major difference is that the stopping angle for the arc is set at π instead of $2 \times \pi$. The rest of the function is the same as drawCircle().

The last function in our canvasBasics app is for displaying text that is passed as a parameter as well as the x and y coordinates of the starting point. We set the fillStyle property to be “black” and font to be sans serif with the 45 pixel height. The actual text is drawn using the fillText() method.

7.3 Adding advice and gauge meter to the Thyroid app using RGraph

After this basic introduction to the canvas element and basic drawing methods, we are going to look at some of the practical applications that use the canvas element. In order to create some really impressive graphics, one tends to use some of the existing packages as we will be doing for our Thyroid and Projectile apps. Clearly, someone has to create these packages from scratch. This book focuses more on using existing packages than creating them.

Let us refresh our memory of where we were in the Thyroid app. Figure 7.17 shows the screenshot of the Records page, where we have entered results of three blood tests. We can go to the menu and press the button for Suggestions, and we see the screen shown in Figure 7.18. The advice is based on the logic provided by the physicians. We will look at this logic later when we see how it is programmed. The drawing methods that we looked at may make it possible for us to draw such a refined figure. However, it will be a fairly complicated program. We will look at a package that will allow us to do this with relative ease, so we can concentrate on the logical aspects of our advice.

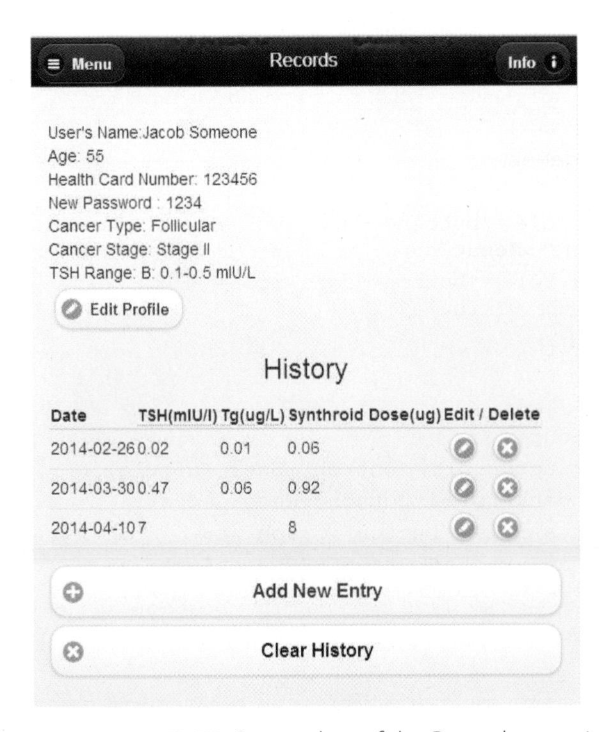

■ **FIGURE 7.17** Screenshot of the Records page in the Thyroid app

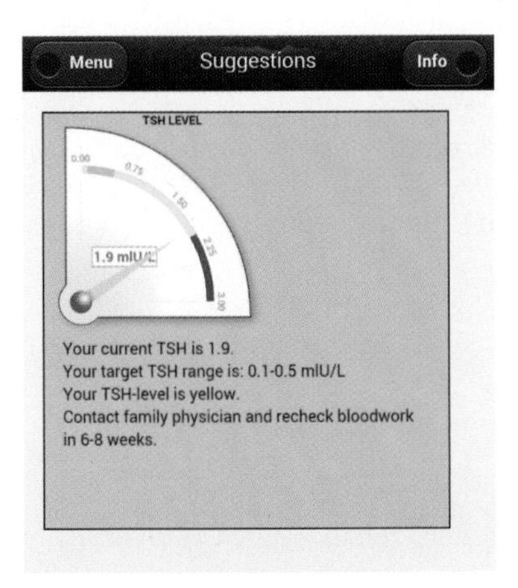

■ **FIGURE 7.18** Screenshot of the Suggestions page in the Thyroid app

Let us also remind ourselves as to how we get to the Suggestions page. Figure 7.19 shows the HTML5 code for Menu page. We have highlighted the button that leads us to the `<div>` section with id "pageAdvice". The HTML5 code for the Suggestions page is reproduced in Figure 7.20. We have studied it in detail previously. The most notable thing about the page for us at this time is the canvas with id "AdviceCanvas". This canvas will be used to render the advice such as the one that is shown in Figure 7.18. This rendering is done by the JavaScript function advicePage(), which is listed in the file Advice.js. Before we look at the function, it may be a good idea to see how the function is triggered. Figure 7.21 shows the function `$(document).on("pageshow", function(){...}`. The highlighted code shows that when the `<div>` section with id "pageAdvice" is the active page, we should call two functions: advicePage() and resizeGraph(). Let us focus on the main part of the function advicePage() that is shown in Figure 7.22.

```
  <!-- Menu page -->
  <div data-role="page" id="pageMenu">
    <div data-role="header">
      <a href="#pageMenu" data-role="button" data-icon="bars" data-
iconpos="left" data-inline="true">Menu</a>
      <a href="#pageAbout" data-role="button" data-icon="info" data-
iconpos="right" data-inline="true">Info</a>
      <h1>Thyroid Cancer Aide</h1>
    </div>
    <div data-role="content">
      <div data-role="controlgroup">
        <a href="#pageUserInfo" data-role="button">User Info</a>
        <a href="#pageRecords" data-role="button">Records</a>
        <a href="#pageGraph" data-role="button">Graph</a>
        <a href="#pageAdvice" data-role="button">Suggestions</a>
      </div>
    </div>
  </div>
```

FIGURE 7.19 HTML5 code for the Menu page in the Thyroid app

```
  <!--Advice Page -->
  <div data-role="page" id="pageAdvice">
    <div data-role="header">
      <a href="#pageMenu" data-role="button" data-icon="bars" data-
iconpos="left" data-inline="true">Menu</a>
      <a href="#pageAbout" data-role="button" data-icon="info" data-
iconpos="right" data-inline="true">Info</a>
      <h1>Suggestions</h1>
    </div>
    <div data-role="content">
      <canvas id="AdviceCanvas" width="550" height="550"
        style="border:1px solid #000000;">
      </canvas>
    </div>
  </div>
```

FIGURE 7.20 HTML5 code for the Suggestions page in the Thyroid app

```
/* Runs the function to display the user information, history, graph or
 * suggestions, every time their div is shown
 */
$(document).on("pageshow", function(){
  if($('.ui-page-active').attr('id')=="pageUserInfo")
  {
    showUserForm();
  }
  else if($('.ui-page-active').attr('id')=="pageRecords")
  {
    loadUserInformation();
    listRecords();
  }
  else if($('.ui-page-active').attr('id')=="pageAdvice")
  {
    advicePage();
    resizeGraph();
  }
  else if($('.ui-page-active').attr('id')=="pageGraph")
  {
    drawGraph();
    resizeGraph();
  }
});
```

FIGURE 7.21 JavaScript code for setting up page shows in the Thyroid app pageLoader.js

```
function advicePage()
{

  if (localStorage.getItem("tbRecords") === null)
  {
    alert("No records exist".);

    $(location).attr("href", "#pageMenu");
  }
  else
  {

    var user=JSON.parse(localStorage.getItem("user"));
    var TSHLevel=user.TSHRange;

    var tbRecords=JSON.parse(localStorage.getItem("tbRecords"));
    tbRecords.sort(compareDates);
    var i=tbRecords.length-1;
    var TSH=tbRecords[i].TSH;
```

```
        var c=document.getElementById("AdviceCanvas");
        var ctx=c.getContext("2d");
        ctx.fillStyle="#c0c0c0";
        ctx.fillRect(0,0,550,550);
        ctx.font="22px Arial";
        drawAdviceCanvas(ctx,TSHLevel,TSH);

    }
}
```

FIGURE 7.22 JavaScript code for the main part of advicePage()—II Advice.js

The function advicePage() checks to make sure that the localStorage has an item with the name "tbRecords". If the item does not exist, an appropriate dialogue box is displayed using the alert() method, and control is returned to the pageMenu as shown in Figure 7.22. Otherwise, we start processing the records. The advice or suggestion is based on the TSH value of the last test entry in comparison to the acceptable TSH range. Therefore, we first get the item with the name "user" from the localStorage, parse it with the JSON.parse() method and save the user. TSHRange in a variable called TSHLevel. Then we retrieve the array of records from the local storage with the name "tbRecords" and parse it with JSON.parse(). We sort it chronologically as before so that the last record will be the most recent one. Since the arrays start with index 0, the last record is in location tbRecords.length–1, which is assigned to a variable i. The TSH value of the last (i^{th}) record is stored in a variable called TSH. As mentioned previously, the physicians told us that the advice will depend only on the suggested TSH range stored in the variable TSHLevel and the last TSH value stored in the variable TSH. Once we have set up the canvas, we will be ready to display suggestions using the logic provided by the physicians. The canvas element with id "AdviceCanvas" is retrieved in a variable called c; its context is then stored in a variable called ctx. We set the fillStyle to be the color "#c0c0c0". It is a hexa-decimal value and is dark gray as shown in Figure 7.18. The fillRect() method actually fills the color in the canvas. The font is set to Arial with a size of 22 pixels. We are now ready to display the suggestions using the advicePage() function with three parameters: ctx, TSHLevel, and TSH.

The function drawAdviceCanvas() is shown in Figure 7.23. We set the font and fillStyle for the text and write the current value of TSH in the x coordinate 25 and row 320. Depending on the TSH range, we use an if/else conditional statement to first write the acceptable TSH range. Then we write the textual advice using functions such as levelAwrite(), levelBwrite(), and levelCwrite(). This is followed by rendering the meter with a dial using functions such as levelAMeter(), levelBMeter(), and levelCMeter(). The meter appears at the top of the page in the app.

The JavaScript code for levelAwrite(), levelBwrite(), and levelCwrite() appears in Figures 7.24, 7.25, and 7.26. These functions contain the if/else statements based on logic provided by the physician. For example, for level A

- TSH between 0.01 and 0.1 means things are going well, so we indicate a green code.
- TSH greater than 0.1 and less than or equal to 0.5 means there is a cause for concern, so we indicate a yellow code.
- For any other TSH level, the situation is critical, so we indicate a red code.

The code is written using the function writeAdvice() shown in Figure 7.27. The function is going to have a total of three lines. These are shown at the bottom of the function. The first line prints the code level at x coordinate 25 and y coordinate 380. The next two lines are advice for the patient. They are determined by the if/else conditions. The first line of advice is written at x coordinate 25 and y coordinate 410, and the second line is written at x coordinate 25 and y coordinate 440.

```
function drawAdviceCanvas(ctx,TSHLevel,TSH)
{
  ctx.font="22px Arial";
  ctx.fillStyle="black";
  ctx.fillText("Your current TSH is " + TSH + "".., 25, 320);

  if (TSHLevel == "LevelA")
  {
    ctx.fillText("Your target TSH range is: 0.01-0.1 mlU/L", 25, 350);
    levelAwrite(ctx,TSH);
    levelAMeter(ctx,TSH);
  }
  else if (TSHLevel == "LevelB")
  {
    ctx.fillText("Your target TSH range is: 0.1-0.5 mlU/L", 25, 350);
    levelBwrite(ctx,TSH);
    levelBMeter(ctx,TSH);
  }
  else if (TSHLevel == "LevelC")
  {
    ctx.fillText("Your target TSH range is: 0.35-2.0 mlU/L", 25, 350);
    levelCwrite(ctx,TSH);
    levelCMeter(ctx,TSH);
  }
}
```

FIGURE 7.23 JavaScript code for rendering the suggestions canvas in the Thyroid app Advice.js

```
function levelAwrite(ctx,TSH)
{
  if ((TSH >= 0.01) && (TSH <= 0.1))
  {
     writeAdvice(ctx",green");
  }
  else if ((TSH > 0.1) && (TSH <= 0.5))
  {
     writeAdvice(ctx",yellow");
  }
  else
  {
     writeAdvice(ctx",red");
  }
}
```

FIGURE 7.24 JavaScript code for textual suggestions for Level A in the Thyroid app Advice.js

```
function levelBwrite(ctx,TSH)
{
  if ((TSH >= 0.1) && (TSH <= 0.5))
  {
    writeAdvice(ctx",green");
  }
  else if ((TSH > 0.5) && (TSH <= 2.0))
  {
    writeAdvice(ctx",yellow");
  }
  else if ((TSH >= 0.01) && (TSH < 0.1))
  {
    writeAdvice(ctx",yellow");
  }
  else
  {
    writeAdvice(ctx",red");
  }
}
```

FIGURE 7.25 JavaScript code for textual suggestions for Level B in the Thyroid app Advice.js

```
function levelCwrite(ctx,TSH)
{
  if ((TSH >= 0.35) && (TSH <= 2.0))
  {
    writeAdvice(ctx",green");
  }
  else if ((TSH > 2) && (TSH <= 10))
  {
    writeAdvice(ctx",yellow");
  }
  else if ((TSH >= 0.1) && (TSH < 0.35))
  {
    writeAdvice(ctx",yellow");
  }
  else
  {
    writeAdvice(ctx",red");
  }
}
```

FIGURE 7.26 JavaScript code for textual suggestions for Level C in the Thyroid app Advice.js

```
function writeAdvice(ctx,level)
{
  var adviceLine1="";
  var adviceLine2="";
  if(level=="red")
  {
    adviceLine1="Please consult with your family";
    adviceLine2="physician urgently".;
  }
  else if(level=="yellow")
  {
    adviceLine1="Contact family physician and recheck bloodwork";
    adviceLine2="in 6-8 weeks".;
  }
  else if(level="green")
  {
    adviceLine1="Repeat bloodwork in 3-6 months".;
  }
  ctx.fillText("Your TSH-level is " + level +""., 25, 380);
  ctx.fillText(adviceLine1, 25, 410);
  ctx.fillText(adviceLine2, 25, 440);
}
```

FIGURE 7.27 JavaScript code for displaying textual suggestions in the Thyroid app Advice.js

The JavaScript code for levelAMeter(), levelBMeter(), and levelCMeter() appears in Figures 7.28, 7.29, and 7.30. These functions use a very useful package called RGraph.

RGraph is a free HTML5 canvas-based JavaScript library written and maintained by Richard Heyes. It can be downloaded from http://www.rgraph.net/. The website also has tutorials and examples. We are not promoting the library but illustrating how one can use such libraries instead of having to start from scratch. RGraph facilitates creation of more than 20 types of web charts including bar charts, line charts, pie charts, meter charts, and gauge charts.

```
function levelAMeter(ctx,TSH)
{
  if (TSH <= 3)
  {
    var cg=new RGraph.CornerGauge("AdviceCanvas", 0, 3, TSH)
      .Set("chart.colors.ranges", [[0.5, 3, "red"], [0.1, 0.5,
"yellow"], [0.01, 0.1, "#0f0"]]);
  }
  else
  {
    var cg=new RGraph.CornerGauge("AdviceCanvas", 0,TSH,TSH)
      .Set("chart.colors.ranges", [[0.5, 3, "red"], [0.1, 0.5,
"yellow"], [0.01, 0.1, "#0f0"], [3.01, TSH, "red"]]);
  }
  drawMeter(cg);
}
```

FIGURE 7.28 JavaScript code for graphical suggestions for Level A in the Thyroid app Advice.js

```
function levelBMeter(ctx,TSH)
{
  if (TSH <= 3)
  {
    var bcg=new RGraph.CornerGauge("AdviceCanvas", 0, 3, TSH)
    .Set("chart.colors.ranges", [[2.01, 3, "red"], [0.51, 2, "yellow"],
[0.1, 0.5, "#0f0"], [0.01, 0.1, "yellow"]]);
  }

  else
  {
    var bcg=new RGraph.CornerGauge("AdviceCanvas", 0, TSH, TSH)
      .Set("chart.colors.ranges", [[2.01, 3, "red"], [0.51, 2",yellow"],
[0.1, 0.5, "#0f0"], [0.01, 0.1, "yellow"], [3, TSH, "red"]]);
  }
  drawMeter(bcg);
}
```

FIGURE 7.29 JavaScript code for graphical suggestions for Level B in the Thyroid app Advice.js

```
function levelCMeter(ctx,TSH)
{
  if (TSH <= 15)
  {
    var ccg=new RGraph.CornerGauge("AdviceCanvas", 0, 15, TSH)
      .Set("chart.colors.ranges", [[10.01, 15, "red"], [2.01, 10",yellow"],
[0.35, 2, "#0f0"], [0.1,.34",yellow"]]);
  }
  else
  {
    var ccg=new RGraph.CornerGauge("AdviceCanvas", 0, TSH, TSH)
      .Set("chart.colors.ranges", [[10.01, 15, "red"], [2.01, 10,
"yellow"], [0.35, 2, "#0f0"], [0.1, 0.34, "yellow"], [15.01, TSH, "red"]]);
  }
  drawMeter(ccg);
}
```

FIGURE 7.30 JavaScript code for graphical suggestions for Level C in the Thyroid app Advice.js

Figures 7.28, 7.29, and 7.30 show the JavaScript code for displaying Corner Gauge meters for three levels of TSH. The structure of the three functions is identical. We will analyze the function levelAMeter() shown in Figure 7.28. We use the value of TSH to decide the range of the meter. The maximum value for level A is expected to be 3. So our meter should go up to 3. If it goes beyond 3, our meter should go up to the TSH value. The statements inside both the if and else parts are the same. They create the CornerGauge meter from the RGraph library using the constructor RGraph.CornerGuage(). The constructor takes four parameters

1. The id of the canvas that will hold the gauge meter ("AdviceCanvas")
2. The lower bound of the value (0 in both cases)
3. The upper bound of the value (3 if TSH is less than or equal to 3; otherwise the same as TSH)
4. The value that will be displayed by the meter

We then set the "chart.colors.ranges" property for level A using an array of arrays (two-dimensional array). We have a total of three arrays: one for code red, the next one for code yellow, and the third for code green. Each one of these arrays is an array itself consisting of three elements. It consists of the lower and upper bound of the range and the color to display for that range. In our case,

- Display red for the TSH in the range [0.5, 3].
- Display yellow for the TSH in the range [0.1,0.5].
- Display green (color with hexadecimal value "0f0") for the TSH in the range [0.01, 0.1].
- If TSH value is greater than 3, additionally display red for the TSH in the range [3.01, TSH].

Once we have set up the CornerGauge object the meter is displayed using the function drawMeter() shown in Figure 7.31. The function drawMeter() receives the CornerGauge object and sets a multitude of properties through a series of calls to the set method. Most of these properties are self-explanatory. Finally, we call the Draw() method, which actually draws the meter.

One last bit of business is left before we move on to the graph page in our Thyroid app. We had seen in Figure 7.21 that after rendering the advice page, we resize the graph. The resize() function is shown in Figure 7.32. It makes sure that our graphs fit properly on smaller devices. The function looks at the width of the window using $(window).width. If the width is less than 700, we need to make adjustments to the canvases in our app. The adjustment is done using the css() method of the canvas element. The css method allows us to dynamically set the display properties of the HTML5 elements. In our case, we are setting the width of each canvas to 50 pixels less than the window width—that is, `$(window).width()- 50`.

Let us now turn our attention to the last page in our Thyroid app. It draws the graph of TSH values, as shown in Figure 7.33.

```
function drawMeter(g)
{
  g.Set("chart.value.text.units.post", " mlU/L")
    .Set("chart.value.text.boxed", false)
    .Set("chart.value.text.size", 14)
    .Set("chart.value.text.font", "Verdana")
    .Set("chart.value.text.bold", true)
    .Set("chart.value.text.decimals", 2)
    .Set("chart.shadow.offsetx", 5)
    .Set("chart.shadow.offsety", 5)
    .Set("chart.scale.decimals", 2)
    .Set("chart.title", "TSH LEVEL")
    .Set("chart.radius", 250)
    .Set("chart.centerx", 50)
    .Set("chart.centery", 250)
    .Draw();
}
```

FIGURE 7.31 JavaScript code for displaying graphical suggestions in the Thyroid app Advice.js

```
function resizeGraph()
{
  if($(window).width() < 700)
  {
    $("#GraphCanvas").css({"width":$(window).width()- 50});
    $("#AdviceCanvas").css({"width":$(window).width()- 50});
  }
}
```

■ **FIGURE 7.32** JavaScript code for resizing graphs in the Thyroid app PageLoader.js

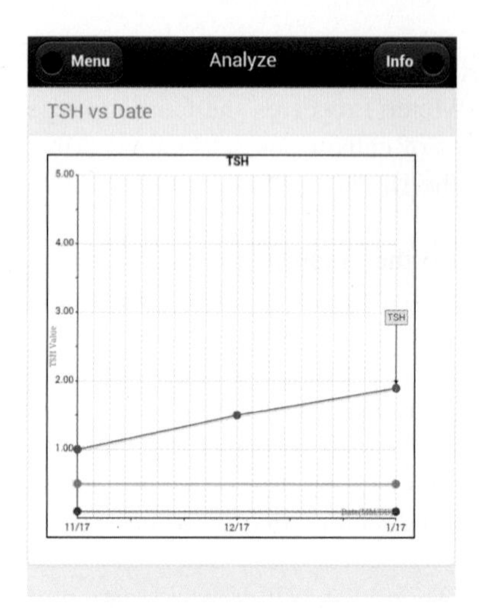

■ **FIGURE 7.33** Screenshot graph page in the Thyroid app Advice.js

7.4 Drawing line graphs in the Thyroid app using RGraph

Let us remind ourselves how we get to the graph page. Figure 7.34 shows the HTML5 code for the menu page. We have highlighted the button that leads us to the `<div>` section with id "pageGraph". The HTML5 code for the graph page is reproduced in Figure 7.35. We have studied it in detail previously. The most notable thing about the page for us at this time is the canvas with id "GraphCanvas". This canvas will be used to render the graph such as the one that is shown in Figure 7.33. This rendering is done by the JavaScript function drawGraph(), which is listed in the file GraphAnimate.js. Let us first see how the function is triggered. Figure 7.36 shows the function `$(document).on("pageshow", function(){...}`. The highlighted code shows that when the `<div>` section with id "pageGraph" is the active page, we should call two functions: drawGraph() and resizeGraph(). We discussed the resizeGraph() function in Figure 7.32. We will now study the graph rendering starting with the function drawGraphic(), which is shown in Figure 7.36.

```
<!-- Menu page -->
<div data-role="page" id="pageMenu">
  <div data-role="header">
    <a href="#pageMenu" data-role="button" data-icon="bars"
data-iconpos="left" data-inline="true">Menu</a>
    <a href="#pageAbout" data-role="button" data-icon="info"
data-iconpos="right" data-inline="true">Info</a>
```

```
        <h1>Thyroid Cancer Aide</h1>
      </div>
      <div data-role="content">
        <div data-role="controlgroup">
          <a href="#pageUserInfo" data-role="button">User Info</a>
          <a href="#pageRecords" data-role="button">Records</a>
          <a href="#pageGraph" data-role="button">Graph</a>
          <a href="#pageAdvice" data-role="button">Suggestions</a>
        </div>
      </div>
    </div>
```

FIGURE 7.34 HTML5 code for the Menu page in the Thyroid app

```
    <!--Graph Page -->
    <div data-role="page" id="pageGraph"  class="test">
      <div data-role="header">
        <a href="#pageMenu" data-role="button" data-icon="bars"
  data-iconpos="left" data-inline="true">Menu</a>
        <a href="#pageAbout" data-role="button" data-icon="info"
  data-iconpos="right" data-inline="true">Info</a>
        <h1>Analyze</h1>
      </div>
      <div class="panel panel-success">
        <div class="panel-heading">
      <h3 class="panel-title">TSH vs Date</h3>
          </div>
      <div class="panel-body">
      <canvas id="GraphCanvas" width="500" height="500"
        style="border:1px solid #000000;">
      </canvas>
        </div>
      </div>
    </div>
```

FIGURE 7.35 HTML5 code for the graph page in the Thyroid app

```
/* Runs the function to display the user information, history,
graph or
 * suggestions, every time their div is shown
 */
$(document).on("pageshow", function(){
  if($('.ui-page-active').attr('id')=="pageUserInfo")
  {
    showUserForm();
  }
  else if($('.ui-page-active').attr('id')=="pageRecords")
  {
    loadUserInformation();
    listRecords();
  }
```

```
    else if($('.ui-page-active').attr('id')=="pageAdvice")
    {
      advicePage();
      resizeGraph();
    }
    else if($('.ui-page-active').attr('id')=="pageGraph")
    {
      drawGraph();
      resizeGraph();
    }
  });
```

FIGURE 7.36 JavaScript code for setting up page shows in the Thyroid app pageLoader.js

The function drawGraph() checks to make sure that the localStorage has an item with the name “tbRecords”. If the item does not exist, an appropriate dialogue box is displayed using the alert() method, and control is returned to the pageMenu as shown in Figure 7.37. Otherwise, we start processing the records for drawing the graph. The else part in the function drawGraph() first sets up the canvas with the function setupCanvas(). It then creates arrays for storing TSH (TSHarr) and Date (Datearr) values. We are going to plot the TSH values, with Date values serving as the x-axis labels. This is the main part of our graph, which is shown

```
function drawGraph()
{
  if(localStorage.getItem("tbRecords") === null)
  {
    alert("No records exist".);

    $(location).attr("href", "#pageMenu");
  }
  else
  {

    setupCanvas();

    var TSHarr=new Array();
    var Datearr=new Array();
    getTSHhistory(TSHarr, Datearr);

    var tshLower = new Array(2);
    var tshUpper = new Array(2);
    getTSHbounds(tshLower,tshUpper);

    drawLines(TSHarr, tshUpper, tshLower, Datearr)
    labelAxes();
  }
}
```

FIGURE 7.37 JavaScript code for the drawGraph() function in the Thyroid app GraphAnimate.js

with blue lines in Figure 7.33. We also note that there is a red horizontal line at the bottom and a green horizontal line above it. These two lines correspond to the lower and upper bounds of the TSH range from the user information. In order to store these values, we use tshLower and tshUpper arrays. Both are of size 2. We use the function getTSHbounds() to populate these arrays. The function drawLines() draws the graphs, and we end with function call labelAxes() to label the axes. We will now go through the functions called from drawGraph in greater detail.

Figure 7.38 shows the relatively straightforward code for setting up the canvas, which includes getting the canvas element using the id, then its context. The fillStyle is set to white (hexadecimal value FFFFFF), and a rectangle of 500 by 500 is filled using the fillRect() method.

The function getTSHhistory(), shown in Figure 7.39, forms the meat of the graph drawing exercise. It populates the arrays of TSH (TSHarr) and Date (Datearr) values. First the item with the name "tbRecords" is retrieved from localStorage and parsed using JSON.parse() and

```
function setupCanvas()
{
    var c=document.getElementById("GraphCanvas");
    var ctx=c.getContext("2d");

    ctx.fillStyle="#FFFFFF";
    ctx.fillRect(0, 0, 500, 500);
}
```

FIGURE 7.38 JavaScript code for the setupCanvas() function in the Thyroid app GraphAnimate.js

```
function getTSHhistory(TSHarr, Datearr)
{
    var tbRecords=JSON.parse(localStorage.getItem("tbRecords"));

    tbRecords.sort(compareDates);

    for (var i=0; i < tbRecords.length; i++)
    {
      var date = new Date(tbRecords[i].Date);

      /*These methods start at 0, must increment
       * by one to compensate
       */
      var m=date.getMonth() + 1;
      var d=date.getDate() + 1;

      //The x-axis label
      Datearr[i]=(m + "/" + d);

      //The point to plot
      TSHarr[i]=parseFloat(tbRecords[i].TSH);
    }
}
```

FIGURE 7.39 JavaScript code for the getTSHhistory() function in the Thyroid app GraphAnimate.js

saved in the variable tbRecords. We chronologically sort the tbRecords array as previously done so that the most recent record is last. The following for loop goes through every element of the array tbRecords using the loop control variable i. The Date field of tbRecords[i] is stored in a variable called date. We then extract the month and date from date using the getMonth() and getDate() methods in variables m and d. We have to add 1 to the values, since the values of month and dates start with 0. The Datearr[i] is set to be a string consisting of month (m) followed by date (d), separated by a slash ("/") character. The TSHarr[i] gets the value from tbRecords[i].TSH. Since this value is in string format, we use the parseFloat() method to convert it to a real number.

The function getTSHbounds() is shown in Figure 7.40. First the item with the name "user" is retrieved from localStorage and parsed using JSON.parse() and saved in the variable user. The item with the name TSHRange is then saved in the variable TSHlevel. Depending on the value of TSHlevel, the values of tshUpper[0], tshUpper[1], tshLower[0], and tshLower[1] are set appropriately. We want to draw the reader's attention to the chained assignment statements such as

```
tshUpper[0] = tshUpper[1] = 0.5;
```

The function drawLines(), shown in Figure 7.41, creates a line chart from the RGraph library using the constructor RGraph.Line(). The constructor takes a minimum of two parameters. The first parameter is the id of the canvas ("GraphCanvas"). The subsequent parameters are the arrays that contain the values that will be plotted. In our case, we have three arrays

1. Array of TSH values from the test results.
2. Upper bound of the TSH—an array with two elements that have the same value that gives us the horizontal line.
3. Lower bound of the TSH—an array with two elements that have the same value that gives us the horizontal line.

```
function getTSHbounds(tshLower, tshUpper)
{
    var user=JSON.parse(localStorage.getItem("user"));
    var TSHLevel=user.TSHRange;

    if (TSHLevel == "StageA")
    {
      tshUpper[0] = tshUpper[1] = 0.1;
      tshLower[0] = tshLower[1] = 0.01;
    }
    else if (TSHLevel == "StageB")
    {
      tshUpper[0] = tshUpper[1] = 0.5;
      tshLower[0] = tshLower[1] = 0.1;
    }
    else
    {
      tshUpper[0] = tshUpper[1] = 2.0;
      tshLower[0] = tshLower[1] = 0.35;
    }
}
```

FIGURE 7.40 JavaScript code for the getTSHbounds() function in the Thyroid app GraphAnimate.js

```
function drawLines(TSHarr, tshUpper, tshLower, Datearr)
{
    var TSHline=new RGraph.Line("GraphCanvas", TSHarr, tshUpper,
tshLower)
        .Set("labels", Datearr)
        .Set("colors", ["blue", "green", "green"])
        .Set("shadow", true)
        .Set("shadow.offsetx", 1)
        .Set("shadow.offsety", 1)
        .Set("linewidth", 1)
        .Set("numxticks", 6)
        .Set("scale.decimals", 2)
        .Set("xaxispos", "bottom")
        .Set("gutter.left", 40)
        .Set("tickmarks", "filledcircle")
        .Set("ticksize", 5)
        .Set("chart.labels.ingraph",
        [,, ["TSH", "blue", "yellow", 1, 80],, ])
        .Set("chart.title", "TSH")
        .Draw();
}
```

FIGURE 7.41 JavaScript code for the drawLines() function in the Thyroid app GraphAnimate.js

We then set a multitude of properties through a series of calls to the set method. Most of these properties are self-explanatory. Let us look at some of the more interesting properties

- "labels" specifies an array that will be used to display values on the x-axis. This array is plotted from the first index to the last index, left to right on the graph.
- "colors" is used to specify line colors. We use blue for the TSH values from the test, green for the upper bound of TSH values, and red for the lower bound of the TSH values.
- chart.labels.ingraph is an array of labels drawn inside the chart. We have just chosen to label the third point. The array can consist of
    - The label text ("TSH")
    - The text color ("blue")
    - The background color ("yellow")
    - Position value: 1 or −1 to denote whether the label should be above or below the line
    - The length of the label pointer (80)

Finally, we call the Draw() method to display the graph.

Once we have displayed the graph we put text for the x and y axes using the function labelAxes() shown in Figure 7.42. After getting the context for the canvas, we set the font and fillStyle. For the x-axis we display "Date (MM/DD)". The y-axis is a little fancier. We rotate the context by –90 degrees or $-\frac{\pi}{2}$ radians. Set the text alignment to be "center" and display the x-axis label to be "TSH values".

Next, we are going to use the knowledge of canvas to enhance our Projectile app. Figure 7.43 reminds us of the input that is accepted in terms of angle and velocity. We have entered 70 degrees for angle and 99 meter/second for velocity. Instead of displaying a table of values, we will draw a graph of height versus distance as shown in Figure 7.44.

```
function labelAxes()
{
     var c=document.getElementById("GraphCanvas");
     var ctx=c.getContext("2d");
     ctx.font="11px Georgia";
     ctx.fillStyle="green";
     ctx.fillText("Date(MM/DD)", 400, 470);
     ctx.rotate(-Math.PI/2);
     ctx.textAlign="center";
     ctx.fillText("TSH Value", -250, 10);
}
```

FIGURE 7.42 JavaScript code for the labelAxes() function in the Thyroid app GraphAnimate.js

Angle:

70

Velocity:

99

Calculate!

FIGURE 7.43 Screenshot of the data entry page for the graphical Projectile app physicsProjectileApp4.html

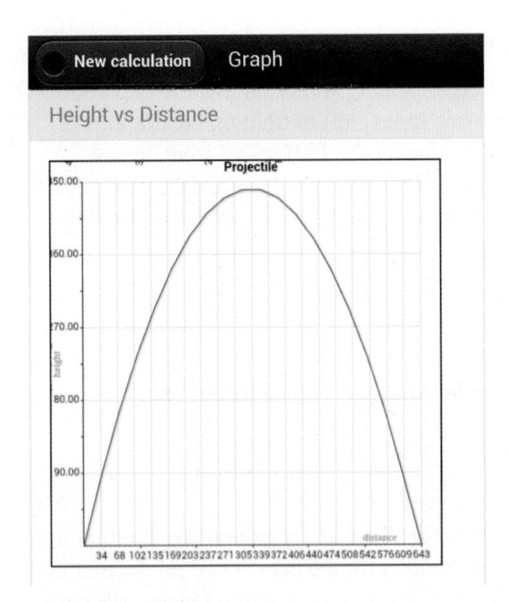

FIGURE 7.44 Screenshot of the Graph page for the graphical Projectile app physicsProjectileApp4.html

We are going to keep as much of the original code as possible and add code that is similar to the graph page from the Thyroid app. One of the objectives of this exercise is to demonstrate how to recycle code from different apps. Figure 7.45 shows the overall structure of the HTML5 code for the graphical Projectile app. We have added RGraph scripts so that we can draw the graph and bootstrap CSS to include a panel. The original HTML5 code from the previous version of the Projectile app is shown in Figure 7.46. We have added one more `<div>` section with data-role of "page" for drawing the graph as shown in Figure 7.47. This page is very similar to the page graph from the Thyroid app. Since we have already studied the essence of both of these pages earlier, we do not need to discuss them in detail again. We will see similar recycled code everywhere needing minimum additional explanation. This is the important feature of the graphical Projectile app: we slap code from two different apps together, make minor adjustments, and voila.

```
<!DOCTYPE html>
<html>
<head>
  <title>Physics Simulator App</title>
  <script type='text/javascript'
    src='scripts/physicsProjectileApp4.js'>
  </script>
  <link rel="stylesheet" href="css/jquery.mobile-1.3.1.min.css">
  <link rel="stylesheet" href="css/bootstrap.css" />
  <script src="scripts/jquery-1.8.3.min.js"> </script>
  <script src="scripts/jquery.mobile-1.3.1.min.js"> </script>
  <script src="scripts/RGraph.common.core.js"></script>
  <script src="scripts/RGraph.common.effects.js"></script>
  <script src="scripts/RGraph.line.js"></script>
  <script src="scripts/RGraph.cornergauge.js"></script>
  <script src="scripts/RGraph.hprogress.js"></script>
  <meta name="viewport"
  content="width=device-width, initial-scale=1.0">
</head>
<body onload='initialize()'>
 <!--  Rest of the body goes here -->
</body>
</html>
```

■ **FIGURE 7.45** Bird's-eye view of HTML5 code for the graphical Projectile app physicsProjectileApp4.html

```
<div data-role="page" id = "pageData">
    <div data-role="content">
      <p>Angle:</p>
      <input type='number' name='angle' id='angle'
        min='0' max='90' placeholder='In degrees'>
      <p>Velocity:</p>
      <input type='number' name='velocity'
        id='velocity'  min='0' max='299792458'
        placeholder='In metres/second'>
      <br/>
      <button onclick='update();'>Calculate!
      <br/>
    </div>
  </div>
```

■ **FIGURE 7.46** HTML5 code for the data entry page of the graphical Projectile app physicsProjectileApp4.html

```
<!--Graph Page -->
  <div data-role="page" id="pageGraph">
    <div data-role="header">
      <a href="#pageData" data-role="button" data-icon="bars" data-
iconpos="left">New calculation</a>
      <h1>Graph</h1>
    </div>
    <div class="panel panel-success">
      <div class="panel-heading">
         <h3 class="panel-title">Height vs Distance</h3>
      </div>
      <div class="panel-body">
        <canvas id="GraphCanvas" width="500" height="500"
          style="border:1px solid #000000;">
        </canvas>
      </div>
    </div>
  </div>
```

FIGURE 7.47 HTML5 code for the graph page of the graphical Projectile app physicsProjectileApp4.html

7.5 Making the output of the Projectile app graphical using RGraph

Let us now turn our attention to the JavaScript code starting with the code for the initialize() function, shown in Figure 7.48, that is called as the onload attribute for the `<body>` element. The only change we have made to the function is the addition of the highlighted statement

```
$(location).attr("href", "#pageData");
```

This statement makes sure that we always go to the data page every time the app is launched or refreshed.

Figure 7.49 shows five functions

- validateAngle() to make sure that the angle is between 1 and 90 degrees
- validateVelocity() to make sure that the velocity is positive and less than the speed of light

```
function initialize()
{
  $(location).attr("href", "#pageData");
  var angleInput=document.getElementById("angle");
  angleInput.addEventListener("blur", validateAngle);

  var velocityInput=document.getElementById("velocity");
  velocityInput.addEventListener("blur", validateVelocity);
}
```

FIGURE 7.48 JavaScript code for the initialize() function of the graphical Projectile app physicsProjectileApp4.js

```
function validateAngle()
{
  var angleInput=document.getElementById("angle");
  if(angleInput.value<1 || angleInput.value>90)
  {
    alert('Angle value must be between 1 and 90');
    angleInput.value="";
  }
}

function validateVelocity()
{
  var velocityInput=document.getElementById("velocity");
  if(velocityInput.value<1)
  {
    alert('Velocity value must be greater than 0')
    velocityInput.value="";
  }
  else if(velocityInput.value>299792458)
  {
    alert('Too fast! The velocity value cannot exceed the speed
of light (299 792 458)!');
    velocityInput.value="";
  }
}
function update()
{
  var angle=document.getElementById("angle").value;
  var velocity=document.getElementById("velocity").value;
  calculate(angle, velocity);
}

function calcDistance(horizontalVelocity, time)
{
  var distance=horizontalVelocity*time;
  return distance;
}
function calcHeight(verticalVelocity, time)
{
  var height=(verticalVelocity*time)-(0.5*9.81*time*time);
  return height;
}
```

FIGURE 7.49 JavaScript code for the previously discussed functions of the graphical Projectile app physicsProjectileApp4.js

- update(), which is specified as the onclick attribute value for the calculate button on the data page
- calcDistance() to calculate distance traveled by the projectile using the horizontal distance traveled based on the horizontal component of the velocity and time
- calcHeight() to calculate height reached by the projectile using the horizontal distance traveled based on the vertical component of the velocity and time

These functions are unchanged from the previous version and reproduced here for completeness.

Figure 7.50 shows the modified version of the function calculate(), which does bulk of the calculations. The changes from the previous version are highlighted. They include the use of the Math.round() function to round off the values of height and distance before pushing them to the end of respective arrays. The rounding off of distances is especially necessary as they will be used as the labels of the x-axis.

Additional modifications to the calculate() are at the end of the function, where instead of updating the table, we make calls to three functions: drawGraph(), labelAxes(), and resizeGraph().

The drawGraph() function is shown in Figure 7.51. It combines the features of setting up the canvas and drawing the line graph from the Thyroid app. We first make graph page active, then retrieve the element with the id "GraphCanvas", set the fillColor to white "FFFFFFF", and fill the 500-by-500 rectangle with the white color. The next statement creates a line chart from the RGraph library using the constructor RGraph.Line(). The constructor takes two parameters. The first parameter is the id of the canvas ("GraphCanvas"). The second parameter is the heightArray. We then set a number of properties through a series of calls to the set() method. These properties are similar to those we saw for the TSH graph in the Thyroid app. Therefore, we will not repeat the explanation.

```
function calculate(angle, velocity)
{
  var horizontalVelocity=velocity*Math.cos((angle*Math.PI)/180);
  var verticalVelocity=velocity*Math.sin((angle*Math.PI)/180);
  var tMaxHeight=verticalVelocity/9.81;
  var tLanding=2*tMaxHeight;
  var heightArray=[];
  var distanceArray=[];
  var timeArray=[];

  if(tLanding<2)
  {
    var interval=0.1;
  }
  else if(tLanding<20)
  {
    var interval=1;
  }
  else
  {
    var interval=10;
  }

  for (var time=0; time<=tLanding+interval; time+=interval)
  {
    timeArray.push(time);

    var height=calcHeight(verticalVelocity, time);

    if(height<0)
```

```
      {
        height=0;
      }
      heightArray.push(Math.round(height));
      var distance=calcDistance(horizontalVelocity, time)

      if(distance<0)
      {
        distance=0;
      }

      distanceArray.push(Math.round(distance));
    }

    drawGraph(distanceArray,heightArray);
    labelAxes();
    resizeGraph();
}
```

■ **FIGURE 7.50** JavaScript code for the calculate() function of the graphical Projectile app physicsProjectileApp4.js

```
function drawGraph(distanceArray, heightArray)
{
    $(location).attr("href", "#pageGraph");
    var c=document.getElementById("GraphCanvas");
    var ctx=c.getContext("2d");

    ctx.fillStyle="#FFFFFF";
    ctx.fillRect(0, 0, 500, 500);
    var TSHline=new RGraph.Line("GraphCanvas", heightArray)
      .Set("labels", distanceArray)
      .Set("colors", ["blue"])
      .Set("shadow", true)
      .Set("shadow.offsetx", 1)
      .Set("shadow.offsety", 1)
      .Set("linewidth", 1)
      .Set("numxticks", 6)
      .Set("scale.decimals", 2)
      .Set("xaxispos", "bottom")
      .Set("gutter.left", 40)
      .Set("ticksize", 5)
      .Set("chart.title", "Projectile")
      .Draw();
}
```

■ **FIGURE 7.51** JavaScript code for the drawGraph() function of the graphical Projectile app physicsProjectileApp4.js

The function labelAxes() shown in Figure 7.52 is essentially the same as the one in the Thyroid app, except for the text used in the label. Finally, the resizeGraph() function (Figure 7.53) adjusts the graphs for smaller screens. The code is similar to the function by the same name in the Thyroid app. Here, we only need to adjust one canvas as opposed to two in the Thyroid app.

In this chapter so far, we have wrapped up our discussion of the Thyroid app by looking at the use of RGraph to display the gauge meter and draw the graph of TSH values. We then recycled the code from the physics Projectile app and the graphics code from the Thyroid app to create a graphical Projectile tracking app. We will now look at how we can access these apps locally.

```
function labelAxes()
{
    var c=document.getElementById("GraphCanvas");
    var ctx=c.getContext("2d");
    ctx.font="11px Georgia";
    ctx.fillStyle="green";
    ctx.fillText("distance", 400, 470);
    ctx.rotate(-Math.PI/2);
    ctx.textAlign="center";
    ctx.fillText("height", -250, 10);
}
```

FIGURE 7.52 JavaScript code for the labelAxes() function of the graphical Projectile app physicsProjectileApp4.js

```
function resizeGraph()
{
  if($(window).width() < 700)
  {
    $("#GraphCanvas").css({"width":$(window).width()- 50});
  }
}
```

FIGURE 7.53 JavaScript code for the resizeGraph() function of the graphical Projectile app physicsProjectileApp4.js

7.6 Creating an icon on the home screen

For the apps that we have looked at so far, we do not necessarily need to be connected to the Internet in order to view them. In this section, we will see how to create an icon on the home screen of our device, so that we do not have to connect to the Internet to view the app (which will be described in the following section). Please note that the process varies with the devices and the version of the operating system. The icon will launch the default browser and load our app with a single tap. We will study the process for iOS. The intent of this section is to demonstrate that every device supports such a creation of icons.

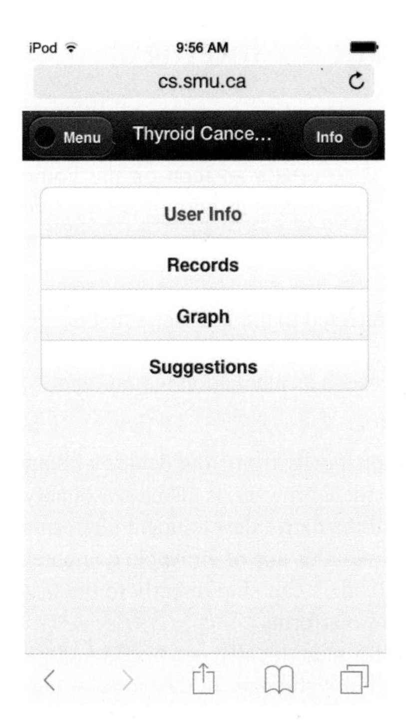

■ **FIGURE 7.54** Locating the page that will appear as an icon on the home screen of the iOS device

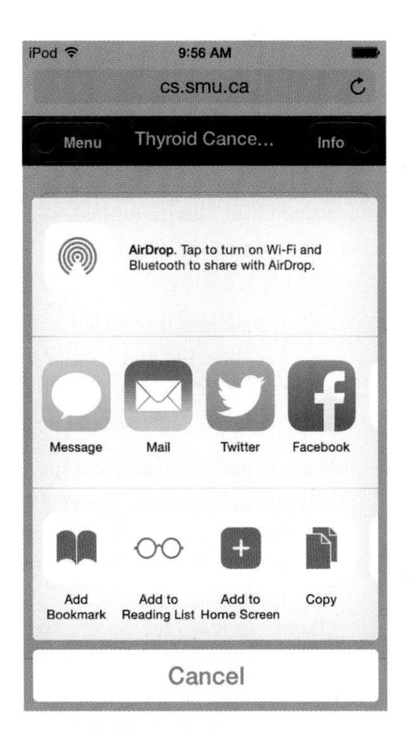

■ **FIGURE 7.55** Tapping on the Add to Home Screen button (iOS)

Let us begin with the iOS. Figure 7.54 shows the page for which we want to create an icon on the home screen. If you tap the center button with an up arrow in the middle, it will bring up the screen shown in Figure 7.55. Pressing on the "Add to Home Screen" button will take us to the screen shown in Figure 7.56. We can change the name of the icon if we wish, and then the icon will appear on our home screen as seen in Figure 7.57.

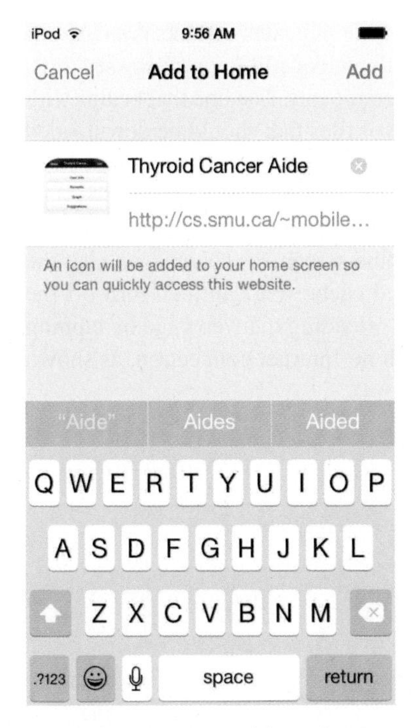

■ **FIGURE 7.56** Adding the icon to the Home Screen (iOS)

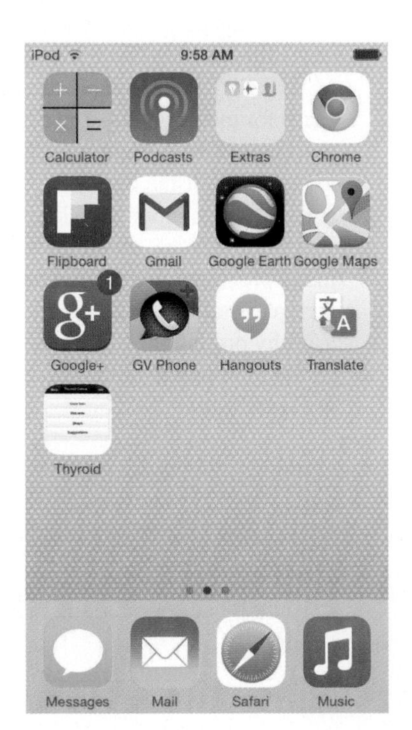

■ **FIGURE 7.57** The icon is now on the Home Screen (iOS)

7.7 Running an app locally without the Internet

So far we have essentially created web pages that can be viewed on a variety of different platforms including mobile devices. We know that readers want to know how this is really mobile computing. In the previous section, we saw how to create an icon on the home screen that can be tapped to directly go to the web page. That makes it more of an app-like experience. However, there are two additional aspects that we need to address in order for this to be a true app

- Access the HTML5 app even when we are not connected to the Internet—that is, store the app locally on our device.
- Make the app full screen without browser controls such as navigation buttons and URL address box.

We will address the first issue in this chapter, running the app locally from your device without Internet connection. The second part, running the app outside a browser, is also a reasonably straightforward process. However, it requires installation of the native development platforms on your computer in order to compile the app code. This includes use of an Apple computer for developing a native iOS app from our HTML5 app. Readers can skip directly to the last chapter to learn the process of converting HTML5 to native platforms.

In order for your device to be able to run the HTML5 page locally, we need to add a manifest file to the website. Figure 7.58 shows the modified web page. In the `<html>` tag, we add an attribute called manifest. Every page with the manifest attribute specified will be cached when the user visits it. It is then available on the device locally, even when the device is not connected to the Internet. Most web pages have multiple files associated with the web page such as images, videos, and css files. In that case, we should specify a manifest file as the value of the manifest attribute as we have done in Figure 7.58. While the manifest file can have any extension, we are using the recommended file extension for manifest files ".appcache" by calling our file manifest.appcache. The web server should be configured to serve the manifest file with the MIME-type "text/cache-manifest". Chances are that your web server is already configured for the manifest files. If you have trouble with the caching and running the app locally, you may want to check with your system administrator.

The manifest.appcache file should be in the same directory as our web page. The content of the file itself is shown in Figure 7.59. The first line of the file indicates that this is a CACHE MANIFEST that contains the list of files that should be downloaded by your browser to its cache for local loading. The character # is used to indicate comments. The line that begins with # is ignored as a comment. The rest of the file is just a list of files that should be downloaded to the cache of the device browser. We have indicated that all the css style files, image files, and scripts along with the html file should be saved in the local cache.

Once we have set up the manifest for our website, we can follow the procedure described in the previous section for creating a bookmark on the home screen. As soon as we visit the web page, all the relevant files will be stored in the local cache. We can then turn off the WiFi by putting the device in the airplane mode and then reloading the web page by tapping on the icon on the home screen. It will run as before with no Internet connection, as shown in Figure 7.60.

```
<!DOCTYPE html>
<html lang=en manifest="manifest.appcache">
<head>
  <title>Thyroid Cancer Aide</title>
    <!-- Rest of the file is the same as previous version  -->
```

FIGURE 7.58 Indicating the location of the manifest file for our web page

```
CACHE MANIFEST
#version 1

index.html

css/bootstrap.css
css/bootstrap.min.css
css/jquery.mobile-1.3.1.min.css

css/images/ajax-loader.gif
css/images/icons-18-white.png
css/images/Thumbs.db

scripts/Advice.js
scripts/bootstrap.js
scripts/GraphAnimate.js
scripts/jquery-1.9.1.min.js
scripts/jquery.mobile-1.3.1.min.js
scripts/Navigation.js
scripts/pageLoader.js
scripts/RGraph.common.core.js
scripts/RGraph.common.dynamic.js
scripts/RGraph.common.effects.js
scripts/RGraph.cornergauge.js
scripts/RGraph.hprogress.js
scripts/RGraph.line.js
scripts/Table.js
scripts/UserForm.js
```

FIGURE 7.59 Manifest file for our web page

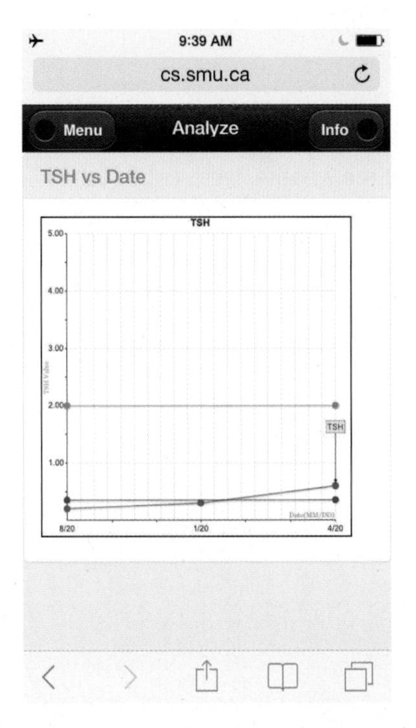

FIGURE 7.60 Thyroid app working with no Internet connection

Readers should experiment with different platforms and our physics Projectile app as well as the temperature converter app so that they can be run locally by tapping on an icon on the home screen.

The apps that we will be looking at in future chapters will be using a server and hence need an Internet connection.

Quick facts/buzzwords

getContext(): A method that must be used with a canvas element before it can be further manipulated.

beginPath(): A canvas method that is used to create a path before starting the rendering of any line or curve.

closePath(): A canvas method that is used to create a path from the current point to the starting point.

moveTo(): A canvas method to take the pointer to a particular x and y coordinate on a canvas.

lineTo(): A canvas method to create a line from the current coordinate to the new coordinate on a canvas.

arc(): A canvas method to create an arc. It is also used to create a circle.

stroke(): A canvas method to actually render the line or curve that has been created by calls such as lineTo() or arc().

fill(): Fills the current drawing/path.

rect(): A canvas method to create a rectangle.

strokeRect(): A canvas method to draw a rectangle with no fill.

fillRect(): A canvas method to draw a filled rectangle.

fillText(): A canvas method to display text.

RGraph: A free JavaScript library to create web charts and graphs.

CornerGauge: An RGraph object to display a corner gauge meter.

Line: An RGraph object to create a line graph.

cache: Memory-on device to store documents for quick access. In our case, to access web pages from local memory even when Internet connection is not on.

manifest: An attribute for the `<html>` tag to indicate that the website needs to be cached.

The chapter also lists a number of properties and methods for canvas context as well as RGraph objects.

CHAPTER 7

Self-test exercises

1. Which method is used to draw a circle on a canvas?
2. Which method is used to draw a rectangle on a canvas?
3. What is a cache in the context of a mobile device?
4. What is a manifest?
5. What does a manifest file contain?

Programming exercises

1. Draw the following figure of a cylinder on a canvas

2. Modify the Projectile app so that it plots a graph of time versus distance.
3. Modify the Projectile app so that it plots a graph of time versus height.
4. Modify the Projectile app so that it calculates vertical velocity over time and creates a plot of vertical velocity against time.
5. Modify the Projectile app so that it calculates vertical velocity over time, and then use calculations of time against height to plot vertical velocity against height.
6. Modify the Projectile app so that it calculates vertical velocity over time, and then use calculations of time against distance to plot vertical velocity against distance.
7. Modify the Projectile app so it can be run locally with no internet connection by tapping on an icon on the home screen.
8. Modify the temperature converter app so it can be run locally with no Internet connection by tapping on an icon on the home screen.

Programming projects

1. **Boiler monitor:** We want to complete the app that monitors temperature and pressure of a boiler. You can model the app based on the Thyroid app.
    - We have already implemented the password-based entry page.
    - We have already implemented the page to get basic information about the boiler such as the boiler ID, date of purchase, maximum allowable values of pressure and temperature, and an ability to change the password.

- We already have a menu page with four choices. We have already implemented the first two choices for data entry and storage
    i. We have already implemented the page to allow you to change the basic information about the boiler.
    ii. We have already implemented the page to enter data—temperature and pressure.
    iii. We should now implement the page to graph the data.
    iv. We should now implement the page to make recommendations based on the values of temperature and pressure.

2. **Blood pressure monitor:** We want to complete the app that monitors blood pressure. You can model the app based on the Thyroid app.
    - We have already implemented the password based entry page.
    - We have already implemented the page to get basic information about the person such as name, date of birth, and health insurance card number, along with an ability to change the password.
    - A menu page with four choices
        i. We have already implemented the page to allow you to change the basic information about the person.
        ii. We have already implemented the page to enter data—blood pressure.
        iii. We should now implement the page to graph the data.
        iv. We should now implement the page to make recommendations based on the values of blood pressure. You should search for acceptable values of blood pressure before implementing this page.

3. **Power consumption monitor:** We want to complete the app that monitors the power consumption of a manufacturing plant. You can model the app based on the Thyroid app.
    - We have already implemented the password-based entry page.
    - We have already implemented the page to get basic information about the plant such as the plant ID and date of installation, along with an ability to change the password.
    - We already have a menu page with four choices. We have already implemented the first two choices for data entry and storage
        i. We have already implemented the page to allow you to change the basic information about the plant.
        ii. We have already implemented the page to enter data—power consumed.
        iii. We should now implement the page to graph the data.
        iv. We should now implement the page to make recommendations based on the values of power consumption. The recommendations should be based on how the recent power consumption compares with the average of the power consumption in the past.

4. **Body mass index (BMI) monitor:** We want to complete the app that monitors the body mass index of a person based on height and weight. You can model the app based on the Thyroid app.
    - We have already implemented the password-based entry page.

- We have already implemented the page to get basic information about the person such as name, date of birth, and health insurance card number, along with an ability to change the password.
- We already have a menu page with four choices. We have already implemented the first two choices for data entry and storage
    i. We have already implemented the page to allow you to change the basic information about the person.
    ii. We have already implemented the page to enter data—height (meter) and weight (kg). Your history should show the value of height and weight, as well as the BMI

$$BMI = \frac{weight}{(height)^2}$$

    iii. We should now implement the page to graph the data, height, and BMI.
    iv. We should now implement the page to make recommendations based on the values of body mass index calculated using the weight and height entered. You should search for acceptable values of BMI before implementing this page.

5. **Managing a line of credit:** We want to complete the app that manages our line of credit at a bank. The line of credit allows us to carry a certain maximum negative balance of p. Negative balance will be charged an interest at a specified fixed rate x. A positive balance can lead to an interest accumulation at a different fixed rate y.
    - We have already implemented the password-based entry page.
    - We have already implemented the page to get basic information about the line of credit such as the name of the bank and date of account opening, along with an ability to change the password.
    - A menu page with four choices: We already have a menu page with four choices. We have already implemented the first two choices for data entry and storage
        i. We have already implemented the page to allow you to change the basic information about the line of credit.
        ii. We have already implemented the page to enter data—credit or debit. The history should show the balance after the debit/credit, which should include interest charges or accumulation.
        iii. We should now implement the page to graph the balance.
        iv. We should now implement the page to make recommendations based on the current balance.
    - If the current balance is positive you should provide a green signal.
    - If the current negative balance is less than 50% of the line of credit, you should provide a yellow alert.
    - If the balance is greater than or equal to 50% but less than 80% of the line of credit, you should provide an amber alert.
    - If the balance is greater than or equal to 80% of the line of credit, you should provide a red alert.

CHAPTER 8

Using servers for sharing and storing information

WHAT WE WILL LEARN IN THIS CHAPTER

1. How to provide facilities for server storage
2. What is Node.js
3. How to sync the device storage with server-side storage

So far all our computations and storage were on the local device. However, one cannot completely rely on the mobile device for longer-term storage. In this chapter, we will set up syncing of information from our Thyroid app to a server. We will learn the server-side version of JavaScript called Node.js. We are going to implement the server storage first using emerging NoSQL databases presented in Chapter 9 based on the JSON format that we have been using. We will then see how the same facility can be implemented using conventional relational databases presented in Chapter 10. Therefore, this chapter does not actually discuss the database management on the server. In this chapter, we will create the web application server using Node.js.

8.1 Introduction

At the end of Chapter 7, we saw how to download the apps to a mobile device using appcache, create an icon on the home screen, and run the apps even when the device is not connected to the Internet. This includes even storing a reasonable amount of data on the mobile device as was the case for our Thyroid app. Running the app and storing the data locally is an important aspect of mobile computing. We cannot always rely on the Internet connection for a number of reasons. The connection may not be available, or the data plans while roaming may be too expensive. However, as apps become more refined, we need to be able to access data from a server. The server is also likely to serve as a more reliable source of data.

8.2 Designing the server-based Thyroid app

The next version of the Thyroid app moves it from storing data locally on the users' devices to using a remote server, so that users can access their data from any device. Although this is largely a change of the architecture of the app, there are some design changes to keep in mind as well.

Primarily, we need to change the design of the login page. In the previous version, the app required only a password to allow users entry, because the assumption was that there would be only one user per device. Now, multiple users can access their own individual data on the same device. To accommodate this, we add a field for the user's email address, which is used as a username. We also add a "Create New User" menu option, which will allow new users to sign up. The "Create New User" button is placed below the "Login" button. Users will only ever create a new user once, but they will log in many times, which is why the "Login" button has higher precedence.

8.3 Signing up to a server-based app

Server-based apps are typically divided into client-side and server-side computing. Client-side scripting includes all code run in the browser, such as all the JavaScript that we have looked at so far (up to and including Chapter 7). Server-side scripting includes any code run outside of a browser on a server, as we will see in this chapter and the following chapters. The server that our apps will run on has all the necessary technologies installed. These technologies include Node.js (server-side JavaScript), mongoDB (a NoSQL database server), and MySQL (SQL/relational database server). The URL for accessing our app on this specialized server is http://140.184.132.239:3001. Here, 140.184.132.239 is the IP address of the server. In most cases, we use an easier to remember name for the IP such as cs.smu.ca that is registered with a domain name server. However, for every such IP name, there is always an IP address as we are using here. Since it was put together for experimentation, the system administrator who configured the server did not bother to register it with a name. The number 3001 that follows the IP number (separated by a colon) is the port for communication used by our app. We will learn more about it when we discuss the server-side computing of our app.

Figure 8.1 shows a modified login screen that now uses the email address as the user identification (ID). Previously we did not need an explicit identification of the user, since the device itself was considered to be the user ID. Now a single server will be dealing with a number of users, so we need a unique ID for each user. This will allow a user to sign in from multiple devices and multiple users to sign in from the same device. We also added a new button to create a new user. The HTML5 code for the login screen is shown in Figure 8.2. The two new elements corresponding to the text box for the email and button for creating a new user are highlighted.

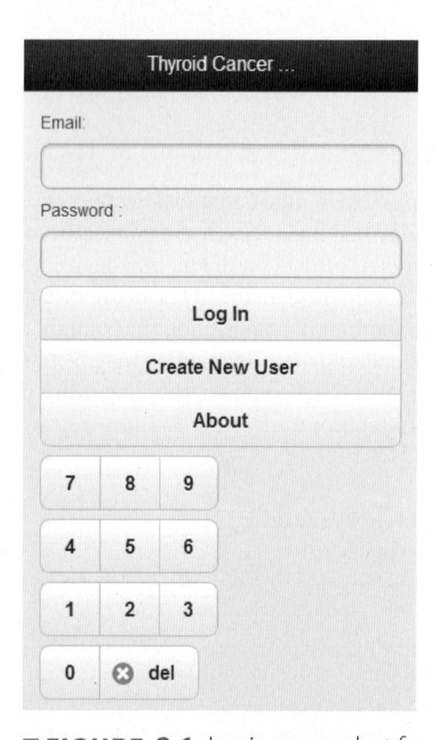

FIGURE 8.1 Login screenshot for server-based Thyroid app

```
<!-- Start of first page -->
<div data-role="page" id="pageHome">
  <div data-role="header">
    <h1>Thyroid Cancer Aide</h1>
  </div>
  <div data-role="content">
    Email: <input type="email" id="email"></input>
    Password : <input type="password" id="passcode"></input>

    <div data-role="controlgroup" id="numKeyPad">
      <a data-role="button" id="btnEnter" type ="submit">Log In</a>
      <a href="#pageUserInfo" data-role="button"
id="btnCreateUser">Create New User</a>
      <a href="#pageAbout" data-role="button">About</a>
    </div>
    <div data-role="controlgroup" data-type="horizontal">
      <a data-role="button" onclick="addValueToPassword(7)">7</a>
      <a data-role="button" onclick="addValueToPassword(8)">8</a>
      <a data-role="button" onclick="addValueToPassword(9)">9</a>
    </div>
    <div data-role="controlgroup" data-type="horizontal">
      <a data-role="button" onclick="addValueToPassword(4)">4</a>
      <a data-role="button" onclick="addValueToPassword(5)">5</a>
      <a data-role="button" onclick="addValueToPassword(6)">6</a>
    </div>
    <div data-role="controlgroup" data-type="horizontal">
      <a data-role="button" onclick="addValueToPassword(1)">1</a>
      <a data-role="button" onclick="addValueToPassword(2)">2</a>
      <a data-role="button" onclick="addValueToPassword(3)">3</a>
    </div>
    <div data-role="controlgroup" data-type="horizontal">
      <a data-role="button" onclick="addValueToPassword(0)">0</a>
      <a data-role="button" onclick="addValueToPassword('bksp')"
data-icon="delete">del</a>
    </div>
  </div>
</div>
```

FIGURE 8.2 HTML5 code for login screen for server-based Thyroid app

Figure 8.3 shows a modified user information screen that additionally accepts the email address as the user identification (ID) that will come up when the user clicks on the "Create New User" button or when the user information option is picked from the menu. The HTML5 code with highlighted text input element for the email address is shown in Figure 8.4.

Once we have entered the user information and submitted the form, the app confirms the creation as shown in Figure 8.5. We can see that the message is coming from the app communicating through port 3001 of the server with IP number 140.184.132.239.

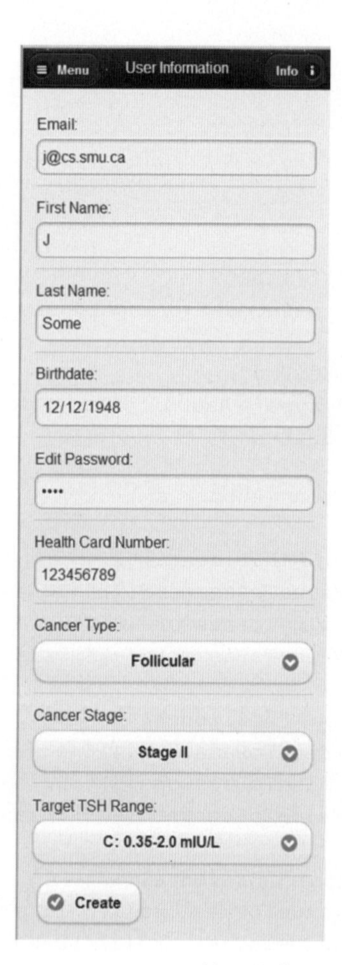

FIGURE 8.3 User Information form screenshot for server-based Thyroid app

```
  <!--User Information Page/Form -->
  <div data-role="page" id="pageUserInfo"  >
    <div data-role="header">
      <a href="#pageMenu" data-role="button" data-icon="bars"
data-iconpos="left" data-inline="true">Menu</a>
      <a href="#pageAbout" data-role="button" data-icon="info"
data-iconpos="right" data-inline="true">Info</a>
      <h1>User Information</h1>
    </div><!-- /header -->
    <div data-role="content">
      <form id="frmUserForm" action="">
          <div data-role="fieldcontain">
            <label for="txtEmail">Email: </label>
            <input type="email" placeholder="name@example.com"
name="txtEmail" data-mini="false" id="txtEmail" value="" required>
          </div>
          <div data-role="fieldcontain">
            <label for="txtFirstName">First Name: </label>
            <input type="text" placeholder="First Name"
name="txtFirstName" data-mini="false" id="txtFirstName" value=""
required>
          </div>
```

```
          <div data-role="fieldcontain">
            <label for="txtLastName">Last Name: </label>
            <input type="text" placeholder="Last Name"
name="txtLastName" data-mini="false" id="txtLastName" value=""
required>
          </div>

  <!—The rest of the code for this page is same as Chapter 5 →
```

FIGURE 8.4 HTML5 code for User Information form for server-based Thyroid app

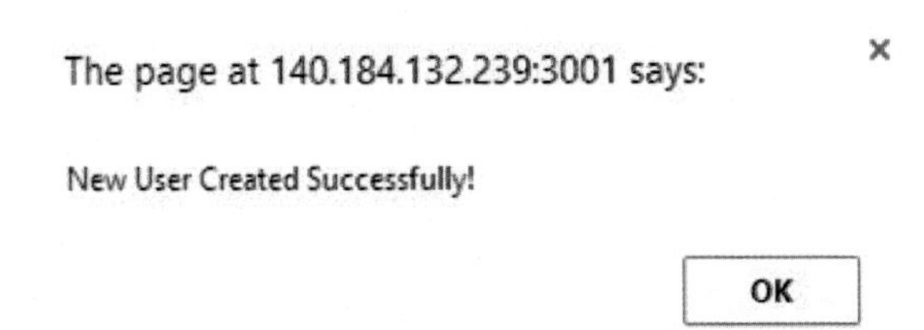

FIGURE 8.5 Confirmation screenshot of user creation for server-based Thyroid app

8.4 Syncing the records between the device and the server

Let us continue with our study of the modified screens for the server-based Thyroid app by looking at the screenshot of the menu page (Figure 8.6), which shows an additional button for syncing the device data with the server database. The HTML5 code in Figure 8.7 highlights the additional button that is linked to the Sync page.

In order to understand how the Sync works, let us add a number of records through the Records page as shown in Figure 8.8. Then we click on the menu from the top left and press the Sync button, which will bring up the Sync page shown in Figure 8.9. The Sync page has three buttons. The first button allows us to upload the records from the device to a database on the server. The remaining two buttons are for downloading the records. The middle button downloads only the records from the dates that do not exist on the device. The last button wipes the existing data on the device and loads the records from the database server. The HTML5 code for the page is shown in Figure 8.10. In addition to the code for the menu and info buttons on the top, we see three more buttons. These buttons will be handled through a script that we will discuss later on in this chapter.

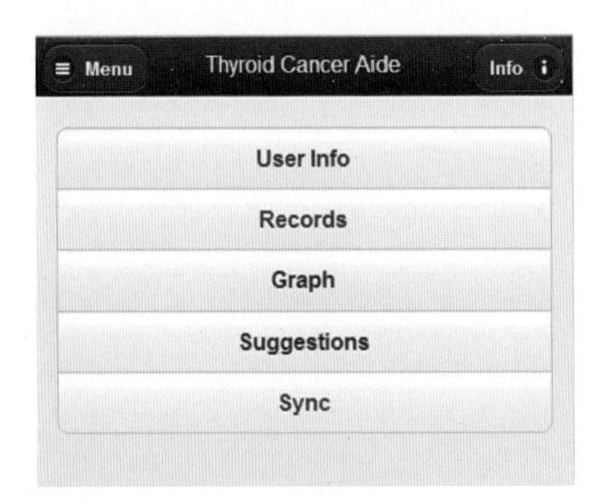

FIGURE 8.6 Screenshot of Menu page for server-based Thyroid app

```
<!-- Menu page -->
<div data-role="page" id="pageMenu">
  <div data-role="header">
    <a href="#pageMenu" data-role="button" data-icon="bars" data-
iconpos="left" data-inline="true">Menu</a>
    <a href="#pageAbout" data-role="button" data-icon="info" data-
iconpos="right" data-inline="true">Info</a>
    <h1>Thyroid Cancer Aide</h1>
  </div>
  <div data-role="content">
    <div data-role="controlgroup">
      <a href="#pageUserInfo" data-role="button">User Info</a>
      <a href="#pageRecords" data-role="button">Records</a>
      <a href="#pageGraph" data-role="button">Graph</a>
      <a href="#pageAdvice" data-role="button">Suggestions</a>
      <a href="#pageSync" data-role="button">Sync</a>
    </div>
  </div>
</div>
```

FIGURE 8.7 HTML5 code for Menu page for server-based Thyroid app

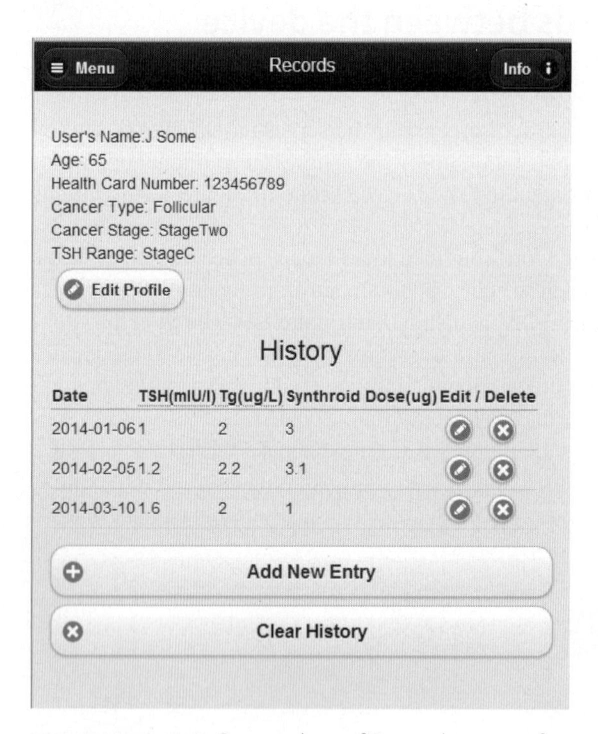

FIGURE 8.8 Screenshot of Records page after adding three records for server-based Thyroid app

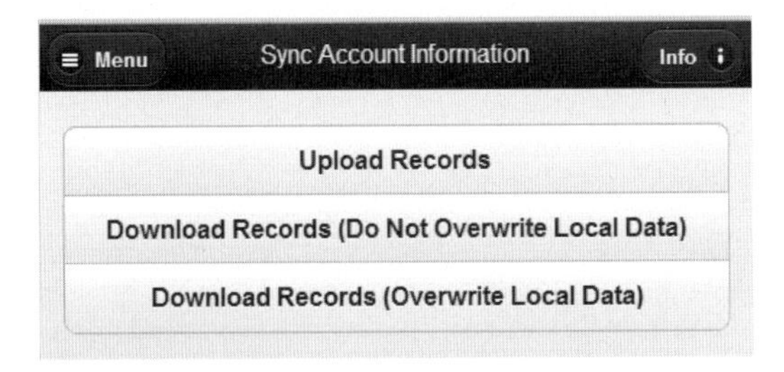

FIGURE 8.9 Screenshot of Sync page for server-based Thyroid app

```
<!--Sync Page -->
<div data-role="page" id="pageSync">
  <div data-role="header">
    <a href="#pageMenu" data-role="button" data-icon="bars" data-
iconpos="left" data-inline="true">Menu</a>
    <a href="#pageAbout" data-role="button" data-icon="info" data-
iconpos="right" data-inline="true">Info</a>
    <h1>Sync Account Information</h1>
  </div>
  <div data-role="content">
    <div data-role="controlgroup">
      <a id="btnUpload" data-role="button">Upload Records</a>
      <a id="btnDownload" data-role="button">Download Records (Do Not
Overwrite Local Data)</a>
      <a id="btnDownloadOverwrite" data-role="button">Download
Records (Overwrite Local Data)</a>
    </div>
  </div>
</div>
```

FIGURE 8.10 HTML5 code for Sync page for server-based Thyroid app

Figure 8.11 shows the confirmation dialogue from the server when the user presses the upload records button. Let us delete one of the records from the device storage as shown in Figure 8.12. We can then go back to the Sync page and download the records (overwrite the local data) as shown by the confirmation in Figure 8.13. We can then go back to the Records page to confirm that the original records have been restored as shown in Figure 8.14. Let us also see how the process of downloading without overwriting local data works. We will first delete one of the records and change the TSH value of the first remaining record as shown in Figure 8.15. We then go back to the sync page and download the records (do not overwrite the local data) as shown by the confirmation in Figure 8.16. We can then go back to the Records page to confirm that the original records are restored as shown in Figure 8.17; here we get the deleted record back, but the changed record stays as it was in Figure 8.15.

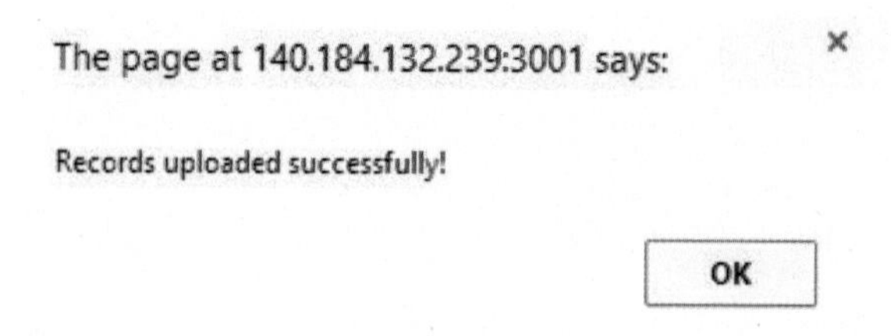

FIGURE 8.11 Confirmation screenshot of record upload for server-based Thyroid app

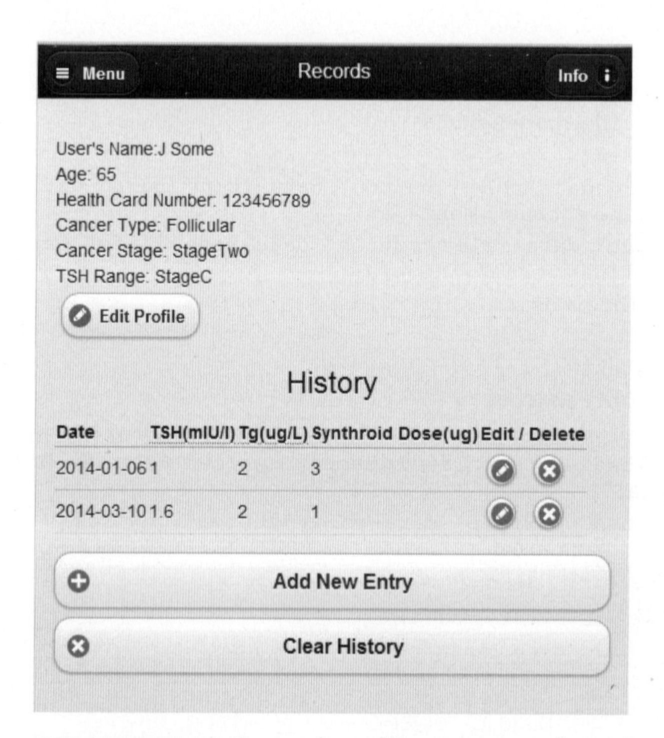

FIGURE 8.12 Screenshot of Records page after deleting one record for server-based Thyroid app

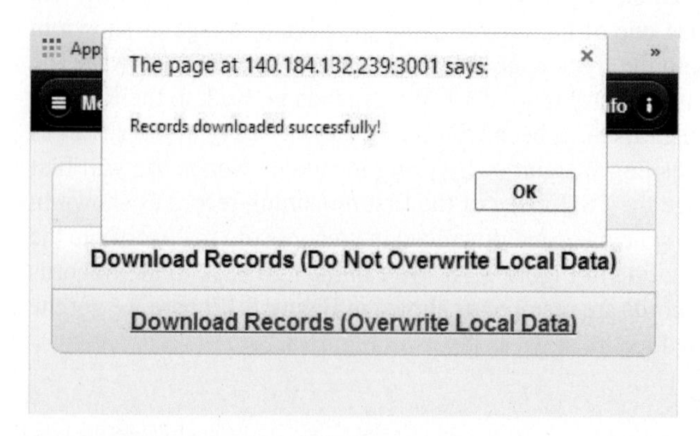

FIGURE 8.13 Confirmation screenshot of record download with overwriting of the local data for server-based Thyroid app

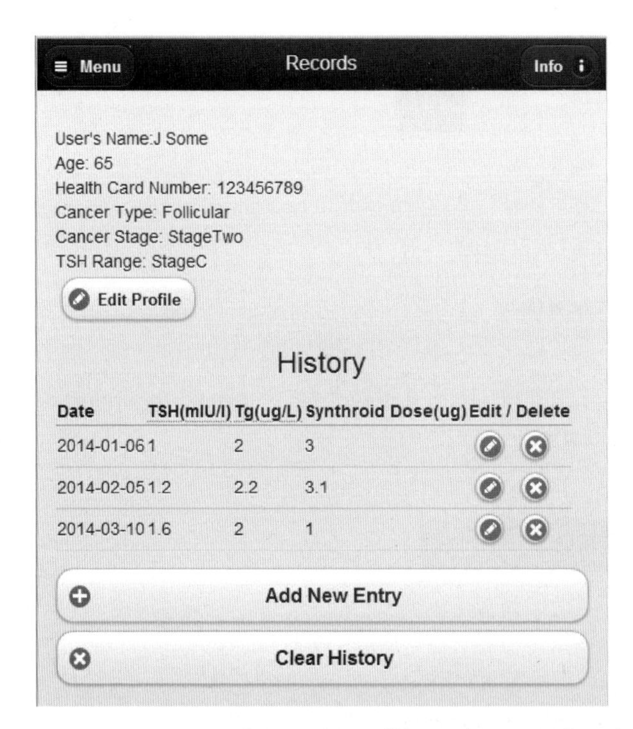

FIGURE 8.14 Screenshot of Records page after downloading records by overwriting the local data for server-based Thyroid app

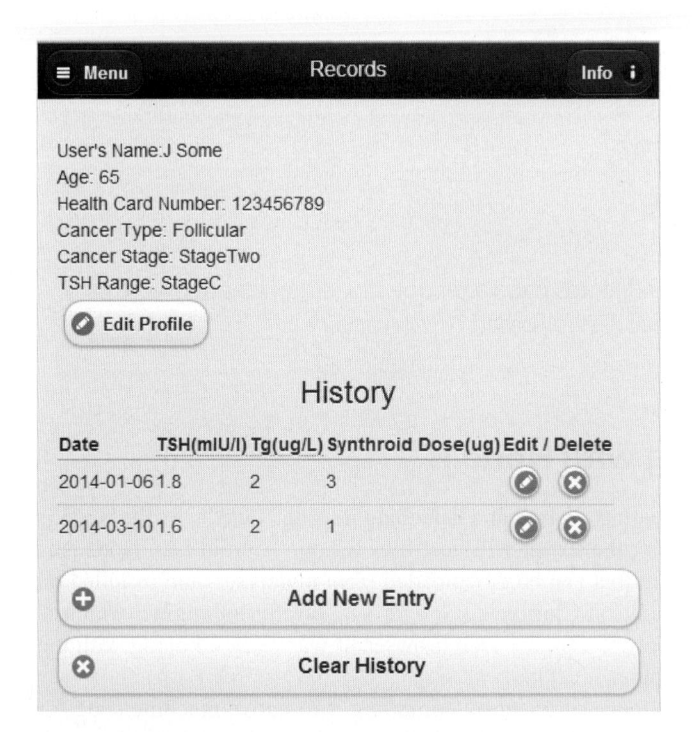

FIGURE 8.15 Screenshot of Records page after deleting one record and changing another one for server-based Thyroid app

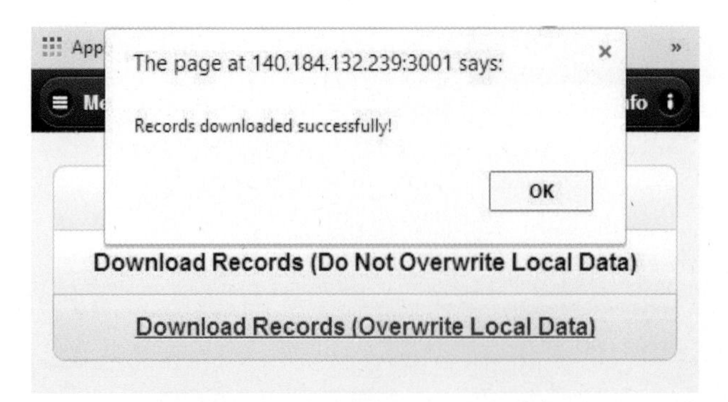

FIGURE 8.16 Confirmation screenshot of a record download without overwriting the local data for server-based Thyroid app

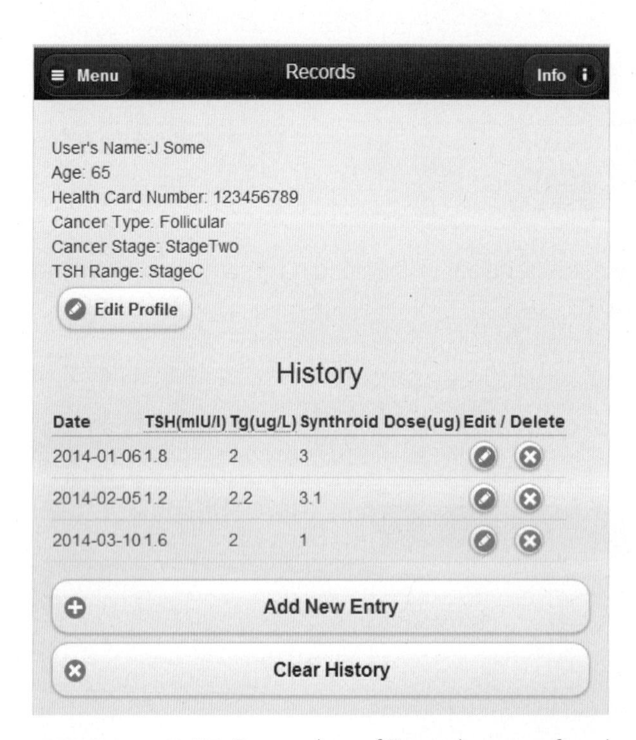

FIGURE 8.17 Screenshot of Records page after downloading without overwriting existing records in the local data for server-based Thyroid app

8.5 Programming with Node.js

Figure 8.18 shows the code for creating a server and launching it using Node.js. As the Node.js file is in our Thyroid app directory, the server will host files for our Thyroid app. Node.js is a platform or runtime environment that can be downloaded from nodejs.org and installed on any computer. The platform is built on Chrome's (Google V8) JavaScript engine, which is open-source developed for the Google Chrome web browser. It was released along with the first version of Chrome on September 2, 2008.

Google V8 makes JavaScript code efficient by compiling it to native machine code for architectures including IA-32, x86-64, ARM, or MIPS ISAs. This compares favorably with traditional techniques such as interpreted bytecode or compiling the entire program to machine code for execution from a filesystem. As conventional programmers know, JavaScript's programming convenience comes at a computational cost. Google V8 further optimizes the code dynamically at runtime using heuristics applied to the profile of the code's typical execution.

Programming in Node.js is based on an asynchronous event-driven, nonblocking I/O model designed to maximize throughput and efficiency. It is a very useful runtime environment for building fast, data-intensive real-time scalable network applications that run across distributed devices. Node.js applications run within the Node.js runtime on a number of popular server operating systems including Mac OS X, Windows, and Linux with no changes. An additional advantage of Node.js applications is that they create multiple parallel threads for file and network events.

So far we have been using a web server such as Apache. With the Node.js platform, we can dispense with the web server, since Node.js contains a built-in asynchronous I/O library for file, socket, and HTTP communication. The HTTP and socket support means that the Node.js platform can act as a web server. We see the launch of such a web server in Figure 8.18. Instead of using a full-blown HTTP service, we are using a minimal web application framework called express (expressjs.com), which provides a robust platform for serving multipage web applications.

```
var express = require('express');
var server = express();

server.use('/scripts', express.static(__dirname + '/scripts'));
server.use('/css', express.static(__dirname + '/css'));
server.use(express.static(__dirname));

server.listen(3001);
```

FIGURE 8.18 Node.js code to create and launch the web server File: app-server.js

8.6 Launching a Node.js app server

We will launch our server running the Thyroid app by running the app-server.js shown in Figure 8.18 with the command

```
node app-server.js
```

The command node invokes the Node.js runtime, and we are passing our file as the script that will be executed by Node.js.

Let us study the code in app-server.js. We first begin by making a function call require(). The function require() is specific to Node.js. It is used to load modules. Programmers with experience in C++ or PHP can think of it as something that is similar to include. Java and C# programmers can liken it to import. However, there is an important difference. The include or import make the modules available in the current scope. The require() function, on the other hand, makes them available through a variable assignment—that is,

```
var express = require('express');
```

as shown in Figure 8.18. We then create a new application with express() and assign it to a variable called server. The Express application is a web server that will serve the web pages of our Thyroid app. There are three calls to the method server.use(). The use() method tells the server to specify various paths for serving static files such as the html, scripts, and css. We are using the JavaScript global variable called __dirname, which corresponds to the current directory.

The three calls thus used essentially tell the server where these files can be found. For example,

```
server.use('/scripts', express.static(__dirname + '/scripts'))
```

is telling the server that the reference to the directory /scripts will be the scripts directory off the current directory (__dirname). The third call, server.use(express.static(__dirname)), is indicating the root directory for the app to be the current directory. Finally the listen() method launches the server, and we start accepting requests through the port 3001.

8.7 Modified navigation in the server-based Thyroid app

Now we are going to look at the rest of the JavaScript code in our server-based Thyroid app. Most of it is inherited from the previous version of the app, except for two changes. We have replaced localStorage with sessionStorage, since we want the longer-term storage to be coming from the server. The sessionStorage behaves much the same as localStorage. The only difference is that when the browser is closed, the information in sessionStorage is lost.

We also simplified the code by not checking to see if the storage is exceeded through a try and catch block. It is unlikely that we will ever come across a storage quota issue. We had introduced it earlier so that users know how the exception handling works. We will study these changes in detail in the two files Navigation.js and UserForm.js, which also involve some communication with the server regarding the user information. In addition, we will be looking at a separate script file called Sync.js, which transfers the table of test records to and from the server.

Let us begin with the Navigation.js file. We will look only at the functions that have code that is different from the previous version. Figure 8.19 shows the main function in the file. There are only two logical parts to this code. We are declaring a variable called SERVER_URL to be 'http://140.184.132.239:3000', which tells that the communication protocol is http, the IP address is 140.184.132.239, and the port is 3000. Readers may recall that the IP address is the same as the one we used for the web server. The port number for the Node.js web server was 3001, and here we are using the port 3000. Port 3000 is used by the Node.js database server, which we will see in the following chapters. While our web server is listening for requests on port 3001, our database server will be listening for requests on port 3000. As mentioned previously, we have switched from localStorage to sessionStorage. Therefore, the if statement in Figure 8.19 makes sure that the browser supports session storage.

```
var SERVER_URL = 'http://140.184.132.239:3000';

if(!sessionStorage) {
  alert('Warning: Your browser does not support features required
for this application, please consider upgrading.');
}
```

FIGURE 8.19 Processing login for server-based Thyroid app File: Navigation.js

When the user clicks the login button (Figure 8.1), the function `$( "#btnEnter" ).click(function() ...)` shown in Figure 8.20 will be called. The function begins by first creating a JSON object called loginCredentials with two name:value pairs. The first one with the name email gets the value of the email element from the form, and the second one with the name password gets the value of the passcode element from the form. We then call the $.post method, which sends a request to the server. The $.post method takes three parameters:

- The first parameter is the URL. In our case it will be a concatenation of the SERVER_URL with the string "/login". That will make it 'http://140.184.132.239:3000/login'.
- The second parameter is the JSON object loginCredentials, which is passed as a parameter.
- The third parameter is an anonymous function that accepts data returned by the server as a parameter.

We will be looking at the details of how the server processes the request 'http://140.184.132.239:3000/login' in the following chapter. At this time, it is sufficient to know that the server verifies that the user password is valid and then returns the user record, which is passed to the anonymous function as a parameter called data. The function makes sure that the data returned is not null and data.email matches the value of the email element from the form. It then assigns the sessionStorage.password field the value of passcode element from the form. The sessionStorage.user is assigned the string version of the JSON element data that is returned by the database server.

We first check to make sure that the user has agreed to the legal notice by looking to see if the agreedToLegal value from data is true. Otherwise, the user is taken to the legalNotice page using the call $.mobile.changePage(). If the user has agreed to the legal notice, we will continue with getting the tbRecords field by posting the URL: 'http://140.184.132.239:3000/getRecords' along with two additional parameters: loginCredentials and another nested anonymous function that receives the data returned by the server. This data is a JSON object; its string equivalent created using JSON.stringify() is assigned to sessionStorage.tbRecords. The control is then transferred to the menu page.

```
$( "#btnEnter" ).click(function()
{
  var loginCredentials = {
    email: $("#email").val(),
    password: $("#passcode").val()
  };
  $.post(SERVER_URL + '/login', loginCredentials, function(data) {
    if(data && data.email==loginCredentials.email) {
      sessionStorage.password = $("#passcode").val();
      sessionStorage.user = JSON.stringify(data);
      if(!data.agreedToLegal) {
        return $.mobile.changePage("#legalNotice");
      }
      $.post(SERVER_URL + '/getRecords', loginCredentials,
function(data) {
        sessionStorage.tbRecords = JSON.stringify(data);
        $.mobile.changePage("#pageMenu");
      }).fail(function(error) {
        alert(error.responseText);
      })
    } else {
      alert('An error occurred logging user in.');
    }
  }).fail(function(error) {
    alert(error.responseText);
  });
});
```

FIGURE 8.20 Verifying login information for server-based Thyroid app File: Navigation.js

We now see a new error-handling feature that is similar to the try and catch block. It is in the form of a fail() function. We have two fail functions in Figure 8.20. The first fail() function corresponds to the second $.post() call corresponding to the request 'http://140.184.132.239:3000/getRecords'. It receives yet another anonymous function as a parameter, which in turn receives the object returned if the $.post() call were to fail. We have named this object appropriately as an error. The function simply displays the responseText field of the error. The else portion shown in Figure 8.20 is executed when either the data was null or the email did not match. We just display a message saying that there was an error in logging the user in. The last piece of code in Figure 8.20 handles the possible failure of the first $.post call corresponding to the request 'http://140.184.132.239:3000/login'. The code for the fail() function is essentially the same as the one that we discussed earlier.

8.8 Modified user form management in server-based Thyroid app

Figure 8.21 shows one of the functions from the UserForm.js that had to be changed to accommodate server storage; the function is saveUserForm(). We have seen the function before when information was stored in the localStorage. We first check to make sure that the information in the form is correct using the function checkUserForm(). We then create the JSON user object using the elements from the form as we had done before. The function saveUserForm() in our new version may be called either to save changed information of an existing user or to create a new user. We distinguish between the two based on the value of the button element btnUserUpdate. If the value is "Create", we need to send the user information to the server. We first create a JSON object called userData with the value of the attribute newUser set to the variable user. We then use the $.post() method with URL 'http://140.184.132.239:3000/saveNewUser' along with two additional parameters, userData and a nested anonymous function, which receives the data returned by the server. This post request sends the new user to the database server. The anonymous function alerts the user that the new user has been created. The sessionStorage.user is assigned the string version of user using JSON.stringify(). The sessionStorage.password is assigned the value of user.newPassword. From this point on the value of the button element btnUserUpdate will be changed to "Update" from the old value of "Create.". The control is then transferred to the menu page. The failure of the $.post() will be handled with the help of the familiar fail() method. The else part in Figure 8.21 corresponds to the situation when the value of btnUserUpdate is not "Create.". In that case, we do not need to go to the server. We simply update the user value in sessionStorage and transfer the control to the menu page.

```
function saveUserForm()
{
  if(checkUserForm())
  {
    var user = {
      "email": $("#txtEmail").val(),
      "firstName": $("#txtFirstName").val(),
      "lastName": $("#txtLastName").val(),
      "healthCardNumber": $("#txtHealthCardNumber").val(),
      "newPassword": $("#changePassword").val(),
      "dateOfBirth": $("#datBirthdate").val(),
      "cancerType": $("#slcCancerType option:selected").val(),
      "cancerStage": $("#slcCancerStage option:selected").val(),
      "tshRange": $("#slcTSHRange option:selected").val()
    };
    if($("#btnUserUpdate").val() == "Create") {
      var userData = {
        newUser: user
      };
      $.post(SERVER_URL + '/saveNewUser', userData, function(data) {
        alert("New User Created Successfully!");
```

```
          sessionStorage.user = JSON.stringify(user);
          sessionStorage.password = user.newPassword;
          $("#btnUserUpdate").val("Update");
          $.mobile.changePage("#pageMenu");
        }).fail(function(error) {
          alert(error.responseText);
        });
      } else {
        user.agreedToLegal =
JSON.parse(sessionStorage.user).agreedToLegal;
        user.password = sessionStorage.password;
        console.log(user.password);
        $.post(SERVER_URL + '/updateUser', user, function(data) {
          sessionStorage.user = JSON.stringify(user);
        }).fail(function(error) {
          alert(error.responseText);
        }).done(function() {
          $.mobile.changePage("#pageMenu");
        });
      }
    }
    else
    {
      alert("Please complete the form properly".);
    }
  }
```

FIGURE 8.21 Saving user information for server-based Thyroid app UserForm.js

8.9 Implementing the syncing of records in the server-based Thyroid app

Figure 8.22 shows Sync.js, which uploads and downloads records to and from the server. As usual, we set the value of SERVER_URL at the top. The function shown in Figure 8.22 is called when the button element btnUpload is pressed. We first check to make sure that the tbRecords field exists in sessionStorage. If the records exist, we create a JSON object called requestBody. It has three attributes: email, password, and newRecords. The email attribute is obtained by using JSON.parse() on the sessionStorage.user object to recover its email value. The password attribute is the same as the password field from the sessionStorage. The newRecords attribute is the JSON object obtained by using JSON.parse() on the sessionStorage.tbRecords. We then use the $.post() method with URL 'http://140.184.132.239:3000/syncRecords' along with two additional parameters: requestBody and a nested anonymous function that receives the data returned by the server. This anonymous function only alerts the user that the records have been uploaded. The failure of the $.post() will be handled with the help of the familiar fail() method. The else part in Figure 8.22 corresponds to the situation when the sessionStorage.tbRecords does not exist. In that case, we alert the user that there are no records to save on the server.

```
var SERVER_URL = "http://140.184.132.239:3000";

// Updates records on the server to match the records on the device
$("#btnUpload").click(function() {
  if(sessionStorage.tbRecords) {
    var requestBody = {
      email: JSON.parse(sessionStorage.user).email,
      password: sessionStorage.password,
      newRecords: JSON.parse(sessionStorage.tbRecords)
    };
    $.post(SERVER_URL + "/syncRecords", requestBody, function(data) {
      alert("Records uploaded successfully!");
    }).fail(function(error) {
      alert(error.responseText);
    });
  } else {
    alert("No records to save!");
  }
});
```

FIGURE 8.22 Uploading records for server-based Thyroid app File: Sync.js

Figures 8.23 and 8.24 show the functions for downloading records. We will discuss them together, since Figure 8.23 is a subset of Figure 8.24. The function shown in Figure 8.23 will be called when the user presses the button element btnDownloadOverwrite. This button corresponds to the simpler overwriting option. The function in Figure 8.24 corresponds to the no overwriting option. That is, if a record with the same date exists, it will not be overwritten.

We will begin with the overwriting option. The function creates a JSON object called credentials. It has two attributes: email and password. The email attribute is obtained by using JSON.parse on the sessionStorage.user object to recover its email value. The password attribute is the same as the password field from the sessionStorage. We then use the $.post() method with URL 'http://140.184.132.239:3000/getRecords' along with two additional parameters: JSON object credentials and a nested anonymous function which receives the data returned by the server. This anonymous function alerts the user that the records have been downloaded. It then assigns sessionStorage.tbRecords, the string version of the data

```
// Downloads records from the server and overwrites all records
// currently on the device
$("#btnDownloadOverwrite").click(function() {
  var credentials = {
    email: JSON.parse(sessionStorage.user).email,
    password: sessionStorage.password
  };
  $.post(SERVER_URL + '/getRecords', credentials, function(data) {
    alert('Records downloaded successfully!');
    sessionStorage.tbRecords = JSON.stringify(data);
  }).fail(function(error) {
    alert(error.responseText);
  })
});
```

FIGURE 8.23 Downloading records (overwriting) for server-based Thyroid app File: Sync.js

```
// Downloads records from the server, only updating records that are
// not already on the device (determined by the date of the record)
$("#btnDownload").click(function() {
  var credentials = {
    email: JSON.parse(sessionStorage.user).email,
    password: sessionStorage.password
  };
  $.post(SERVER_URL + '/getRecords', credentials, function(data) {
    alert('Records downloaded successfully!');
    var tbRecords = JSON.parse(sessionStorage.tbRecords);
    for(var i=0; i<data.length; i++) {
      var exists = false;
      for(var j=0; j<tbRecords.length; j++) {
        if(tbRecords[j].date==data[i].date) {
          exists = true;
        }
      }
      if (!exists) {
        tbRecords.push(data[i]);
      }
    }
    sessionStorage.tbRecords = JSON.stringify(tbRecords);
  }).fail(function(error) {
    alert(error.responseText);
  })
});
```

FIGURE 8.24 Downloading records (no overwrite) for server-based Thyroid app File: Sync.js

created using JSON.stringify(). The failure of the $.post() will be handled with the help of the familiar fail() method. The highlighted code in the "no overwrite" version shown in Figure 8.24 is how the function called when the button element btnDownload is pressed differs from the one in Figure 8.23.

Let us just focus on the highlighted code. We first create a JSON object called tbRecords from sessionStorage.tbRecords using JSON.parse(). That will make tbRecords an array of JSON objects. The data returned by the server is also an array of JSON objects. We iterate through this array with a for loop using a loop control variable i. We assume that data[i] (i.e., the array of records returned by the server) does not exist in the sessionStorage by setting a Boolean variable exists to false. We then iterate through the array tbRecords with a nested for loop using a loop control variable j. We compare the date field of data[i] and tbRecords[j]. If they match, then a record with the same date exists in the sessionStorage, so we set exists to true. At the end of the inner for loop, we check to see if a record that matches the date from data[i] exists. If no record with the same date as data[i] exists, we push data[i] on the tbRecords array. If it does exist, we do not do anything. That is why the if has no else part. When the outside loop is over, we simply set sessionStorage.tbRecords to the tbRecords array.

In this chapter, we revisited our Thyroid app, which was originally storing data on the local device. This allowed us to use the app even when the device was not connected to the Internet. However, it also means that if the device is damaged, we will lose all our data. Therefore, we created a server-based version, where the user can upload the information from the sessionStorage to a server. We only looked at the web server computing of this app. The database server computing is implemented using two competing database technologies. In the following chapter, we will look at a NoSQL database implementation. The NoSQL databases are natural for JSON objects from our sessionStorage. However, the conventional relational database will continue to play a major role in modern-day computing. Therefore, we will reimplement the database server computations using a relational database in Chapter 10.

Quick facts/buzzwords

Node.js: A JavaScript-based platform for server-side computing. It uses an asynchronous and event-driven model to maximize throughput and efficiency.

Relational databases: A structured model for organizing records in a collection of related tables called relations.

NoSQL: An alternative to relational databases for storing unstructured data.

Google V8: An efficient JavaScript engine developed by Google for the Chrome browser. Node.js is built on Google V8.

require(): A method to include/import to load Node.js modules.

express: A robust minimal web application framework for serving web apps with Node.js.

sessionStorage: Local storage on the device that is retained through a browsing session.

post(): A Node.js method for sending requests to the server.

fail(): A method that is invoked when a function call fails.

CHAPTER 8

Self-test exercises

1. What are the advantages of client-side computing?
2. What are the advantages of server-side computing?
3. What is Node.js?

Programming projects

1. **Boiler monitor server-side version:** We want to change the app that monitors the temperature and pressure of a boiler so that the data is stored on a server instead of in localStorage. You can model the app based on the server-side Thyroid app. In this chapter, we will only modify the web pages to provide the server-side update functionality. We will defer the node.js programming until the following chapter.
    - Modify the password-based entry page to accept a unique ID of the boiler and an ability to create a new boiler ID.
    - We already have a menu page with four choices. Add a fifth option to sync the records to the server.
2. **Blood pressure monitor server-side version:** We want to change the app that monitors blood pressure so that the data is stored on a server instead of in localStorage. You can model the app based on the server-side Thyroid app. In this chapter, we will only modify the web pages to provide the server-side update functionality. We will defer the node.js programming until the following chapter.
    - Modify the password-based entry page to accept a unique ID of the person and an ability to create a new personal ID.
    - We already have a menu page with four choices. Add a fifth option to sync the records to the server.
3. **Power consumption monitor server-side version:** We want to change the app that monitors the power consumption of a manufacturing plant so that the data is stored on a server instead of in localStorage. You can model the app based on the server-side Thyroid app. In this chapter, we will only modify the web pages to provide the server-side update functionality. We will defer the node.js programming until the following chapter.
    - Modify the password-based entry page to accept a unique ID of the plant and an ability to create a new plant ID.
    - We already have a menu page with four choices. Add a fifth option to sync the records to the server.
4. **Body mass index (BMI) monitor server-side version:** We want to change the app that monitors the body mass index of a person based on height and weight so that the data is stored on a server instead of in localStorage. You can model the app based on the server-side Thyroid app. In this chapter, we will only modify the web pages to provide the server-side update functionality. We will defer the node.js programming until the following chapter.
    - Modify the password-based entry page to accept a unique ID of the person and an ability to create a new personal ID.
    - We already have a menu page with four choices. Add a fifth option to sync the records to the server.

5. **Managing a line of credit server-side version:** We want to change the app that manages our line of credit at a bank so that the data is stored on a server instead of in localStorage. You can model the app based on the server-side Thyroid app. In this chapter, we will only modify the web pages to provide the server-side update functionality. We will defer the node.js programming until the following chapter.
   - Modify the password-based entry page to accept a unique ID of the organization and an ability to create a new ID for the organization.
   - We already have a menu page with four choices. Add a fifth option to sync the records to the server.

CHAPTER 9

Using MongoDB server for sharing and storing information

WHAT WE WILL LEARN IN THIS CHAPTER:

1. What are non-relational or NoSQL databases in JSON format
2. How to create collections in a NoSQL database
3. How to insert/delete/update records in a NoSQL collection
4. How to connect to a NoSQL database server using node.js
5. How to manipulate NoSQL database using node.js

In this chapter, we will study a NoSQL based server-side database management for our server-based Thyroid app.

9.1 Emergence of NoSQL database models

Modern-day computing began with scientific computations. Fortran was one of the primary languages that led scientific computing in 1950s, 1960s, and even the 1970s. Business applications programming with the advent of COBOL was not too far behind. Databases played a major role in business applications of computers. COBOL had a number of built-in file operations that made it easy to store, retrieve, and modify the datasets. It soon became obvious that careful design and planning were required to store all the data used in business applications. One of the primary design issues in creating a database is the representation model. A database model defines the structure and operations that will be performed on a database. Early databases used somewhat unstructured hierarchical and network models that were programmed with COBOL's file systems. In 1970, E. F. Codd from IBM introduced the relational database model that revolutionized the database world. The data in relational databases is structured in a number of related tables. We will discuss the relational model in more detail in Chapter 10. The use of tables in relational databases seemed to be the most natural and structured representation of databases. It should be noted that a large amount of data is still maintained in relational databases, and that is unlikely to change. The structured format of relational databases is very important and useful in creating applications that guarantee less error-prone data management with the use of database normalization. However, with the proliferation of text, images, videos, and other unstructured data flowing through the information grid, the relational database model is not always the best option. In this chapter, we will look at NoSQL databases. The name NoSQL is derived from the SQL (structured query language) that is used in a relational database model. We will introduce NoSQL databases using its most popular implementation, MongoDB.

9.2 Introduction to MongoDB

Data in a MongoDB database is stored in a collection. A collection is a set of documents. Each document is a set of { name : value } pairs in the familiar JSON format discussed in Chapter 6 while studying localStorage and used throughout our Thyroid app. Readers who are familiar with relational databases will find collections to be analogous to tables and documents to be analogous to records. Since most readers are unlikely to be familiar with MongoDB, we will get a brief glimpse of how it works. We will assume that your system administrator has installed MongoDB on your server. Unlike relational databases, there is no intuitive graphical user interface (GUI) for MongoDB. This lack of a GUI is partly due to the fact that NoSQL databases do not have structured objects such as tables in a relational database. This lack of structure makes it difficult to create an intuitive layout. However, these are early days in the NoSQL world. As the field matures, we may be able to manipulate a NoSQL database through a GUI. For now, let us work with the command line interface (CLI). Figure 9.1 shows a sample session using MongoDB on our departmental UNIX server. The first line is the prompt from the UNIX command line. We type mongo as the command and enter the MongoDB environment. On our server, we immediately get an error saying that we are not authorized to execute some commands that give startup warnings. Let us not worry about this error message and move on. The system administrator has created a database for us with the name "mobileappbook". We will access this database using the command

```
> use mobileappbook
```

```
cs/pawan 3 >mongo
MongoDB shell version: 2.6.1
connecting to: test
Error while trying to show server startup warnings: not authorized on
admin to execute command { getLog: "startupWarnings" }
> use mobileappbook
switched to db mobileappbook
> show collections
2014-06-06T10:19:13.768-0300 error: {
        "$err" : "not authorized for query on
mobileappbook.system.namespaces",
        "code" : 13
} at src/mongo/shell/query.js:131
> db.auth('mobileappbook','xxxxxxx')
1
> show collections
system.indexes
> db.intro.insert({ message : "Hello world of MongoDB" })
WriteResult({ "nInserted" : 1 })
> db.intro.insert({ messge : "Hello again world of MongoDB" })
WriteResult({ "nInserted" : 1 })
> db.intro.find()
{ "_id" : ObjectId("5391c1e23fd5fbceeb4626d7"),
... "message" : "Hello world of MongoDB" }
{ "_id" : ObjectId("5391c2093fd5fbceeb4626d8"),
... "messge" : "Hello again world of MongoDB" }
```

FIGURE 9.1 Adding collections and documents in a MongoDB session

Note: ">" is a MongoDB prompt; we do not type it.

The command

```
> show collections
```

can be used to see the collections in our database. However, as Figure 9.1 shows, we are slapped with an error message saying that we are not authorized to look at the collections in the database "mobileappbook". We need authorization. The system administrator, Andrew, has given us a password for the database. We execute the command

```
> db.auth('mobileappbook', 'xxxxxxx')
```

to get the necessary authorization using the method db.auth(). Now we can re-execute the command to see all the collections we have.

```
> show collections
system.indexes
```

Any collection with the name system.*name* is a special collection maintained by the system. Users are strongly discouraged to create any collection that begins with "system".
We can create a new collection by simply inserting a document such as

```
> db.intro.insert({ message : "Hello world of MongoDB" })
WriteResult({ "nInserted" : 1 })
```

The above command creates a collection called intro and inserts a document with the name "message" and value "Hello world of MongoDB" in that collection. We can see the unstructured and unfettered world of a NoSQL database with this command. In a relational database we would have had to first create a table (collection in MongoDB) and specify the number of fields and the type of data that will be stored in that fields. Then we execute an insert command to insert values into pre-specified fields. In MongoDB, we can make up collection names and field names as we go. This convenience comes at a price as we can see with a spelling mistake in the following insert command in the same collection

```
> db.intro.insert({ messge : "Hello again world of MongoDB" })
WriteResult({ "nInserted" : 1 })
```

We ended up creating a document with the name "messge", when we meant to create a document with the name "message". This is one of the risks of using NoSQL databases such as MongoDB. You have significant flexibility and power, but the power comes with responsibility.

If you wanted to retrieve documents from a collection you would use the method find(), which is similar to the select command from SQL. An example of the find() method is shown below

```
> db.intro.find()
{ "_id" : ObjectId("5391c1e23fd5fbceeb4626d7"),
... "message" : "Hello world of MongoDB" }
{ "_id" : ObjectId("5391c2093fd5fbceeb4626d8"),
... "messge" : "Hello again world of MongoDB" }
```

We see that there is an additional field in the document called "_id". This is automatically added by MongoDB as a unique identifier.

Note that the three dots (...) at the beginning of a line just show a continuation of the previous line. They are used to wrap the text for improved readability and are not part of the input or output.

The find() command can be used to restrict the number of documents retrieved. It can also be used to retrieve only part of a document. We will look at some of these more advanced features later on this chapter.

If you want to delete documents from a collection you use the method remove(), which is similar to the delete command in SQL. The session for deleting documents and collections is shown in Figure 9.2. Let us remove the document with spelling error

```
> db.intro.remove({messge: "Hello again world of MongoDB"})
WriteResult({ "nRemoved" : 1 })
```

We indicated that we wanted to delete a particular document by specifying it as {messge: "Hello again world of MongoDB"}.

We can confirm that the document is gone from the collection

```
> db.intro.find()
{ "_id" : ObjectId("5391c1e23fd5fbceeb4626d7"),
... "message" : "Hello world of MongoDB" }
```

```
> db.intro.remove({messge: "Hello again world of MongoDB"})
WriteResult({ "nRemoved" : 1 })
> db.intro.find()
{ "_id" : ObjectId("5391c1e23fd5fbceeb4626d7"),
... "message" : "Hello world of MongoDB" }
> db.intro.insert({ messge : "Hello again world of MongoDB" })
WriteResult({ "nInserted" : 1 })
> db.intro.find()
{ "_id" : ObjectId("5391c1e23fd5fbceeb4626d7"),
... "message" : "Hello world of MongoDB" }
{ "_id" : ObjectId("5391c2093fd5fbceeb4626d8"),
... "messge" : "Hello again world of MongoDB" }
> db.intro.remove({messge:{$regex:''}})
WriteResult({ "nRemoved" : 1 })
> db.intro.remove({})
WriteResult({ "nRemoved" : 1 })
> db.intro.find()
> show collections
intro
system.indexes
> db.inro.insert({message : "Another hello"})
WriteResult({ "nInserted" : 1 })
> show collections
inro
intro
system.indexes
> db.inro.drop()
true
> db.random.find()
> quit()
cs/pawan 4 >
```

■ FIGURE 9.2 Deleting collections and documents from MongoDB session

If we did not want to type the entire value, we will have to use regular expressions. Figure 9.2 adds that document with the spelling error back in the collection and confirms that the document has been inserted

```
> db.intro.insert({ messge : "Hello again world of MongoDB" })
WriteResult({ "nInserted" : 1 })
> db.intro.find()
{ "_id" : ObjectId("5391c1e23fd5fbceeb4626d7"),
... "message" : "Hello world of MongoDB" }
{ "_id" : ObjectId("5391c2093fd5fbceeb4626d8"),
... "messge" : "Hello again world of MongoDB" }
```

Now, let us try a different version of the remove command with the help of a regular expression referred to as $regex in MongoDB.

```
> db.intro.remove({messge:{$regex:''}})
WriteResult({ "nRemoved" : 1 })
```

Instead of specifying the value, we are specifying the $regex with a pattern that is an empty string ''. It will match any value of document with the name messge. That is, we will delete all the documents with the name messge. You can specify a partial pattern for $regex, or a more refined regular expression. We are not going to discuss regular expressions in greater detail at this point. We confirm that the documents are gone from the collection

```
> db.intro.find()
{ "_id" : ObjectId("5391c1e23fd5fbceeb4626d7"),
... "message" : "Hello world of MongoDB" }
```

If we wanted to empty an entire collection—that is, delete all the documents—we can use empty braces {} in the remove method as

```
> db.intro.remove({})
WriteResult({ "nRemoved" : 1 })
> db.intro.find()
```

The find() method above produces no output indicating an empty collection. The collection is still there as confirmed by the following command

```
> show collections
intro
system.indexes
```

Our spelling mistakes can also add unintended collections as we can see in the following command, which adds a collection "inro"

```
> db.inro.insert({message : "Another hello"})
WriteResult({ "nInserted" : 1 })
```

If we look at the collections, we have added a new collection called inro

```
> show collections
inro
intro
system.indexes
```

We can completely delete an entire collection using the drop() method, which is equivalent to the drop command from SQL, as

```
> db.inro.drop()
true
```

Finally, let us look at the forgiving nature of MongoDB and search for documents in a collection called random

```
> db.random.find()
```

MongoDB does not give any indication that such a collection does not exist. We terminate our MongoDB session with a call to the method quit()

```
> quit()
```

9.3 Modeling a NoSQL database

Let us now create a more refined MongoDB collection to explore more features of NoSQL data management. Figure 9.3 shows a library collection. It consists of two documents. Each document has two attributes: author and books. The author attribute is just a string that stores author name. The books attribute is an array of subdocuments. Each subdocument has two attributes: title and year. The title attribute stores strings while the year attribute stores an integer.

Readers familiar with relational database models will see an immediate divergence model since the NoSQL model is not in the first normal form. The first normal form (1NF) in relational databases was necessitated by the technological limitations of the era. It mandated that an attribute can store only one atomic value. For example, if an author "Pawan" writes three books "Web mining", "Web programming", and "Mobile app development", we cannot have a

```
{
  {
    author : "Pawan",
    books :
    [
       { title : "Web mining", year : 2007 },
       { title : "Web programming", year : 2012 },
       { title : "Mobile app development", year : 2015 }
    ]
  }

  {
    author : "Walter",
    books :
    [
      { title : "Intro programming", year : 2000 },
      { title : "Data structures", year : 2002 }
    ]
  }
}
```

FIGURE 9.3 A library collection in NoSQL

single record with "Author = Pawan" and "Product = {"Web mining", "Web programming", "Mobile app development"}". Instead, we need three records

- "Author = Pawan" and "Product = Web mining"
- "Author = Pawan" and "Product = Web programming"
- "Author = Pawan" and "Product = Mobile app development"

In the NoSQL model shown in Figure 9.3, we do not repeat the author's name three times.

Let us try to implement the database model from Figure 9.3 using MongoDB. Unlike relational databases, we do not need to create the entire structure of the collection. We can build it as we go. Figure 9.4 shows our MongoDB session, which starts by inserting a document {author: "Pawan"} in a collection called library. This insert creates the library collection as well. Now we want to update this record by adding the books field with an array of two books.

We update an existing document with a method called update(). The method needs at least two parameters: a query to identify the document and the updated document. Let us look at the update method call in Figure 9.4.

```
> db.library.insert({author:"Pawan"})
WriteResult({ "nInserted" : 1 })

> db.library.update(
... {author:"Pawan"},
... {author:"Pawan",
...  books:
...  [
...     {title:"Web mining", year : 2007},
...     {title:"Web programming", year : 2012}
...  ]
... }
... )
WriteResult({ "nMatched" : 1, "nUpserted" : 0, "nModified" : 1 })
> db.library.find()
{ "_id" : ObjectId("53920e9e410b268e7c85cbc8"),
... "author" : "Pawan",
... "books" :
... [
... { "title" : "Web mining",
... "year" : 2007 },
... { "title" : "Web programming", "year" : 2012 }
... ] }
```

FIGURE 9.4 Inserting nested documents with an array in MongoDB

Here the query is `{author:"Pawan"}`, which identifies the document. We then specify the updated document as

```
{author:"Pawan",
... books:
... [
...    {title:"Web mining", year : 2007},
...    {title:"Web programming", year : 2012}
... ]
... }
```

The new document now has an array called books with two elements. Each element has a title and a year attribute. We use the find() method to verify that the document is as we wanted it to be. The update() command can build on the existing record instead of specifying a complete new copy of the record as we did in Figure 9.4. Let us say we wanted to add a third book to the books array. Figure 9.5 shows a more sophisticated version of the update. Our query part is the same as before—that is, `{author:"Pawan"}` to identify the document. The update part uses $push operator as

```
{$push: {books: {title:"Mobile app development",year:2015}}}
```

to append a third element to the array. Figure 9.5 also has a find() method call that confirms that the book's array now indeed has three elements.

```
> db.library.update(
... {author:"Pawan"},
... {$push:
... {books:
... {title:"Mobile app development",year:2015}
...}})
WriteResult({ "nMatched" : 1, "nUpserted" : 0, "nModified" : 1 })
> db.library.find()
{ "_id" : ObjectId("53920e9e410b268e7c85cbc8"),
... "author" : "Pawan",
... "books" : [
... { "title" : "Web mining", "year" : 2007 },
... { "title" : "Web programming", "year" : 2012 },
... { "title" : "Mobile app development", "year" : 2015 }
... ] }
>
```

■ **FIGURE 9.5** Pushing documents in an array in MongoDB

Figure 9.6 shows the selection and projection aspects of the find() method. Selection restricts the number of records that will be retrieved while projection limits the number of fields from the selected records that will be retrieved. We add one more record to our library collection for an author called "Walter" who has two books. A basic call to the find() method shows that now we have two records in the library collection: one for author "Pawan" with three books and another for author "Walter" with two books. Let us say that we only want to see the record(s) for author "Walter". This is the selection operation. The find() method can accept a `<query>` such as shown in the following call

```
db.library.find({author:"Walter"})
```

As the output shows, only the record with author "Walter" is retrieved. If we only wanted to see the books attribute of author "Walter", we can also add the projection to the find() method call as

```
db.library.find({author:"Walter"},{books:1})
```

Our projection {books:1} specifies that we want the attribute books by putting the digit 1 against it. We can use 0 to exclude fields in the projection. We have barely scratched the surface of what MongoDB has to offer. Given the number of technologies that we are covering in this book, we refer the readers to http://www.mongodb.org/ for more sophisticated usage of MongoDB.

```
> db.library.insert(
... {author:"Walter",
... books: [
... {title:"Intro programming", year : 2000},
... {title:"Data structures", year : 2002} ] } )
WriteResult({ "nInserted" : 1 })
> db.library.find()
{ "_id" : ObjectId("53920e9e410b268e7c85cbc8"),
... "author" : "Pawan",
... "books" : [
... { "title" : "Web mining", "year" : 2007 },
... { "title" : "Web programming", "year" : 2012 },
... { "title" : "Mobile app development", "year" : 2015 }
... ] }

...{ "_id" : ObjectId("5392fc90b17e85f5823eab6c"),
... "author" : "Walter",
... "books" : [
... { "title" : "Intro programming", "year" : 2000 },
... { "title" : "Data structures", "year" : 2002 }
... ] }
> db.library.find({author:"Walter"})
{ "_id" : ObjectId("5392fc90b17e85f5823eab6c"),
... "author" : "Walter",
... "books" : [
... { "title" : "Intro programming", "year" : 2000 },
... { "title" : "Data structures", "year" : 2002 }
... ] }
> db.library.find({author:"Walter"},{books:1})
{ "_id" : ObjectId("5392fc90b17e85f5823eab6c"),
... "books" : [
... { "title" : "Intro programming", "year" : 2000 },
... { "title" : "Data structures", "year" : 2002 }
... ] }
>
```

FIGURE 9.6 Finding projections of documents in MongoDB

9.4 Modeling a NoSQL database for the Thyroid app

Armed with adequate knowledge of MongoDB, we are ready to create a database model for our server version of the Thyroid app. We have already discussed the client side of the server-based Thyroid app in the previous chapter. Based on all the information that is collected from the user and records form, we have to store the following fourteen attributes

1. email address
2. First name
3. Last name
4. Date of birth
5. Health insurance card number
6. Password
7. Cancer type
8. Cancer stage
9. TSH range

10. legal waiver
11. Date of test
12. TSH
13. Tg
14. Synthroid

The first step in creating a database is to design a data model. In a NoSQL database, the data modeling will consist of designing the structure of the collections. The easiest way to design a database would be to create one single collection consisting of all fourteen attributes.

We can then have a record for each blood test done by the patients, in which case we will be repeating user information from attributes 1 to 10 for each test result. This will lead to redundancy in the database. Moreover, if the user information were to change, we will have to update every single test record. This will lead to additional overhead. Moreover, if these updates were not programmed properly, we may have inconsistency about the user information in the test records, destroying the integrity of the database. A solution is to separate the collection into two collections.

One collection is called users with the following ten fields

1. email address
2. First name
3. Last name
4. Date of birth
5. Health insurance card number
6. Password
7. Cancer type
8. Cancer stage
9. TSH range
10. legal waiver

Another is called records with the following five fields

1. email address
2. Date of test
3. TSH
4. Tg
5. Synthroid

These two collections are linked together through the email attribute that serves as user-id. The division of collections as described here will turn our database into second normal form in relational database terminology. We will discuss it in more detail in the following chapter.

Figure 9.7 shows the users collection stored in our MongoDB database. As before, MongoDB adds a unique _id attribute for each record. Another thing that we notice is the fact that the password field is encrypted, so the system administrator cannot see the actual password used by the user. We will look at the encryption of the password when we study the node.js scripts that are used to maintain the database. The records collection from MongoDB is shown in Figure 9.8.

```
{
    "_id" : ObjectId("539841e7533eec66206ccfaa"),
    "email" : "j@cs.smu.ca",
    "firstName" : "J",
    "lastName" : "Some",
    "healthCardNumber" : "123456789",
    "newPassword" : "1234",
    "dateOfBirth" : "1948-12-12",
    "cancerType" : "Follicular",
    "cancerStage" : "StageTwo",
```

```
    "tshRange" : "StageC",
    "agreedToLegal" : "1",
    "password" : "$2a$10$UbNqzmIWTFQoiEoWEGOefVUz6354j6f9d3HV9U/K"
}
{
    "_id" : ObjectId("539fbd478dea969a5d220f1e"),
    "email" : "user@cs.smu.ca",
    "firstName" : "X".,
    "lastName" : "User",
    "healthCardNumber" : "1111222233334444",
    "newPassword" : "1234",
    "dateOfBirth" : "1951-12-09",
    "cancerType" : "Follicular",
    "cancerStage" : "StageTwo",
    "tshRange" : "StageB",
    "agreedToLegal" : "1",
    "password" : "$2a$10$rCR2ZWfXSbO2Q9HiGfE4wu4lXPcxCDBL.
dvouVbxK"
}
```


■ **FIGURE 9.7** Users collection in MongoDB for the server-based Thyroid app


```
{
    "date" : "2014-05-25",
    "tsh" : "0.8",
    "tg" : "0.7",
    "synthroidDose" : "0.6",
    "user" : "user@cs.smu.ca",
    "_id" : ObjectId("539fbeec8dea969a5d220f1f")
}
{
    "date" : "2014-06-02",
    "tsh" : "0.7",
    "tg" : "0.6",
    "synthroidDose" : "0.5",
    "user" : "user@cs.smu.ca",
    "_id" : ObjectId("539fbeec8dea969a5d220f20")
}{
    "date" : "2014-06-09",
    "tsh" : "0.6",
    "tg" : "0.5",
    "synthroidDose" : "0.4",
    "user" : "user@cs.smu.ca",
    "_id" : ObjectId("539fbeec8dea969a5d220f21")
}
{
    "date" : "2014-01-06",
    "tsh" : "1",
    "tg" : "2",
    "synthroidDose" : "3",
    "user" : "j@cs.smu.ca",
```

```
        "_id" : ObjectId("53a039258dea969a5d220f22")
}
{
        "date" : "2014-03-10",
        "tsh" : "1.6",
        "tg" : "2",
        "synthroidDose" : "1",
        "user" : "j@cs.smu.ca",
        "_id" : ObjectId("53a039258dea969a5d220f24")
}
{
        "date" : "2014-02-05",
        "tsh" : "1.2",
        "tg" : "2.2",
        "synthroidDose" : "3.1",
        "user" : "j@cs.smu.ca",
        "_id" : ObjectId("53a039258dea969a5d220f23")
}
```

FIGURE 9.8 Records collection in MongoDB for the server-based Thyroid app

9.5 Launching the MongoDB server for the Thyroid app

The database server of the server-side Thyroid app is launched with the following command

```
node server.js
```

The command node is used to invoke the Node.js runtime environment, and we are passing the name of the file containing the script as a parameter. This file contains all of our database server program that will handle all the requests. We begin by looking at the main part of the program shown in Figures 9.9 and 9.10. It begins with setting up variables using the method require(). We had looked at the method require() in the previous chapter, which introduced the web server of our app. The method require() essentially imports the specified modules and makes them available to us through the variables we have assigned them to. Here we are importing the bcrypt module for encrypting the password. We have already seen the express module in the previous chapter. It is a minimal web application framework called express (expressjs.com) that provides a robust platform for serving multi-page web applications. Express is required on the database server to handle POST and GET requests from the browser. We also import the MongoDB module so that the database server can connect to the MongoDB database. The following lines set up values of various variables needed to connect to the MongoDB

- user
- password
- host—IP address of the machine hosting MongoDB server. Since we are on the same server as the machine hosting MongoDB, we use the IP address 127.0.0.1. It usually corresponds to the local host.
- port—we are using 27017, which is the default port used by the MongoDB server.
- database

All of the above information is used to construct a string called connectionString. It will be passed to the connect function.

```
var bcrypt = require('bcrypt');
var express = require('express');
var mongodb = require('mongodb');

var user = 'mobileappbook';
var password = 'xxxxxxxx';
var host = '127.0.0.1';
var port = '27017'; // Default mongodb port
var database = 'mobileappbook';
// var connectionString = 'mongodb://' + user + ':' + password + '@' +
var connectionString = 'mongodb://' +
   host + ':' + port + '/' + database;

// These will be set once connected, used by other functions below
var usersCollection;
var recordsCollection;
```

FIGURE 9.9 Main part of the server.js in the MongoDB server-based Thyroid app File: server.js

In addition, we have declared two variables called usersCollection and recordsCollection. These will be initialized to the appropriate MongoDB collections when we connect to the database.

The variable allowCrossDomain is related to Cross-Origin Resource Sharing (CORS), which is a World Wide Web Consortium (W3C)—a body that establishes standards for the web specification. Normally, the browser only accepts data from a single origin. CORS allows data to come from multiple domains. The variable allowCrossDomain is in fact a function that enables any origin ('*' – wild card), methods `'GET,PUT,POST,DELETE'`, and headers `'CONTENT-Type'`.

We then create a new application with express() and assign it to a variable called app. There are two calls to the method app.use(). The first enables us to parse JSON objects. The second one enables CORS support.

```
//CORS Middleware, causes Express to allow Cross-Origin Requests
var allowCrossDomain = function(req, res, next) {
  res.header('Access-Control-Allow-Origin', '*');
  res.header('Access-Control-Allow-Methods', 'GET,PUT,POST,DELETE');
  res.header('Access-Control-Allow-Headers', 'Content-Type');

  next();
}

var app = express();
app.use(express.bodyParser());
app.use(allowCrossDomain);
```

FIGURE 9.10 Initializing and setting up the MongoDB server-based Thyroid app with Cross Origin Resource Sharing

Before we start processing the requests to our MongoDB server we call the method mongodb.connect() as the last part of the main program shown. The method accepts two parameters. The first one is the connectionString that contains information for connecting to the MongoDB server such as the server IP address, port, username, and password. The second parameter is the function that will be executed when we connect to MongoDB. The function checks to see if the object error exists. If it does, it will throw the error object. We had seen this type of exception handling, which includes try, catch, and throw while studying local storage in Chapter 6.

The function then assigns the users collection to the variable usersCollection and records collection to the variable recordsCollection. This will be available for the rest of the functions since they were declared in the main.

The call to method process.on() is a signal handling method. Readers who are familiar with C on UNIX-based systems will find it similar to the function signal(). The object process is a global object in Node.js that is available from anywhere, and it allows us to manipulate the existing process. In our case, the process is the MongoDB connection. The method process.on() allows us to specify what to do when some of the familiar standard UNIX system signals. Readers who are not acquainted with signal handling may want to just focus on the method shown in Figure 9.11. We are defining a function that will be called when the process receives a signal to terminate (SIGTERM). In that case, we will write a message to the system console saying that we are shutting the server down. We should then close the express server using the call app.close() and the MongoDB server using the call mongodb.close() before terminating.

The last thing we do as part of the mongodb.connect() call is launch the server using the method app.listen(). It takes two parameters. The first parameter is the port number where we will be listening to the server requests. We have chosen the port number to be 3000. The second parameter is an anonymous function. It writes a message to the system console saying that the database server is now listening on the specified port.

```
mongodb.connect(connectionString, function(error, db) {
  if (error) {
    throw error;
  }

  usersCollection = db.collection('users');
  recordsCollection = db.collection('records');

  // Close the database connection and server when the application ends
  process.on('SIGTERM', function() {
    console.log("Shutting server down".);
    db.close();
    app.close();
  });

  var server = app.listen(3000, function() {
    console.log('Listening on port %d', server.address().port);
  });
});
```

FIGURE 9.11 Call to the connect() method for MongoDB in the server-based Thyroid App
File: server.js

The rest of the functions in the server.js handle various requests to the website including saveNewUser, login, updateUser, getRecords, and syncRecords. We will look at the Node.js code for each one of these requests in this chapter.

9.6 Saving a new user on the MongoDB server for the Thyroid app

Figure 9.12 shows the code for processing a saveNewUser request. We are running the method app.post(), which receives two parameters. The first one is the request '/saveNewUser' that will be handled by the server, and the second parameter is an anonymous function that contains the rest of the code. We log the fact that we are sending the records on the system console. We create a variable called user that is the newUser field from the body of the parameter request.

```
// Creates a new user in the database
app.post('/saveNewUser', function(request, response) {
  console.log('New user being created.');
  var user = request.body.newUser;
  console.log(request.body);
  if (!user.email || !user.newPassword || !user.firstName ||
!user.lastName || !user.healthCardNumber || !user.dateOfBirth ||
!user.cancerType || !user.cancerStage || !user.tshRange) {
    return response.json(400, 'Missing a parameter.');
  }

  var salt = bcrypt.genSaltSync(10);
  var passwordString = '' + user.newPassword;
  user.password = bcrypt.hashSync(passwordString, salt);
  user.agreedToLegal = false;

  usersCollection.find({
    email: user.email
  }, function(err, result) {
    if (err) {
      return response.send(400, 'An error occurred creating this
user.');
    }
    if (result.length) {
      return response.send(400, 'A user with this email address already
exists.');
    }
    usersCollection.insert(user, function(err, result) {
      if (err) {
        return response.send(400, 'An error occurred creating this
user.');
      }
      return response.json(200, 'User created successfully!');
    });
  });
});
```

FIGURE 9.12 node.js method to process saveNewUser request in MongoDB File: server.js

If any of the data that is required to create a new user such as the email, new password, first name, or last name in the request body is null, we return the response with error code 400 and a message saying that login information is missing. The next step in the process is to encrypt the password before storing it in the database. The encryption is a one-way encryption. The encrypted password cannot be decrypted to obtain the original password. That means even the database administrator does not know what your password is. The program will only encrypt the password you supply and compare it with the stored password. We have created the object bcrypt by including the bcrypt library. We use this object for our encryption.

The first thing we do is generate a string called salt. The salt string is added to the password string before encryption making it difficult for attacks that could try a list of password guesses such as dictionary words. The salt is generated using the method bcrypt.genSaltSync(). The parameter 10 is optional and tells how many rounds should be used for random generation. More rounds means more computations. We then declare a variable called passwordString that is a concatenation of an empty string and the newPassword field of user. We then generate the user.password using the bcrypt.hashSync(). The method returns the encrypted string based on the passwordString and newly generated salt. We also set the agreedToLegal field to be false, so that the user will be taken to the disclaimer page next.

We look at the MongoDB usersCollection using the find() method. This call to the find method is a little more complex than the one we have studied so far. The find receives two parameters. The first parameter is a JSON object with an email field equal to the email value we received through the request.body. The second parameter is an anonymous function that contains the rest of the code in Figure 9.12. The anonymous function receives two objects. The first one, called err, is the possible error that could be thrown by the find() method. If err is not null, we return the response with error code 400 and a message saying that there was an error creating the user. The second object, called result, is not the actual document retrieved from the usersCollection, but a cursor (or pointer) that points to the retrieved records. We are really not interested in any records returned by the find() method. In fact, if we do get a record back, we have an erroneous situation, since it means that the user with the same email is already in the database. Since we are using the email address as the unique ID, we cannot add another user with the same email. This is checked by looking at the length of result. If the length is not zero, we have a user with the same email address in the database. Therefore, we return the response with error code 400 and a message saying that a user with the same email is already in the database. Otherwise, we call the insert() method for the usersCollection. It again receives two parameters: the user object and an anonymous function. The first one, called err, is the possible error that could be thrown by the insert() method. If err is not null we return the response with error code 400 and a message saying that there was an error creating the user. The second object called result is not very useful to us for the insert. If we did not return error code 400 until this point, we can send a code 200 and a message confirming that the user was successfully created.

9.7 Processing login with the MongoDB server for the Thyroid app

Figure 9.13 shows the code for processing a login request. We are running the method app.post(), which receives two parameters. The first one is the request '/login', which will be handled by the server, and the second parameter is an anonymous function that contains the rest of the code. The anonymous function takes two parameters. The request is received by the post() method and the response that will be returned. We log the fact that we are logging in the user on the system console. If either the email or password in the request body is empty, we return the response with error code 400 and a message saying that login information is missing.

If we have email and password, we proceed further. We look at the MongoDB usersCollection using the find() method. The find receives two parameters. The first parameter is a JSON object with the email field equal to the email value we received through the request.body. The second parameter is an anonymous function that contains the rest of the code in Figure 9.13. The anonymous function receives two objects. The first one, called err, is the possible error that could be thrown by the find() method. Earlier we had sent returned a code

```
// Uses a email address and password to retrieve a user from the
database
app.post('/login', function(request, response) {
  console.log('Logging user in');

  if (!request.body.email || !request.body.password) {
    return response.send(400, 'Missing log in information.');
  }

  usersCollection.find({
    email: request.body.email
  }, function(err, result) {
    if (err) {
      throw err;
    }
    result.next(function(err, foundUser) {
      if (err) {
        throw err;
      }
      if (!foundUser) {
        return response.send(400, 'User not found.');
      }
      if (!bcrypt.compareSync('' + request.body.password,
foundUser.password)) {
        return response.send(400, 'Invalid password');
      }
      delete foundUser.password;
      return response.send(200, foundUser);
    });
  });
});
```

FIGURE 9.13 node.js method to process login request in MongoDB File: server.js

of 400 and an error message when err was not null. This time we just throw the object err. It will terminate the method call and can be handled by the browser. The second object is called result. As mentioned previously, the object result is not the actual document retrieved from the usersCollection but a cursor (or pointer) that points to the retrieved records. In order to get the records that satisfy the query (i.e., first parameter in the find() method), we use the method result.next(), which receives an anonymous function that contains the rest of the code in Figure 9.13. This anonymous function receives two parameters. The first one, called err, is the possible error that could be thrown by the next() method. The second object, called foundUser, will be the retrieved result from the usersCollection in MongoDB. If err is not null, we just throw it and are done with the processing of the request. If err is null, we continue with the rest of the processing. We then check to see if the foundUser is null. If it is null, we return the response with error code 400 and a message saying that the user is not found. Otherwise, we have retrieved the user document and are ready to process the information about the user.

The processing involves comparing the password in the body of the request with the user password. We do not need to encrypt the entered password. The method bcrypt.compareSync() will compare the entered password with the password field of foundUser. We had to append the entered password to an empty string to convert it to a string as in ''+request.body.password. If the call to bcrypt.compareSync() returns false, we return the response with error code 400 and a message saying that the password is invalid. Otherwise, the user is successfully logged in. We delete the password from the foundUser document (it does not delete it from the database). That way, when we return the foundUser, we are not sending back the encrypted password. Finally, we return the code 200 and foundUser as an indication of success.

9.8 Updating user data in the MongoDB server for the Thyroid app

Figure 9.14 shows the code for processing an updateUser request. A lot of code is similar to the previous post methods. However, we explain all the steps for completeness. We are running the method app.post(), which receives two parameters. The first one is the request '/updateUser', which will be handled by the server, and the second parameter is an anonymous function that contains the rest of the code. The anonymous function takes two parameters: the request that is received by the post() method and the response that will be returned. We log the fact that we

```
// Update user information
app.post('/updateUser', function(request, response) {
  console.log('User being updated.');
  var user = request.body;
  usersCollection.find({
    email: request.body.email
  }, function(err, result) {
    if (err) {
      throw err;
    }
    result.next(function(err, foundUser) {
      if (err) {
        return response.send(400, 'An error occurred finding a user to
update.');
      }
      if (!foundUser) {
        return response.send(400, 'User not found.');
      }
      if (!bcrypt.compareSync('' + request.body.password, foundUser.
password)) {
        return response.send(400, 'Invalid password.');
      }
      if (user.newPassword) {
        var salt = bcrypt.genSaltSync(10);
        var passwordString = '' + user.newPassword;
        user.password = bcrypt.hashSync(passwordString, salt);
      }
      delete user._id;
      usersCollection.update({
        email: user.email
      }, user, null, function(err) {
        if (err) {
          console.log(err);
          return response.send(400, 'An error occurred updating users
information.');
        }
        return response.send(200, foundUser);
      });
    });
  });
});
```

FIGURE 9.14 node.js method to process updateUser request in the MongoDB File server.js

are updating the user on the system console. We are saving the body of the request in a variable called user for easy reference.

We look at the MongoDB usersCollection using the find() method. The find receives two parameters. The first parameter is a JSON object with the email field equal to the email value we received through the request.body or user. The second parameter is an anonymous function that contains the rest of the code in Figure 9.14. The anonymous function receives two objects. The first one, called err, is the possible error that could be thrown by the find() method. If err is not null, we throw the object err. It will terminate the method call and can be handled by the browser. The second object, called result, is a cursor (or pointer) that points to the retrieved records. In order to get the records, we use the method result.next(), which receives an anonymous function that contains the rest of the code in Figure 9.14. This anonymous function receives two parameters. The first one, called err, is the possible error that could be thrown by the next() method. The second object called foundUser will be the retrieved result from the usersCollection in MongoDB. If err is not null, we return the response with error code 400 and a message saying that the user could not be updated. If err is null, we continue with the rest of the processing. We then check to see if the foundUser is null. If it is null, we return the response with error code 400 and a message saying that the user is not found. Otherwise, we have retrieved the user document and are ready to process the information about the user. The processing involves comparing the password in the body of the request with the user password.

We do not need to encrypt the entered password. The method bcrypt.compareSync() will compare the entered password with the password field of foundUser. We had to append the entered password to an empty string to convert it to a string as in ''+request.body.password. If the call to bcrypt.compareSync() returns false, we return the response with error code 400 and a message saying that the password is invalid. If the updating includes a new password for the user, user.newPassword will not be null. In that case, we encrypt the password as we did while creating the new user and save it in user.password. We delete the system-generated _id field before updating the user with the usersCollection.update() method. The update function takes four parameters. The first parameter is a JSON object with an email field equal to the email value we received through the request.body or user. It serves as a query to locate the document in MongoDB. The second parameter is the entire user object, which will be inserted as the updated document in MongoDB. The third parameter can be used to specify a number of options that can be specified, such as whether we should insert the user if it does not exist, or can we have multiple copies with the same email ID. We have set the third parameter to null, so we accept default values that do not insert the user if it does not exist and does not keep multiple copies. Finally, the fourth parameter is an anonymous function that takes one parameter called err. If err is not null, the update did not work out. We log the err object on the system console and return the response with error code 400 and a message saying that user could not be updated. If err is null, we continue with the rest of the processing, where we return the code 200 and foundUser as an indication of success.

9.9 Downloading records from the MongoDB server for the Thyroid app

Figure 9.15 shows the code for processing a getRecords request. We are running the method app.post(), which receives two parameters. The first one is the request '/getRecords' that will be handled by the server, and the second parameter is an anonymous function that contains the rest of the code. The anonymous function takes two parameters: the request that is received by the post() method and the response that will be returned. We log the fact that we are sending the records on the system console. If either the email or password in the request body is empty, we return the response with error code 400 and a message saying that login information is missing.

If we have an email and password, we proceed further. We look at the MongoDB usersCollection using the find() method. The find receives two parameters. The first parameter is a JSON object with an email field equal to the email value we received through the request. body. The second parameter is an anonymous function that contains the rest of the code

```
// // Get all records associated with the given user
app.post('/getRecords', function(request, response) {
  console.log('Sending records');
  if (!request.body.email || !request.body.password) {
    return response.send(400, 'Missing log in information.');
  }
  usersCollection.find({
    email: request.body.email
  }, function(err, result) {
    if (err) {
      throw err;
    }
    result.next(function(err, foundUser) {
      if (err) {
        throw err;
      }
      if (!foundUser) {
        return response.send(400, 'User not found.');
      }
      if (!bcrypt.compareSync('' + request.body.password,
foundUser.password)) {
        return response.send(400, 'Invalid password');
      }
      getRecords(request, function(err, result) {
        if (err) {
          return response.send(400, 'An error occurred retrieving
records.');
        }
        result.toArray(function(err, resultArray) {
          if (err) {
            return response.send(400, 'An error occurred processing
your records.');
          }
          return response.send(200, resultArray);
        });
      });
    });
  });
});
```

FIGURE 9.15 node.js method to process getRecords request in MongoDB File: server.js

in Figure 9.15. The anonymous function receives two objects. The first one, called err, is the possible error that could be thrown by the find() method. The second object is called result. The object result is a cursor (or pointer) that points to the retrieved records. In order to get the records, we use the method result.next(), which receives an anonymous function that contains the rest of the code in Figure 9.15. It receives two parameters. The first one, called err, is the possible error that could be thrown by the next() method. The second object, called foundUser, will be the retrieved result from the usersCollection in MongoDB. If err is not null, we just throw it and are done with the processing of the request. If err is null, we continue with the rest of the processing.

We then check to see if the foundUser is null. If it is null, we return the response with error code 400 and a message saying that the user is not found. Otherwise, we have retrieved the user document. Next, we compare the password in the body of the request with the user password. The method bcrypt.compareSync() compares the entered password with the password field of foundUser. We had to append the entered password to an empty string to convert it to a string as in ''+request.body.password. If the call to bcrypt.compareSync() returns false, we return the response with error code 400 and a message saying that the password is invalid.

After all of the checks, we finally get the records for the user using the method getRecords() shown in Figure 9.16. We will discuss the method as soon as we have finished with the rest of the code in Figure 9.15. We pass two parameters to getRecords(). The first one, called err, is the possible error that could be thrown by the next() method. The second object, called result, will be the retrieved result from the recordsCollection in MongoDB. If err is not null, we return the response with error code 400 and a message saying that there was an error in retrieving records. If err is null, we continue with the rest of the processing. The result is expected to be an array of records. Therefore, we call a JavaScript method toArray(). It will return the retrieved documents pointed at by result in the form of an array. We pass an anonymous function as a parameter to the toArray() method, which in turn receives two objects. The first one, called err, is the possible error that could be thrown by the toArray() method. The second object called resultArray will be the array returned by the toArray() method. If err is not null, we return the response with error code 400 and a message saying an error occurred while processing the records. If err is null, we return the resultArray() with code 200 indicating success.

Let us look at the getRecords() method (Figure 9.16) that was used by the post() method that serviced /getRecords request. The method is relatively brief. It takes the request as the first parameter. An interesting aspect of the method is the fact that we are receiving a function as the second parameter. The function logs the fact that we are retrieving records on the system console, and makes a call to the recordsCollection.find() method. It sends a JSON object with an email field equal to the email value we received. The second parameter is the function that we received as the second parameter to getRecords() itself. We see how the results from getRecords() will be processed in Figure 9.15.

```
// Helper function to get all records for a given user
function getRecords(request, callback) {
  console.log('Retrieving records.')
  recordsCollection.find({
    user: request.body.email
  }, callback);
}
```

FIGURE 9.16 node.js method to get records from MongoDB File: server.js

9.10 Uploading records to the MongoDB server for the Thyroid app

Figure 9.17 shows the code for processing a syncRecords request. We are running the method app.post(), which receives two parameters. The first one is the request '/syncRecords' that will be handled by the server, and the second parameter is an anonymous function that contains the rest of the code. The anonymous function takes two parameters: the request that is received by the post() method and the response that will be returned. We log the fact that we are sending the records on the system console. If either the email or password in the request body is empty, we return the response with error code 400 and a message saying that login information is missing.

If we have an email and password, we proceed further. We look at the MongoDB usersCollection using the find() method. The find receives two parameters. The first parameter is a JSON object with an email field equal to the email value we received through the request.body. The second parameter is an anonymous function that contains the rest of the code in Figure 9.15. The anonymous function receives two objects. The first one, called err, is the possible error that could be thrown by the find() method. The second object is called result.

```
// Updates the records in the database with the provided records
app.post('/syncRecords', function(request, response) {
  console.log('Save Records Request Received.');

  if (!request.body.email || !request.body.password) {
    return response.send(400, 'Missing log in information.');
  }

  usersCollection.find({
    email: request.body.email
  }, function(err, result) {
    if (err) {
      throw err;
    }
    result.next(function(err, foundUser) {
      if (err) {
        throw err;
      }
      if (!foundUser) {
        return response.send(400, 'User not found.');
      }
      if (!bcrypt.compareSync('' + request.body.password,
foundUser.password)) {
        return response.send(400, 'Invalid password');
      }
      var newRecords = request.body.newRecords;
      for (var i = 0; i < newRecords.length; i++) {
        newRecords[i].user = request.body.email;
        delete newRecords[i]._id;
      }
      syncRecords(request, response, newRecords);
    });
  });
});
```

FIGURE 9.17 node.js method to process syncRecords request in MongoDB File server.js

The object result is a cursor (or pointer) that points to the retrieved records. In order to get the records, we use the method result.next(), which receives an anonymous function that contains the rest of the code in Figure 9.15. It receives two parameters. The first one, called err, is the possible error that could be thrown by the next() method. The second object called foundUser will be the retrieved result from the usersCollection in MongoDB. If err is not null, we just throw it and are done with the processing of the request. If err is null, we continue with the rest of the processing.

We then check to see if the foundUser is null. If it is null, we return the response with error code 400 and a message saying that the user is not found. Otherwise, we have retrieved the user document. Next, we compare the password in the body of the request with the user password. The method bcrypt.compareSync() compares the entered password with the password field of foundUser. We had to append the entered password to an empty string to convert it to a string as in ''+request.body.password. If the call to bcrypt.compareSync() returned false, we return the response with error code 400 and a message saying that the password is invalid.

We do a little bit of cleanup of the records that we received through the request body by first assigning request.body.newRecords to a variable called newRecords. We go through the array newRecords and set its user field to the email that we have received through the request.body. We delete the __id field. Once we have finished the cleanup we call the method syncRecords() (Figure 9.18), which receives the request, the response, and newRecords as the parameters.

The function syncRecords(), shown in Figure 9.18, performs two operations on the recordsCollection. It removes all the records for the user and then adds the new records passed by the third parameter. Let us go through the code line by line to understand the processing. We first make a call to the recordsCollection.remove() method. The remove method takes three parameters. The first parameter is a JSON object with an email field equal to the email value we received through the request.body or user. It serves as a query to locate the document in MongoDB. The second parameter can be used to specify a number of options including whether to delete one document or all the documents matching the query. We have set the second parameter to null, which will delete all the copies. Finally, the third parameter is an anonymous function that takes one parameter called err. If err is not null, the remove did not work out. We log the err object on the system console. If err is not null, we return the response with error code 400 and a message saying that the records could not be synced. If err is null, we continue with the rest of the processing.

If we have successfully deleted the existing records for the user, we add the new records received through the parameter recordsToSave using the recordsCollection.insert() method. The insert() method receives two parameters: the array recordsToSave and an anonymous function. The first one, called err, is the possible error that could be thrown by the insert() method. If err is not null, we return the response with error code 400 and a message saying that there was an error syncing the records. The second object called result is not very useful to us for the insert. If we did not return error code 400 until this point, we can send a code 200 and a message confirming that the records were successfully synced.

In this chapter, we discussed the database server computing in the server-based Thyroid app using MongoDB, a NoSQL database management system. In the following chapter, we will see how exactly the same support can be provided using the traditional relational databases.

```
// Helper function to remove all old records and insert all new records
// We do this instead of a combination of UPDATE and
// INSERT statements to simplify this process
function syncRecords(request, response, recordsToSave) {
  recordsCollection.remove({
    user: request.body.email
  }, null, function(err) {
    if (err) {
      return response.send(400, 'Error occurred syncing records');
    }
    recordsCollection.insert(recordsToSave, function(err, result) {
      if (err) {
        console.log(err);
        return response.send(400, 'Error occurred syncing records');
      }
      return response.send(200, 'Records synced.');
    });
  });
}
```

FIGURE 9.18 node.js method to sync records in MongoDB File: server.js

Quick facts/buzzwords

NoSQL: An alternative to relational databases for storing unstructured data.

MongoDB: One of the most popular NoSQL database management systems.

mongo: A command to launch an interactive MongoDB session.

use: A command to switch to a desired database.

show collections: A command to look at collections in the current database.

insert(): A method to add documents to a MongoDB collection.

update(): A method to update documents in a MongoDB collection.

remove(): A method to delete documents from a MongoDB collection.

find(): A method to retrieve documents from a MongoDB collection.

drop(): A method to delete an entire MongoDB collection.

db.auth(): A method to authenticate a user in MongoDB.

Database modeling: A process of determining the structure of the database.

bcrypt: A module for encrypting strings in node.js.

CORS: Cross-Origin Resource Sharing.

process.on(): A method for processing signals in node.js.

listen(): A method to listen for server requests in node.js.

genSaltSync(): A method to generate salt in bcrypt to increase the number of possibilities.

hashSync(): A method to encrypt in bcrypt module.

compareSync(): A method to compare a string with an encrypted one in the bcrypt module.

throw(): A method to throw error/exception objects in JavaScript.

Self-test exercises

1. Name three database models.
2. Who introduced the relational database model?
3. What is MongoDB?
4. What is the command used to view existing collections in MongoDB?
5. What is the command used to retrieve records in MongoDB?
6. What is the command used to add records in MongoDB?
7. What is the command used to delete records in MongoDB?

Programming exercises

1. Create the MongoDB collection on your MongoDB server as shown in the text and add the library collection.
2. Write a JavaScript function to add the following books in the library collection for author John Grisham. The number in the parentheses is the year of publication.

   A Time to Kill (1989)
   The Firm (1991)
   The Pelican Brief (1992)
   The Client (1993)
   The Chamber (1994)
   The Rainmaker (1995)
   The Runaway Jury (1996)
   The Partner (1997)
   The Street Lawyer (1998)
   The Testament (1999)
   The Brethren (2000)
   A Painted House (2001)
   Skipping Christmas (2001)
   The Summons (2002)
   The King of Torts (2003)
   Bleachers† (2003)
   The Last Juror (2004)
   The Broker (2005)
   Playing for Pizza (2007)
   The Appeal (2008)
   The Associate (2009)
   The Confession (2010)
   The Litigators (2011)
   Calico Joe (2012)
   The Racketeer (2012)
   Sycamore Row (2013)
   Gray Mountain (2014)

3. Write queries to find the following
   - All the books published before 2000 (by any author)
   - All the books published by John Grisham
   - All the books not published by John Grisham
   - All the books published between years 2000 and 2009
   - The number of books published after 2009

4. Write queries to delete the following
   - All the books published before 2000 (by any author)
   - All the books published by John Grisham
   - All the books not published by John Grisham
   - All the books published between years 2000 and 2009 (by any author)

Programming projects

1. **Boiler monitor server-side version:** We want to change the app that monitors the temperature and pressure of a boiler so that the data is stored on a MongoDB server instead of in localStorage. You can model the app based on the server-side Thyroid app. We have modified the webpages in the previous chapter. In this chapter, you need to design the MongoDB database so that the records can be stored and retrieved using the boiler ID as the key, and to create the routes in your server for the app to access.
2. **Blood pressure monitor server-side version:** We want to change the app that monitors blood pressure so that the data is stored on a MongoDB server instead of in localStorage. You can model the app based on the server-side Thyroid app. We have modified the webpages in the previous chapter. In this chapter, you need to design the MongoDB database so that the records can be stored and retrieved using the person's ID as the key, and to create the routes in your server for the app to access.
3. **Power consumption monitor server-side version:** We want to change the app that monitors the power consumption of a manufacturing plant so that the data is stored on a MongoDB server instead of in localStorage. You can model the app based on the server-side Thyroid app. We have modified the webpages in the previous chapter. In this chapter, you need to design the MongoDB database so that the records can be stored and retrieved using the plant ID as the key, and to create the routes in your server for the app to access.
4. **Body mass index (BMI) monitor server-side version:** We want to change the app that monitors the body mass index of a person based on height and weight so that the data is stored on a MongoDB server instead of in localStorage. You can model the app based on the server-side Thyroid app. We have modified the webpages in the previous chapter. In this chapter, you need to design the MongoDB database so that the records can be stored and retrieved using the person's ID as the key, and to create the routes in your server for the app to access.
5. **Managing a line of credit server-side version:** We want to change the app that manages our line of credit at a bank so that the data is stored on a server instead of in localStorage. You can model the app based on the server-side Thyroid app. We have modified the webpages in the previous chapter. In this chapter, you need to design the MongoDB database so that the records can be stored and retrieved using the organization's ID as the key, and to create the routes in your server for the app to access.

CHAPTER 10

Using a relational database server for sharing and storing information

WHAT WE WILL LEARN IN THIS CHAPTER

1. What are relational databases
2. How to create tables in a relational database
3. How to insert/delete/update records in a relational table
4. How to connect to a relational database server using node.js
5. How to manipulate relational database using node.js

In the previous chapter, we used the NoSQL database server to store data from the Thyroid app users. NoSQL databases use JSON objects to structure the data. Traditional relational databases will continue to play an important part in the modern computing world. Therefore, we implement the same server-side data support using relational databases.

10.1 Relational databases

As mentioned in Chapter 9, modern-day computing began with scientific computations. Fortran was one of the primary languages that led scientific computing in the 1950s, 1960s, and even the 1970s. Business applications programming with the advent of COBOL was not too far behind. Databases played a major role in business applications used on computers. COBOL had a number of built-in file operations that made it easy to store, retrieve, and modify the datasets. It soon became obvious that careful design and planning was required to store all the data used in business applications. One of the primary design issues in creating a database is the representation model. A database model defines the structure and operations that will be performed on a database. Early databases used somewhat unstructured hierarchical and network models that were programmed with COBOL's file systems. In 1970, E. F. Codd from IBM introduced the relational database model that revolutionized the database world. In Chapter 9, we looked at an alternative database model called NoSQL that is popular for working with unstructured data. In the following chapters, we will see another app that uses multimedia and unstructured data that has exposed the limitations of the relational model. However, the relational database remains the most widely used database model. Therefore, it is important to study the relational model. We will study the use of a relational database to support our Thyroid app.

The term "relational database" comes from the fact that the data is represented as mathematical relations. This allows us to manipulate the data using relational calculus or algebra. The mathematical foundations for the relational database model are used to minimize data redundancy and verify data integrity. While some readers may find this mathematical treatment a little intimidating, the relational database actually makes it easier to visualize the database model. Using more visual interpretation, one can say that the data in a relational database is stored in a table. Each row of the table corresponds to a record, which is called "tuple" in relational algebra. A field in the record corresponds to a column in a table, also called an "attribute".

10.2 Modeling a relational database

The easiest way to design a database would be to create one single database as shown in Figure 10.1. The table consists of fourteen attributes/columns

1. email address
2. Password
3. First name
4. Last name
5. Date of birth
6. Health insurance card number
7. Cancer type
8. Cancer stage
9. TSH range
10. legal waiver
11. Date of test
12. TSH
13. Tg
14. Synthroid

We have a row/record for each blood test completed for the patients. The problem with this table is the fact that we are repeating information about the user for every blood test. This leaves redundancy in the database and will increase the storage requirements. More importantly, it

email	Password	First name	Last name	Date of birth	Health insurance card number	Cancer type
j@cs.smu.ca	1953	J.	Some	1953/02/03	123456	Medular
j@cs.smu.ca	1953	J.	Some	1953/02/03	123456	Medular
k@cs.smu.ca	1950	K.	Other	1950/02/23	183456	Folicular
j@cs.smu.ca	1953	J.	Some	1953/02/03	123456	Medular
k@cs.smu.ca	1950	K.	Other	1950/02/23	183456	Folicular
j@cs.smu.ca	1953	J.	Some	1953/02/03	123456	Medular

Cancer stage	TSH range	Agreed to legal waiver	Date of test	TSH	Tg	Synthroid
I	0.1–0.3	True	2011/12/05	0.20	0.6	1.7
I	0.1–0.3	True	2011/12/05	0.20	0.9	0.7
II	0.01–0.1	True	2012/12/05	0.20	0.4	1.1
I	0.1–0.3	True	2012/11/15	0.20	0.5	1.8
II	0.01–0.1	True	2010/01/25	0.02	0.2	1.9
I	0.1–0.3	True	2013/11/02	0.20	0.16	1.7

FIGURE 10.1 A non-normalized relational database

leaves our database open to possible errors—for example, if the user information were to change, such as the "TSH range" or the "Cancer stage". Our programmer will have to make sure that the records are updated for every blood test by the user. Even if the programmer took proper care to make these changes, we will be performing a large number of unnecessary updates, causing too many disk accesses (one of the major bottlenecks in system performance). This is where Codd's master stroke of using relational algebra provides an elegant solution by defining normal forms. A database is in a normal form if the relationships between attributes are defined with mathematical precision. In the original marquee paper in 1970, Codd specified three normal forms. The first normal form (1NF) in relational databases was necessitated by the technological limitations of the era. It mandated that an attribute can store only one atomic value. For example, if an author "Pawan" writes three books—*Web Mining, Web Programming*, and *Mobile App Development*—we cannot have a single record with "Author = Pawan" and "Product = {"Web mining", "Web programming", "Mobile app development"}". Instead, we need three records

- "Author = Pawan" and "Product = Web mining"
- "Author = Pawan" and "Product = Web programming"
- "Author = Pawan" and "Product = Mobile app development"

The second and third normal forms use the mathematical definition of functions to eliminate the functional dependency of nonkey attributes on the primary key. A primary key uniquely identifies a record in a table. In our database from Figure 10.1, the health insurance card number uniquely determines

1. email address
2. Password
3. First name
4. Last name
5. Date of birth
6. Health insurance card number
7. Cancer type
8. Cancer stage
9. TSH range
10. legal waiver

Then we can define a function, f(email address) that will give us a unique set of values for the above ten attributes. We can use this functional dependency to separate these ten attributes from the other four attributes. We need to repeat the email address for the other set to link the two sets together

1. User (email address)
2. Date of test
3. TSH
4. Tg
5. Synthroid

These two sets of attributes are connected with a common attribute. We call this attribute "email address" in the first table and "User" resulting in the database shown in Figure 10.2. The names of the common attributes do not have to match as long as they have the same type of information.

These two tables can be joined together with the help of the common attribute to recover the original table. Therefore, the decomposition shown in Figure 10.2 is said to be lossless.

The third normal form eliminates a little more complexity as it uses transitive functional dependency. Database researchers exploited the relational algebra to propose a number of additional normal forms including a Boyce-Codd normal form (BCNF) and the fourth and fifth normal forms (4NF and 5NF). The sixth normal form (6NF) was shown to be useful for temporal databases. Many programmers ensure that the database is at least in second normal form as shown in Figure 10.2.

email	Password	First name	Last name	Date of birth	Health insurance card number	Cancer type
j@cs.smu.ca	1953	J.	Some	1953/02/03	123456	Medular
j@cs.smu.ca	1953	J.	Some	1953/02/03	123456	Medular
k@cs.smu.ca	1950	K.	Other	1950/02/23	183456	Folicular
j@cs.smu.ca	1953	J.	Some	1953/02/03	123456	Medular
k@cs.smu.ca	1950	K.	Other	1950/02/23	183456	Folicular
j@cs.smu.ca	1953	J.	Some	1953/02/03	123456	Medular

Cancer stage	Agreed to legal waiver	TSH range	User	Date of test	TSH	Tg	Synthroid
I	True	0.1–0.3	j@cs.smu.ca	2011/12/05	0.20	0.6	1.7
I	True	0.1–0.3	j@cs.smu.ca	2011/12/05	0.20	0.9	0.7
II	True	0.01–0.1	k@cs.smu.ca	2012/12/05	0.20	0.4	1.1
I	True	0.1–0.3	j@cs.smu.ca	2012/11/15	0.20	0.5	1.8
II	True	0.01–0.1	k@cs.smu.ca	2010/01/25	0.02	0.2	1.9
I	True	0.1–0.3	j@cs.smu.ca	2013/11/02	0.20	0.16	1.7

FIGURE 10.2 A relational database in second normal form

10.3 SQL

Structured Query Language (SQL) is the most commonly used language for manipulation and retrieval of data in the relational database world. SQL commands can be grouped into various categories

- Definition: For managing the structure of the database
- Retrieval: Most frequently used function to extract data from the database
- Manipulation: Adding, deleting or modifying data in the database
- Control: To manage access to the database

In this chapter, we will focus on the first three sets of commands. The control commands are the prerogative of a database or system administrator.

There are two types of interfaces for working with a database management system: graphical user interface (GUI) and command line interface (CLI). GUI tends to be easier to use, while command line provides more flexibility. We will look at GUI for the Definition commands. These commands tend to be used sparingly and usually at the setup time. We are unlikely to use these commands from our programs. Therefore, a graphical treatment may be sufficient. The retrieval and manipulation commands, on the other hand, will be an integral part of our JavaScript programming. It is imperative for us to understand these commands at a reasonable depth, so we will use the CLI to explore them.

We will use the database structure shown in Figure 10.2 to implement the database using MySQL (www.mysql.com). MySQL is one of the most popular open source relational database management systems. In order to follow along, you need to either install MySQL or have access to a MySQL platform through your course instructor. Your system administrator will create an account and a database for you. Please note the distinction between an account and

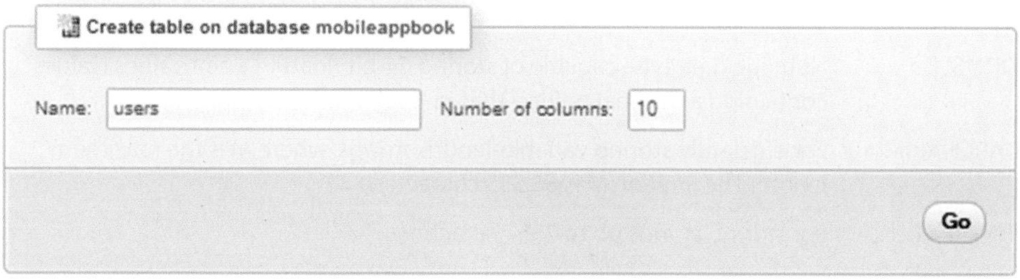

Create table on database mobileappbook

Name: users Number of columns: 10

Go

■ **FIGURE 10.3** Creating the User table using phpMyAdmin

a database, even though many times in a classroom setting they may have the same name. The account is used for logging into the MySQL system. The database is where you store all your tables. A single account may have access to multiple databases. For example, in your case you will likely have your own database as well as a common course database maintained by your instructor. Similarly, a database can be accessed by multiple users. For example, your common course database will be accessible to all the students in the course as well as the course instructor. The control commands are used by the system administrator to control access to these databases. Your instructor will have read and write access to the common course database, while all the students may only have the read access to the database. Once you have logged into the system, you can choose an appropriate database and begin your implementation. Our discussion will begin with the graphical interface for MySQL called phpMyAdmin. The graphical user interface is particular to MySQL. However, the SQL commands are more or less the same for any relational database management system.

We will create two tables as shown in Figure 10.2. Figure 10.3 shows a dialogue box once we have chosen the database that we are building. We indicate that the table name is Users with 10 attributes. Pressing the button "Go" will bring up a screen like the one shown in Figure 10.4. We now have to enter the attribute names and types. MySQL supports a wide range of data types. Figure 10.5 shows some of the commonly used data types. A more detailed list can be found at www.mysql.com. Entering names of attributes and types as shown in

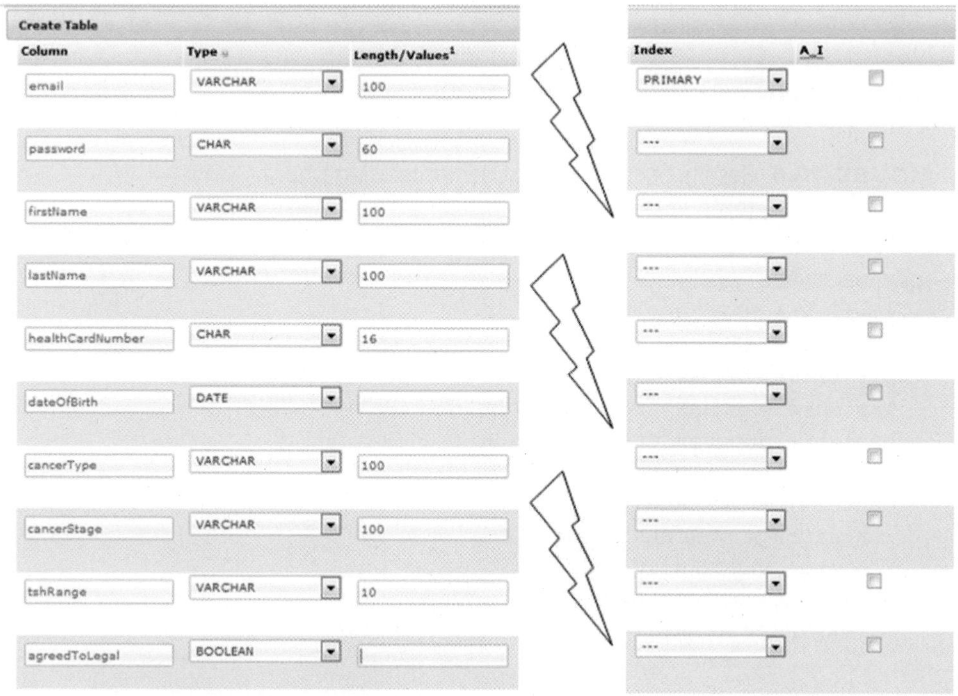

Create Table

Column	Type	Length/Values[1]	Index	A_I
email	VARCHAR	100	PRIMARY	
password	CHAR	60	---	
firstName	VARCHAR	100	---	
lastName	VARCHAR	100	---	
healthCardNumber	CHAR	16	---	
dateOfBirth	DATE		---	
cancerType	VARCHAR	100	---	
cancerStage	VARCHAR	100	---	
tshRange	VARCHAR	10	---	
agreedToLegal	BOOLEAN		---	

■ **FIGURE 10.4** Specifying attributes for the User table using phpMyAdmin

INTEGER	Numeric data type capable of storing 32-bit integer values
DOUBLE	Numeric data type capable of storing 64-bit floating point values (values containing a decimal point—that is, "real numbers")
VARCHAR(M)	For efficiently storing variable-length strings, where M is the maximum length. The limit on M is 65,535 characters.
TINYTEXT	For storing strings up to 255 characters
TEXT	For storing large amounts of text (65,535 characters)
DATE	For storing date values

■ **FIGURE 10.5** Common data types in SQL

Figure 10.3 and pressing the "Save" button creates the table shown in Figure 10.6. This figure shows the complete structure of the table. It also describes the primary and unique keys as well as other indexed attributes. The figure also shows the query in SQL that was used to create the table. Since it is a little difficult to read the query, it is shown in a more readable form in Figure 10.7. The same query can work with any database management system other than MySQL. The query uses one of the data definition SQL commands called **CREATE**. The data definition commands are used to create, delete, and modify structures of tables in the database. Figure 10.8 shows some of the other data definition commands that can be useful to us. We have seen the CREATE command in action. Discussion of all the commands is beyond the scope of this chapter. We will leave the exploration of the DROP and ALTER commands as exercises for the readers.

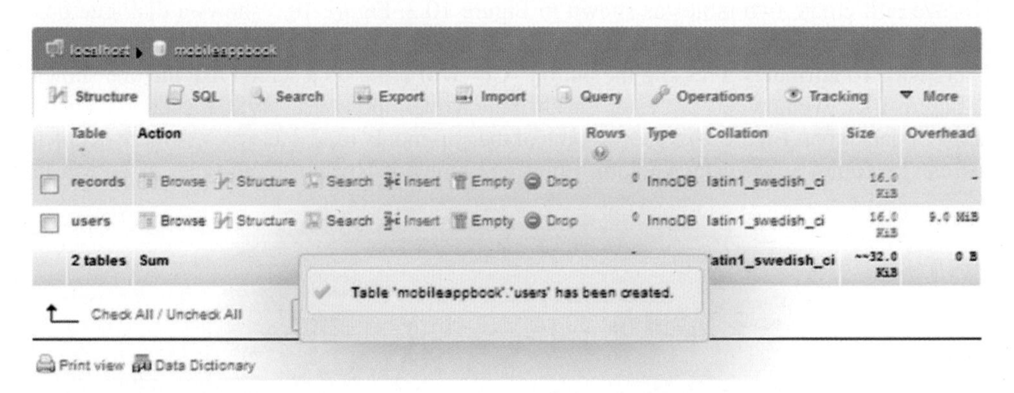

■ **FIGURE 10.6** Results of creating the User table using phpMyAdmin

```
CREATE TABLE users (
  email VARCHAR(100) NOT NULL,
  password CHAR(60) NOT NULL,
  firstName VARCHAR(100) NOT NULL,
  lastName VARCHAR(100) NOT NULL,
  healthCardNumber CHAR(16),
  dateOfBirth DATE NOT NULL,
  cancerType VARCHAR(100),
  cancerStage VARCHAR(100),
  tshRange VARCHAR(10),
  agreedToLegal BOOLEAN,
  PRIMARY KEY (email)
);
```

■ **FIGURE 10.7** SQL command for creating the User table

CREATE	For creating an object such as a table in the database.
DROP	For permanently deleting an object (column or table) from the database.
ALTER	For modification of an existing object (column or table).

■ **FIGURE 10.8** Data definition SQL commands

We will add one more table to store our Records consisting of the following attributes

1. User (email address)
2. Date of test
3. TSH
4. Tg
5. Synthroid

Figure 10.9 shows our database consisting of two tables. There are a number of icons next to each table. The first icon allows us to browse the table, the second icon can be used to look at the structure of the table, the third icon can be used to search the table, the fourth icon is used to insert records in the table, the fifth icon empties the table, while the sixth icon deletes the table. These operations correspond to the data manipulation statements to add, update, and delete data shown in Figure 10.10. This set of statements is important to us from the programming point of view. We will need to use them from our programs for modifying our database. We will first look at them with the help of a GUI and later on use them in our programs. Clicking on the insert icon next to the Records table will bring up a dialogue box for inserting records. Figure 10.11 shows the insertion of a record in the Records table from our database using the phpMyAdmin interface. Figure 10.12 confirms that the rows were added to the table and also shows the corresponding SQL query. Since we will be using the INSERT command in our programs, we should study its syntax in a little more detail. The general format of the INSERT command is as follows

```
        INSERT INTO table_name
        ( list of attributes )
  VALUES
        ( list of values for record 1),
        ( list of values for record 2),
        ( list of values for record 3),
        ...
        ( list of values for last record)
```

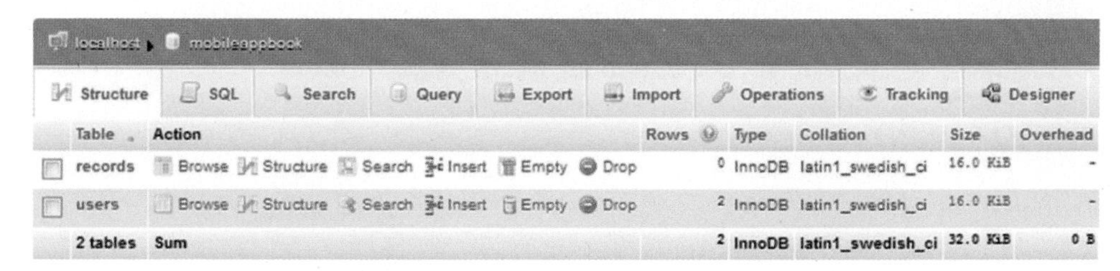

■ **FIGURE 10.9** Database consisting of two tables

INSERT	Adds rows or tuples to an existing table.
UPDATE	Used to modify the values of a set of existing table rows.
TRUNCATE	Deletes all data from a table.
DELETE	Removes existing rows from a table.

■ **FIGURE 10.10** Data manipulation SQL commands

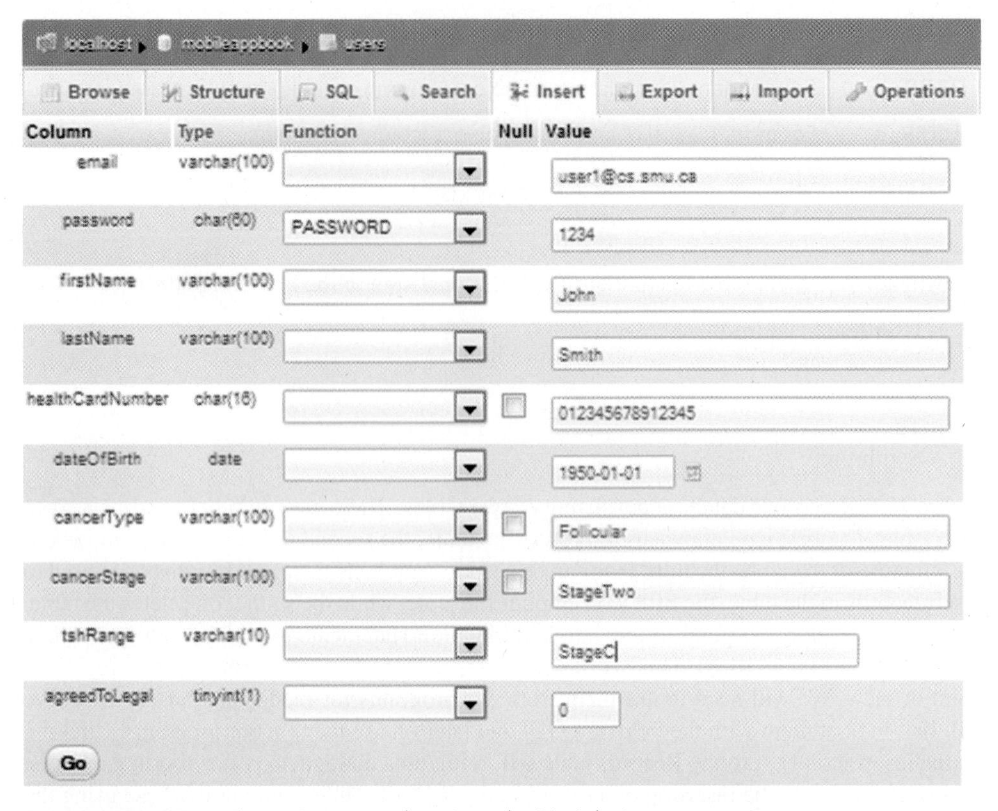

■ **FIGURE 10.11** Inserting records using phpMyAdmin

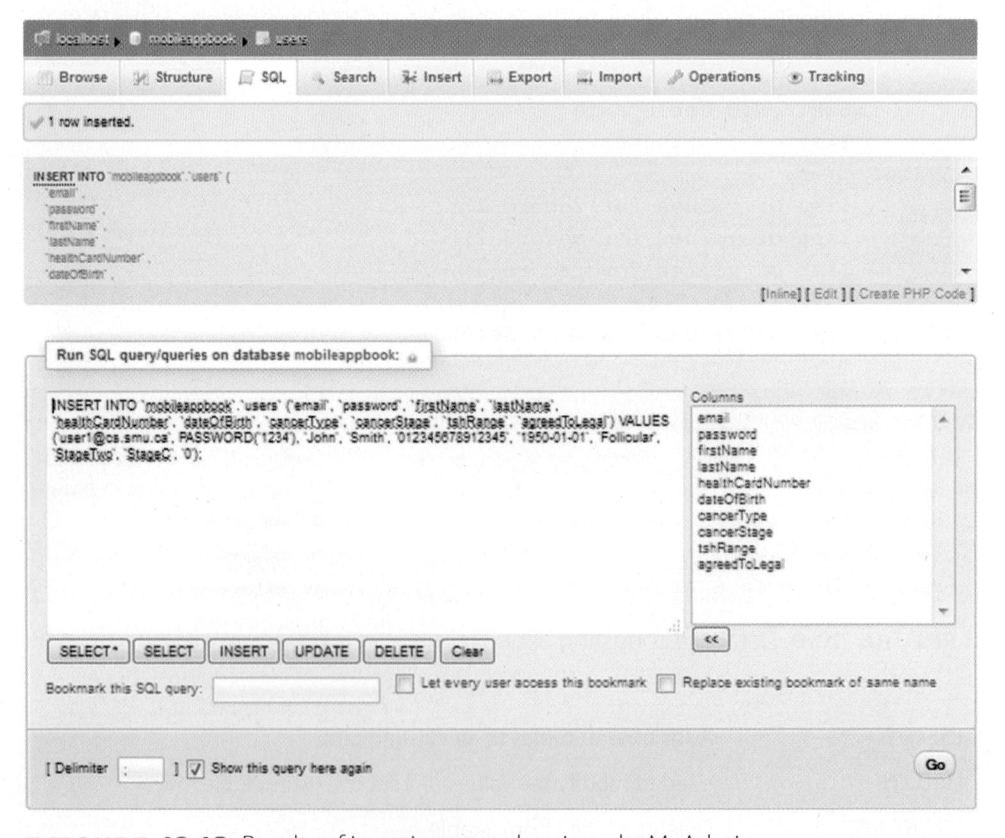

■ **FIGURE 10.12** Results of inserting records using phpMyAdmin

In our query shown in Figure 10.13 we are indicating we want to enter the values of attributes

1. User (email address)
2. Date of test
3. TSH
4. Tg
5. Synthroid

The first record has the value

1. user@mobileappbook.com
2. 2014-01-13
3. 0.5
4. 0.5
5. 0.5

The second record has the value

1. user2@mobileappbook.com
2. 2014-02-28
3. 0.4
4. 0.3
5. 1.2

```
mysql> INSERT INTO records
    ->    (user, date, tsh, tg, synthroidDose)
    ->    VALUES
    ->    ("user@mobileappbook.com", "2014-01-13", 0.5, 0.5, 0.5),
    ->    ("user2@mobileappbook.com", "2014-02-28", 0.4, 0.3, 1.2);
Query OK, 2 rows affected (0.00 sec)
Records: 2 Duplicates: 0 Warnings: 0
```

FIGURE 10.13 SQL command for inserting records

The INSERT command can be used to insert a large number of records in our database even from a text file in comma separated value (CSV) format. LOAD DATA is another SQL command that will provide you with an efficient mechanism for adding records from a text file. We will not discuss these commands in any more detail in this chapter.

Let us now go back to the database and click on the first icon to browse our records table. It will bring up a screen as shown in Figure 10.14. In addition to displaying the record, we have two icons on the left of the icon. The first icon allows us to edit/update the record. Clicking on the first icon will lead us to the dialogue box shown in Figure 10.15. We can change the values of any field we want and press the "Go" button. Figure 10.16 shows the results from our update operation along with the SQL UPDATE command used to carry out the operation. The general format of the UPDATE operation is as follows

```
UPDATE table_reference
SET col_name1=value
col_name2=value ...
WHERE where_condition
ORDER BY ...
LIMIT row_count
```

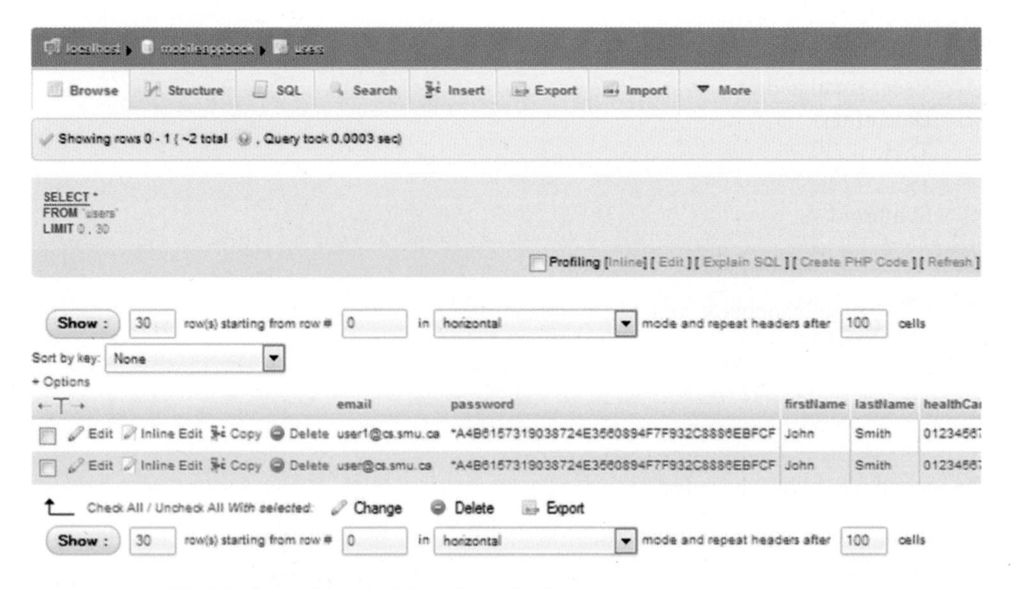

■ **FIGURE 10.14** Browsing a table using phpMyAdmin

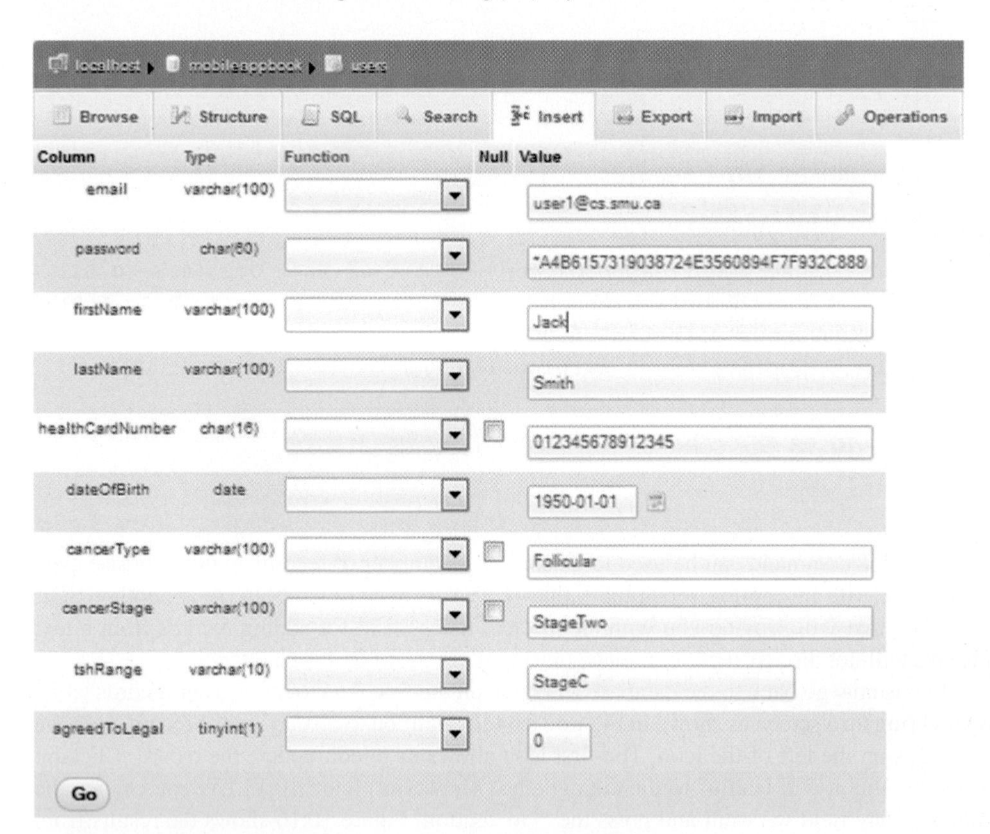

■ **FIGURE 10.15** Updating records using phpMyAdmin

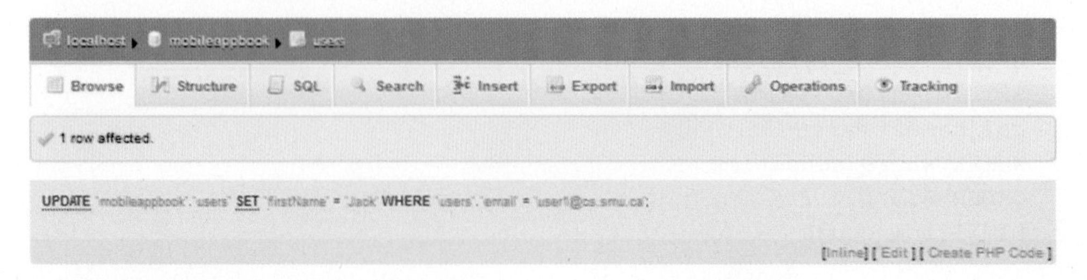

■ **FIGURE 10.16** Results of updating records using phpMyAdmin

Only the SET (to specify the new values) clause is mandatory. However, you will probably want to specify a WHERE clause to update only certain records. Otherwise, all records will be updated. You can change values of one or more columns. In Figure 10.17, we are replacing the value of TSH to 0.6 for the record with the following attribute values: user=user2@mobileappbook.com and date="2014-02-28";

Let us go back to browsing our records table as shown in Figure 10.14. The second icon next to a record allows us to delete the record. Clicking on the delete icon will lead us to the warning dialogue box shown in Figure 10.18 along with the SQL DELETE command that will be used to carry out the operation. We have reproduced it in Figure 10.19 for better readability.

```
DELETE  FROM table_reference
WHERE where_condition
ORDER BY ...
LIMIT row_count
```

All clauses are optional; however, you will probably want to specify a WHERE clause. If no WHERE clause is specified, this command deletes all records. In Figure 10.19, we deleted a record with user=user2@mobileappbook.com.

```
mysql> UPDATE records
    ->     SET tsh=0.6
    ->     WHERE user="user2@mobileappbook.com"
    ->     AND date="2014-02-28";
Query OK, 1 row affected (0.00 sec)
Rows matched: 1  Changed: 1  Warnings: 0
```

FIGURE 10.17 SQL command for updating a record

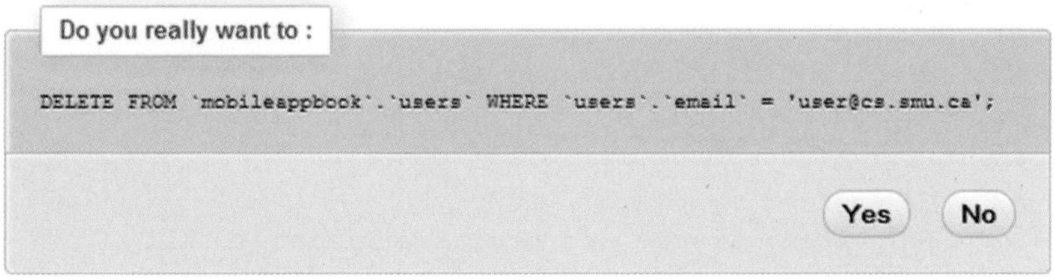

```
Do you really want to :

DELETE FROM `mobileappbook`.`users` WHERE `users`.`email` = 'user@cs.smu.ca';

Yes   No
```

FIGURE 10.18 Deleting records using phpMyAdmin

```
mysql> DELETE FROM records
    ->    WHERE user="user2@mobileappbook.com";
Query OK, 1 row affected (0.00 sec)
```

FIGURE 10.19 SQL command for deleting a record

The DELETE command can also be used for deleting multiple records by stating an appropriate condition in the WHERE clause. If you wanted to empty the entire table, it might be better to use the TRUNCATE command, which can be invoked by clicking on the fifth icon next to the table name. It will bring up a dialogue box such as the one shown in Figure 10.20. The corresponding query is TRUNCATE users. There is not much more in the syntax for the TRUNCATE command. It is one of the simplest and most destructive SQL commands.

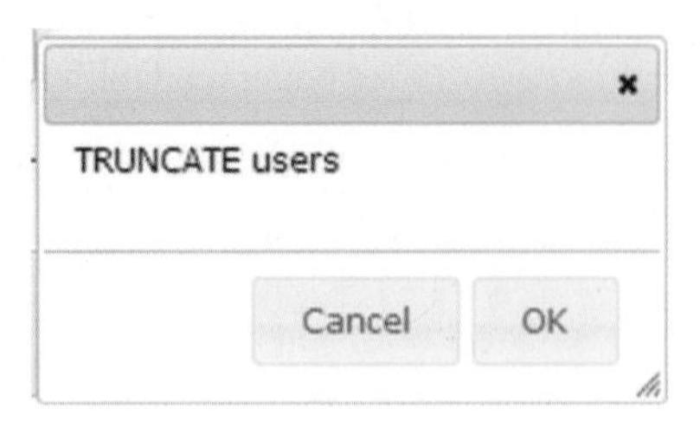

■ **FIGURE 10.20** Emptying a table using phpMyAdmin

We will now turn our attention to data retrieval, which is the most frequently used function from SQL. The records are retrieved from a database using a SELECT statement. In a SELECT query, the user only describes the desired result set. A database management system such as MySQL optimizes the query into an efficient query plan. The following are commonly used keywords with a SELECT command

- FROM indicates from which tables the data is to be taken.
- WHERE specifies which rows are to be retrieved.
- GROUP BY is used to combine rows with related values.
- HAVING acts much like a WHERE, but it operates on the results of the GROUP BY.
- ORDER BY is used to identify which columns are used to sort the resulting data.

Figure 10.21 is our first exposure to the command line interface for MySQL and the SELECT command in action. The SELECT command can be used through the GUI interface as well. However, there is very little assistance from the interface in creating the SQL queries. It is

```
$ mysql -u mobileappbook -p
Enter password:
Welcome to the MySQL monitor.  Commands end with ; or \g.
Your MySQL connection id is 308
Server version: 5.5.37-0ubuntu0.12.04.1 (Ubuntu)

Copyright (c) 2000, 2014, Oracle and/or its affiliates. All
rights reserved.

Oracle is a registered trademark of Oracle Corporation and/or its
affiliates. Other names may be trademarks of their respective
owners.

Type 'help;' or '\h' for help. Type '\c' to clear the current
input statement.

mysql> use mobileappbook
Reading table information for completion of table and column names
You can turn off this feature to get a quicker startup with -A
```

```
Database changed
mysql> show tables;
+------------------------+
| Tables_in_mobileappbook |
+------------------------+
| records                |
| users                  |
+------------------------+
2 rows in set (0.00 sec)

mysql> SELECT * FROM records;
+---+----------------------+------------+-----+-----+--------------+
|id | user                 | date       | tsh | tg  | synthroidDose |
+---+----------------------+------------+-----+-----+--------------+
| 1 | user@mobileappbook.com | 2014-01-13 | 0.5 | 0.5 |          0.5 |
+---+----------------------+------------+-----+-----+--------------+
1 row in set (0.00 sec)

mysql> quit
Bye
$
```

■ **FIGURE 10.21** First look at the command line SQL interface and the SELECT command

mostly a textual exercise. Therefore, we are going to study them through the command line interface (CLI). You can execute the same commands by clicking on the SQL tab in the GUI. Your MySQL commands can also work from your UNIX/Linux command line by typing

```
mysql -u username -p
```

The option -u means that the next word will be the username. The option -p means the password will be typed at the prompt. After typing the password, we are in the MySQL command line interface. We need to type the USE command as shown in Figure 10.21 to change to the database. We then use the SHOW TABLES command to get the list of the tables in the database. We use a couple of simple executions of the SELECT command

```
SELECT * FROM records;
```

Here, the asterisk (*) is used to indicate we want to see all the fields. Figure 10.22 shows a more complex SELECT command. It uses the WHERE and ORDER BY clauses.

```
mysql> SELECT user, tsh
    ->   FROM records
    ->   WHERE tsh > 1
    ->   ORDER BY tsh DESC;
+------------------------+------+
| user                   | tsh  |
+------------------------+------+
| user2@mobileappbook.com |  1.4 |
| user@mobileappbook.com  | 1.35 |
| user2@mobileappbook.com | 1.22 |
+------------------------+------+
4 rows in set (0.00 sec)
```

■ **FIGURE 10.22** The SELECT command with WHERE and ORDER BY clauses

Figure 10.23 shows the use of a built-in function COUNT() that is used along with GROUP BY. There are a number of such built in aggregate functions including SUM() and AVERAGE().

As we mentioned earlier, a normalized database tends to span across multiple tables. They usually share a common attribute such as "email/user" in our database. When we need to retrieve information that is spread across multiple tables, we need to join these tables as shown by the query in Figure 10.24.

```
mysql> SELECT user, count(*)
    ->     FROM records
    ->     WHERE tsh > 1
    ->     GROUP BY user;
+------------------------+----------+
| user                   | count(*) |
+------------------------+----------+
| user2@mobileappbook.com |        2 |
| user@mobileappbook.com  |        1 |
+------------------------+----------+
2 rows in set (0.01 sec)
```

FIGURE 10.23 The SELECT command with GROUP BY clause and COUNT function

```
mysql> SELECT user, firstName AS first, lastName AS last, date, tsh
    ->     FROM users AS u, records AS r
    ->     WHERE u.email = r.user
    ->     AND tsh > 1
    ->     ORDER BY date;
+------------------------+--------+-------+------------+------+
| user                   | first  | last  | date       | tsh  |
+------------------------+--------+-------+------------+------+
| user2@mobileappbook.com | Johnny | Smith | 2014-03-05 |  1.4 |
| user@mobileappbook.com  | Sara   | Cohen | 2014-04-21 | 1.35 |
| user2@mobileappbook.com | Johnny | Smith | 2014-05-04 | 1.22 |
+------------------------+--------+-------+------------+------+
3 rows in set (0.01 sec)
```

FIGURE 10.24 The SELECT command used to retrieve data from multiple files

10.4 Launching the MySQL server for the Thyroid app

Similar to the MongoDB-based application, the database server of the MySQL-based Thyroid app is launched with the following command

node server.js

The command node is used to invoke the Node.js runtime environment, and we are passing the name of the file containing the script as a parameter. This file contains all of our database server program that will handle all the requests. We begin by looking at the main part of the program shown in Figure 10.25. It begins with setting up variables using the method require(). We had looked at the method require() in previous chapters that introduced the client side

```
var express = require('express');
var mysql = require('mysql');
var database = mysql.createConnection({
  host: '127.0.0.1',
  user: 'mobileappbook',
  password: 'xxxxxxxxxx',
  database: 'mobileappbook',
  connectTimeout: 10000
});
database.connect();

//CORS middleware
var allowCrossDomain = function(req, res, next) {
    res.header('Access-Control-Allow-Origin', '*');
    res.header('Access-Control-Allow-Methods',
'GET,PUT,POST,DELETE');
    res.header('Access-Control-Allow-Headers', 'Content-Type');

    next();
}

var app = express();
app.use(express.bodyParser());
app.use(allowCrossDomain);
```

FIGURE 10.25 Main part of the server.js in the MySQL server-based Thyroid app File: server.js

of our app. The method require() essentially imports the specified modules and makes them available to us through the variables we have assigned them to. Here we are importing the express and mysql modules. We have already seen the express module in a previous chapter. It is a minimal web application framework called express (expressjs.com) that provides a robust platform for serving multi-page web applications. Express is required on the database server to handle POST and GET requests from the browser. The mysql module will help us communicate with the mysql server. We then create an object called database that is created using the call to the method mysql.createConnection(). The method receives a JSON object with the following fields

- host—IP address of the machine hosting the MySQL server. Since we are on the same server as the machine hosting MySQL, we use the IP address 127.0.0.1. It usually corresponds to the local host.
- user
- password
- database
- connectTimeout—time in milliseconds before a timeout occurs during the initial connection to the MySQL server. If we do not specify, two minutes is used.

We then call database.connect() to establish the connection.

The variable allowCrossDomain is related to Cross-Origin Resource Sharing (CORS). This is a World Wide Web Consortium (W3C)—a body that establishes standards for the web specification. Normally, the browser only accepts data from a single origin. CORS allows data to come from multiple domains. The variable allowCrossDomain is in fact a function that enables any origin ('*'—wild card), methods

```
'GET,PUT,POST,DELETE', and headers 'CONTENT-Type'.
```

We then create a new application with express() and assign it to a variable called app. There are two calls to the method app.use(). The first enables us to parse JSON objects. The second one enables CORS support.

The call to method process.on() is a signal-handling method. Readers who are familiar with C on UNIX-based systems will find it similar to the function signal(). The object process is a global object in node.js that is available from anywhere, and it allows us to manipulate the existing process. In our case, the process is the MySQL connection. The method process.on() allows us to specify what to do with some of the familiar standard UNIX system signals. Readers who are not acquainted with signal handling may want to just focus on the method shown in Figure 10.26. We are defining a function that will be called when the process receives a signal to terminate (SIGTERM). In that case, we will write a message to the system console saying that we are shutting the server down. We should then shutdown the MySQL server using the call database.end() and close the express server using the call app.close() before terminating.

The last thing we do is launch the server using the method app.listen() as shown in Figure 10.27. It takes two parameters. The first parameter is the port number where we will be listening to the server requests. We have chosen the port number to be 3000. The second parameter is an anonymous function. It writes a message to the system console saying that the server is now listening on the specified port.

The rest of the functions in the server.js handle various requests to the website including saveNewUser, login, updateUser, getRecords, and syncRecords. We will look at the node.js code for each one of these requests in this chapter.

```
process.on('SIGTERM', function() {
  console.log("Shutting server down".);
  database.end();
  app.close();
});
```

FIGURE 10.26 Signal processing in the MySQL server-based Thyroid app File: server.js

```
var server = app.listen(3000, function() {
  console.log('Listening on port %d', server.address().port);
});
```

FIGURE 10.27 Starting the MySQL server-based Thyroid app File: server.js

10.5 Saving a new user on the MySQL server for the Thyroid app

Figure 10.28 shows the code for processing saveNewUser request. We are running the method app.post(), which receives two parameters. The first one is the request '/saveNewUser' that will be handled by the server, and the second parameter is the anonymous function that contains the rest of the code. We log the fact that we are sending the records on the system console. We

```
// Creates a new user in the database
app.post('/saveNewUser', function(request, response) {
  console.log('New user being created.');
  var user = request.body.newUser;
  var query = database.query('INSERT INTO user VALUES (?, PASSWORD(?),
?, ?, ?, ?, ?, ?, ?, ?)', [user.email,
    user.newPassword,
    user.firstName,
    user.lastName,
    user.healthCardNumber,
    user.dateOfBirth,
    user.cancerType,
    user.cancerStage,
    user.tshRange,
    false // agreedToLegal
  ], function(err, result) {
    if (err) {
      console.log(err);
      return response.send(400, 'An error occurred creating this
user.');
    }
    return response.json(200, 'User created successfully!');
  });
});
```

FIGURE 10.28 node.js method to process saveNewUser request for MySQL File: server.js

create a variable called user that is the newUser field from the body of the parameter request. We then create an SQL INSERT query using the method database.query(). The method receives three parameters. The first parameter is a string with the INSERT query. Instead of using the values of the attributes, we use the question mark (?). The values will be provided by an array of values from the appropriate fields for the user object as the second parameter to the database.query() method. We use the PASSWORD() function on the second parameter. That will make sure that an encrypted value of the password will be stored in the database. The third parameter to database.query() is an anonymous function that receives two parameters. The first one, called err, is the possible error that could be thrown by the INSERT query. If err is not null, we return the response with error code 400 and a message saying that there was an error creating the user. The second object called result is not very useful to us for the insert. If we did not return error code 400 until this point, we can send a code 200 and a message confirming that the user was successfully created.

10.6 Processing login with the MySQL server for the Thyroid app

Figure 10.29 shows the code for processing a login request. We are running the method app.post(), which receives two parameters. The first one is the request '/login' that will be handled by the server, and the second parameter is the anonymous function that contains the rest of the code. The anonymous function takes two parameters: the request that is

```
// Uses a email address and password to retrieve a user from the
database
app.post('/login', function(request, response) {
  console.log('Logging user in');
  if (!request.body.email || !request.body.password) {
    return response.send(400, 'Missing log in information.');
  }
  var query = database.query('SELECT * '
    + ' FROM user '
    + 'WHERE user.email=? AND user.password=PASSWORD(?)',
[request.body.email,
    request.body.password
  ], function(err, result) {
    if (err) {
      return response.send(400, 'Error logging user in.');
    }
    if(!result.length) {
      return response.send(400, 'Invalid password and email
provided.');
    }
    delete result[0].password;
    return response.send(200, result[0]);
  });

});
```

FIGURE 10.29 node.js method to process login request for MySQL File: server.js

received by the post() method and the response that will be returned. We log the fact that we are logging in the user on the system console. If either the email or password in the request body is empty, we return the response with error code 400 and a message saying that login information is missing.

We then create an SQL SELECT query using the method database.query(). The method receives three parameters. The first parameter is a string with the SELECT query. Instead of using the values of the attributes, we use the question mark (?). The values will be provided by an array of values from the appropriate fields for the user object as the second parameter to the database.query() method. We use the PASSWORD() function on the password value. That will make sure that an encrypted value of the password that is entered will be compared with the encrypted value stored in the database. The select will return a record only if both email and encrypted passwords matched. The third parameter to database.query() is an anonymous function that receives two parameters. The first one, called err, is the possible error that could be thrown by the SELECT query. If err is not null, we return code 400 with the message saying that there was an error logging the user in. The second parameter, called result, is the array of records. If the length of result (result.length) is zero, we did not retrieve any records that matched both email and password. In that case, we return the response with error code 400 and a message saying that the password is invalid. Otherwise, the user is successfully logged in.

We delete the password from the result[0] (first record) (it does not delete it from the database). That way, when we return the result[0], we are not sending back the encrypted password. Finally, we return the code 200 and result[0] as an indication of success.

10.7 Updating user data in the MySQL server for the Thyroid app

Figure 10.30 shows the code for processing updateUser request. A lot of code is similar to the previous post methods. However, we explain all the steps for completeness. We are running the method app.post(), which receives two parameters. The first one is the request '/updateUser', which will be handled by the server, and the second parameter is the anonymous function that contains the rest of the code. The anonymous function takes two parameters: the request that is received by the post() method and the response that will be returned. We log the fact that we are updating the user on the system console. We are saving the body of the request in a variable called user for easy reference. If the newPassword field is null, we set it to be the same as the password field.

We then create an SQL UPDATE query using the method database.query(). The method receives three parameters. The first parameter is a string with the UPDATE query. Instead of using the values of the attributes, we use the question mark (?). The values will be provided by an array of values from the appropriate fields for the user object as the second parameter to the database.query() method. We use the PASSWORD() function on the password value. That will make sure that an encrypted value of the password that is entered will be compared with the encrypted value stored in the database. We also use the formatDateOfBirth() function shown in Figure 10.31 to format the dateString. We will discuss this function after we have finished studying Figure 10.30. All we need to know is that our date now will be a string of the form yyyy/mm/dd.

```
// Update user information
app.post('/updateUser', function(request, response) {
  console.log('User being updated.');
  var user = request.body;
  if(!user.newPassword) {
    user.newPassword = user.password;
  }
  var query = database.query('UPDATE user SET password=PASSWORD(?),
firstName=?, lastName=?, healthCardNumber=?, dateOfBirth=?,
cancerType=?, cancerStage=?, tshRange=?, agreedToLegal=? '
    + 'WHERE email=? and password=PASSWORD(?)', [user.newPassword,
    user.firstName,
    user.lastName,
    user.healthCardNumber,
    formatDateOfBirth(user.dateOfBirth),
    user.cancerType,
    user.cancerStage,
    user.tshRange,
    user.agreedToLegal,
    user.email,
    user.password
  ], function(err, result) {
    if (err) {
      return response.send(400, 'An error occurred creating this
user.');
    }
    return response.send(200, 'User created successfully!');
  });
});
```

FIGURE 10.30 node.js method to process updateUser request for MySQL File: server.js

```
function formatDateOfBirth(dateString) {
  var date = new Date(dateString);
  var month = date.getMonth()+1;
  var day = date.getDate();
  var year = date.getFullYear();
  return year + '-' +
    ((''+month).length<2 ? '0' : '') + month + '-' +
    ((''+day).length<2 ? '0' : '') + day;
}
```

FIGURE 10.31 node.js method to format date of birth string for MySQL

The update will be successful if both email and encrypted password match as shown in the WHERE clause. The third parameter to database.query() is an anonymous function that receives two parameters. The first one, called err, is the possible error that could be thrown by the UPDATE query. If err is not null, we return the response with error code 400 and a message saying that there was an error creating the user. The second object called result is not very useful to us for the update. If we did not return error code 400 until this point, we can send a code 200 and a message confirming that the user was successfully created.

We have seen the code used in the formatDateOfBirth() function in Chapter 6. We reproduce the explanation here again with an apology to the readers for the complexity of the code. We do not know its original author, but it can be found often on various websites. We think it is necessary that the readers become familiar with it. If the explanation is difficult to follow, just remember that it creates a string of the form yyyy/mm/dd such as 2015/01/09 for January 9, 2015.

Let us go through it one piece at a time. These pieces are concatenated together with the help of the + operator. We will not worry about some of the trivial parts of the string such as '/', which just adds a slash between values of year, month, and date. The problem is that the length of month and date can be one character (e.g., '5') or two characters (e.g., '12'). We want to make sure that the length is always two characters by preceding the value with '0' for single character values. This is done with the help of a conditional expression. A conditional expression has three parts. The part before the question mark (?) is a condition. If the condition is true, then the expression takes the value before the colon (:). If the condition is false, the expression takes the value after the colon (:).

Let us see how ((''+month).length<2 ? '0' : '') works. We are first creating a string ''+month by concatenating month to a blank character. Next, we look at the length of our string. If the length is less than 2, the expression evaluates to '0'. If the length is not less than 2, the expression evaluates to ''.

This conditional expression can also be written with an if-else construct

```
if(''+month).length<2)
{
    currentDate = currentDate + '0'
}
else
{
    currentDate = currentDate + ''
}
```

As mentioned before, conditional expressions are not everyone's cup of tea. All we need to know is that our currentDate now will be a string of the form yyyy/mm/dd.

10.8 Downloading records from the MySQL server for the Thyroid app

Figure 10.32 shows the code for processing a getRecords request. We are running the method app.post(), which receives two parameters. The first one is the request '/getRecords' that will be handled by the server, and the second parameter is the anonymous function that contains the rest of the code. The anonymous function takes two parameters: the request that is received by the post() method and the response that will be returned. We log the fact that we are sending the records on the system console. If either the email or password in the request body is empty, we return the response with error code 400 and a message saying that login information is missing.

We then create an SQL SELECT query using the method database.query(). The method receives three parameters. The first parameter is a string with the SELECT query. Instead of using the values of the attributes, we use the question mark (?). The values will be provided by an array of values from the appropriate fields for the user object as the second parameter to the database.query() method. We use the PASSWORD() function on the password value. That will make sure that an encrypted value of the password that is entered will be compared with the encrypted value stored in the database. It should be noted that we are joining both the users and the records table in this query using user.email and records.user as the join attributes. The select will return a record only if both email and encrypted password matched. The third parameter to database.query() is an anonymous function that receives two parameters. The first one, called err, is the possible error that could be thrown by the SELECT query. The second parameter, called result, is the array of records. If err is not null, we return code 400 with the message saying that there was an error retrieving the records. Otherwise, we return code 200 and the result.

```
// // Get all records associated with the given user
app.post('/getRecords', function(request, response) {
  console.log('Sending records');

  if (!request.body.email || !request.body.password) {
    return response.send(400, 'Missing log in information.');
  }

  var query = database.query('SELECT records.date, records.tsh,
records.tg, records.synthroidDose ' + ' FROM records, user ' + 'WHERE
records.user=user.email' + ' AND user.email=? AND
user.password=PASSWORD(?)' + ' ORDER BY records.date DESC',
[request.body.email,
    request.body.password
  ], function(err, result) {
    if (err) {
      return response.send(400, 'An error occurred in retrieving
records.');
    }
    return response.send(200, result);
  });

});
```

FIGURE 10.32 node.js method to process a getRecords request in MySQL File: server.js

10.9 Uploading records to the MySQL server for the Thyroid app

Figure 10.33 shows the code for processing a syncRecords request. We are running the method app.post(), which receives two parameters. The first one is the request '/syncRecords' that will be handled by the server, and the second parameter is the anonymous function that contains the rest of the code. The anonymous function takes two parameters: the request that is received by the post() method and the response that will be returned. We log the fact that we are sending

```
// Updates the records in the database with the provided records
app.post('/syncRecords', function(request, response) {
  console.log('Save Record Request Received!');

  if (!request.body.email || !request.body.password) {
    return response.send(400, 'Missing log in information.');
  }

  var query = database.query('SELECT email FROM user WHERE email=? AND
password=PASSWORD(?)', [request.body.email,
    request.body.password
  ], function(err, result) {
    if (err) {
      return response.send(400, 'Error logging user in.');
    }
    if (!result) {
      return response.send(400, 'Invalid user credentials.');
    }
  });
  var deleteQuery = database.query('DELETE FROM records WHERE user=?',
[request.body.email], function(err, result) {
    if(err) {
      return response.send(400, 'Error occurred syncing records');
    }
  });
  var newRecords = request.body.newRecords;
  var recordsToSave = [];
  for (var i = 0; i < newRecords.length; i++) {
    var newRecord = [request.body.email, newRecords[i].date,
newRecords[i].tsh, newRecords[i].tg, newRecords[i].synthroidDose];
    recordsToSave.push(newRecord);
  }
  var insertQuery = database.query('INSERT INTO records (user, date,
tsh, tg, synthroidDose) VALUES ?', [recordsToSave], function(err,
result) {
    if(err) {
      return response.send(400, 'Error occurred syncing records');
    }
    return response.send(200, 'Records synced.');
  });
});
```

FIGURE 10.33 node.js method to process a syncRecords request using MySQL—I
File: server.js

the records on system console. If either the email or password in the request body is empty, we return the response with error code 400 and a message saying that login information is missing.

Before inserting the records, we should make sure that the user with the given email actually exists in the database. The code for checking this is similar to the one used for processing a /login request. We create an SQL SELECT query using the method database.query(). The method receives three parameters. The first parameter is a string with the SELECT query. Instead of using the values of the attributes, we use the question mark (?). The values will be provided by an array of values from the appropriate fields for the user object as the second parameter to the database.query() method. We use the PASSWORD() function on the password value. That will make sure that an encrypted value of the password that is entered will be compared with the encrypted value stored in the database. The select will return a record only if both email and encrypted password matched. The third parameter to database.query() is an anonymous function that receives two parameters. The first one, called err, is the possible error that could be thrown by the SELECT query. If err is not null, we return code 400 with the message saying that there was an error logging the user in. The second parameter, called result, is the array of records. If the length of result (result.length) is zero, we did not retrieve any records that matched both email and password. In that case, we return the response with error code 400 and a message saying that the password is invalid. Otherwise, the user is a valid user.

We then delete all the existing records for the user using the method database.query(). The method receives three parameters. The first parameter is a string with the DELETE query. Instead of using the values of the attributes, we use the question mark (?). The values will be provided by an array of values from the appropriate fields for the user object as the second parameter to the database.query() method. The third parameter to database.query() is an anonymous function that receives two parameters. The first one, called err, is the possible error that could be thrown by the DELETE query. If err is not null, we return the response with error code 400 and a message saying that there was an error syncing the user. The second object called result is not very useful to us for the delete.

If we have successfully deleted the existing records for the user, we add the new records received through the array recordsToSave. We first save the newRecords from the body of the request in a variable called newRecords. We then initialize the recordsToSave array. The array is then built in a for loop. We create individual elements with the date, tsh, tg, and synthroidDose information from the newRecords array, and the email from the body of the record and store it in a variable called newRecord. We then push newRecord on the recordsToSave array.

Once we have created the recordsToSave array, we create an SQL INSERT query using the method database.query(). The method receives three parameters. The first parameter is a string with the INSERT query. Instead of using the values of the attributes, we use the question mark (?). The values will be provided by an array of values from the appropriate fields for the user object as the second parameter to the database.query() method. The third parameter to the database.query() is an anonymous function that receives two parameters. The first one, called err, is the possible error that could be thrown by the INSERT query. If err is not null, we return the response with error code 400 and a message saying that there was an error syncing the records. The second object called result is not very useful to us for the insert. If we did not return error code 400 until this point, we can send a code 200 and a message confirming that the records were synced.

In this chapter, we concluded the discussion on database server computing of our server-based Thyroid app using the conventional relational databases. The next two chapters will go back to the NoSQL/JSON databases and explain how to store and retrieve multimedia information. We will also study how to use the maps and geographic locations.

Quick facts/buzzwords

Relational database model: The most popular database model for storing structured data.
MySQL: A popular open source relational database management systems.
mysql: A command to launch an interactive command line session on MySQL.
use: A command to switch to a desired database.
show tables: A command to look at the tables in the current database.
describe: A command to look at the structure of a database.
INSERT: A command to add records to a MySQL table.
UPDATE: A command to update records in a MySQL table.
DELETE: A command to delete records from a MySQL table.
SELECT: A command to retrieve records from a MySQL table.
TRUNCATE: A command to delete records from a MySQL table.
DROP: A command to delete an entire MySQL table.
CORS: Cross-Origin Resource Sharing.
process.on(): A method for processing signals in node.js.
listen(): A method to listen for server requests in node.js.

CHAPTER 10

Self-test exercises

1. Name three database models.
2. Who introduced the relational database model?
3. What is MySQL?
4. What is the command used to retrieve records in MySQL?
5. What is the command used to add records in MySQL?
6. What is the command used to delete records in MySQL?

Programming exercises

1. Create the MySQL database on your MySQL server as shown in the text and add the library table.
2. Write a JavaScript function to add the following books to the library table for author John Grisham. The number in the parentheses is the year of publication.

   A Time to Kill (1989)
   The Firm (1991)
   The Pelican Brief (1992)
   The Client (1993)
   The Chamber (1994)
   The Rainmaker (1995)
   The Runaway Jury (1996)
   The Partner (1997)
   The Street Lawyer (1998)
   The Testament (1999)
   The Brethren (2000)
   A Painted House (2001)
   Skipping Christmas (2001)
   The Summons (2002)
   The King of Torts (2003)
   Bleachers (2003)
   The Last Juror (2004)
   The Broker (2005)
   Playing for Pizza (2007)
   The Appeal (2008)
   The Associate (2009)
   The Confession (2010)
   The Litigators (2011)
   Calico Joe (2012)
   The Racketeer (2012)
   Sycamore Row (2013)
   Gray Mountain (2014)

3. Write queries to find the following
   - All the books published before 2000 by any author
   - All the books published by John Grisham
   - All the books not published by John Grisham
   - All the books published between years 2000 and 2009 by any author
   - The number of books published after 2009 by any author

4. Write queries to delete the following
    - All the books published before 2000 by any author
    - All the books published by John Grisham
    - All the books not published by John Grisham
    - All the books published between years 2000 and 2009 by any author

Programming projects

1. **Boiler monitor server-side version:** We want to change the app that monitors the temperature and pressure of a boiler so that the data is stored on a MySQL server instead of in localStorage. You can model the app based on the server-side Thyroid app. We have modified the webpages in the previous chapter. In this chapter, you need to design the MySQL database so that the records can be stored and retrieved using the boiler ID as the key and to create the routes in your server for the app to access.
2. **Blood pressure monitor server-side version:** We want to change the app that monitors blood pressure so that the data is stored on a MySQL server instead of in localStorage. You can model the app based on the server-side Thyroid app. We have modified the webpages in the previous chapter. In this chapter, you need to design the MySQL database so that the records can be stored and retrieved using the person's ID as the key and to create the routes in your server for the app to access.
3. **Power consumption monitor server-side version:** We want to change the app that monitors the power consumption of a manufacturing plant so that the data is stored on a MySQL server instead of in localStorage. You can model the app based on the server-side Thyroid app. We have modified the webpages in the previous chapter. In this chapter, you need to design the MySQL database so that the records can be stored and retrieved using the plant ID as the key and to create the routes in your server for the app to access.
4. **Body mass index (BMI) monitor server-side version:** We want to change the app that monitors the body mass index of a person based on height and weight so that the data is stored on a MySQL server instead of in localStorage. You can model the app based on the server-side Thyroid app. We have modified the webpages in the previous chapter. In this chapter, you need to design the MySQL database so that the records can be stored and retrieved using the person's ID as the key and to create the routes in your server for the app to access.
5. **Managing a line of credit server-side version:** We want to change the app that manages our line of credit at a bank so that the data is stored on a server instead of in localStorage. You can model the app based on the server-side Thyroid app. We have modified the webpages in the previous chapter. In this chapter, you need to design the MySQL database so that the records can be stored and retrieved using the organization's ID as the key and to create the routes in your server for the app to access.

CHAPTER 11

JavaScript templating

WHAT WE WILL LEARN IN THIS CHAPTER

1. What is JavaScript templating
2. What is Handlebars
3. How to create apps using Handlebars

11.1 JavaScript templating

So far we have been using HTML5 along with CSS3 to control the content and display of our pages and screens. In recent years, a number of content management systems have made it easier for people with little or no knowledge of HTML to build web pages. These content management systems are easy to use and allow users to simply type in the necessary text, choosing a variety of display options. However, they provide limited flexibility. An alternative to the content management systems is templating.

Templating usually consists of code similar to HTML5 with some additional syntax. It provides a consistent look and feel that is useful when the layout is the same throughout a website, but the content of each page may change. A template is also helpful when a lot of data is being shown, such as displaying a large array of objects. There are often many elements of a web app that are used on every page; coordinating changes to these elements between every page can be difficult for individuals to manage. A simpler alternative is to use server-side templating. We will look at JavaScript templating using a popular library called Handlebars.js for an app called Explorador that will help users explore parks in Halifax. Our study will span two chapters, the first looking at the presentation and Handlebars templates. The JavaScript programming for both the client and server code will be discussed in the following chapter.

11.2 Explorador app

The app that will help us search for parks in the Halifax region uses a MongoDB database installed on the same server (140.184.132.239) as the one that we saw with the server-based version of the Thyroid app. As before we need to start the app as shown in Figure 11.1 by running the app.js script using the command

```
node app.js
```

We will look at the JavaScript in app.js in the following chapter. At this time, all we need to know is that the express server is listening for requests on port 8080. We can access the app using the URL: http://140.184.132.239:8080, which displays a welcome and tutorial screen as shown in Figure 11.2.

```
Running the server:
pawan@HRM-ParkApps:~/explorador$ node app.js
========================================================================
=  Please ensure that you set the default write concern for the database by setting    =
=   one of the options                                                                  =
=                                                                                       =
=     w: (value of > -1 or the string 'majority'), where < 1 means                      =
=        no write acknowledgement                                                        =
=     journal: true/false, wait for flush to journal before acknowledgement              =
=     fsync: true/false, wait for flush to file system before acknowledgement            =
=                                                                                       =
=  For backward compatibility safe is still supported and                               =
=   allows values of [true | false | {j:true} | {w:n, wtimeout:n} | {fsync:true}]       =
=   the default value is false which means the driver receives does not                 =
=   return the information of the success/error of the insert/update/remove             =
=                                                                                       =
=   ex: new Db(new Server('localhost', 27017), {safe:false})                            =
=                                                                                       =
=   http://www.mongodb.org/display/DOCS/getLastError+Command                            =
=                                                                                       =
=  The default of no acknowledgement will change in the very near future                 =
=                                                                                       =
=  This message will disappear when the default safe is set on the driver Db            =
========================================================================
Express server listening on port 8080
```

FIGURE 11.1 Launching the Explorador server

FIGURE 11.2 Welcome screen for the iOS Explorador app

The main use of this app is to make selections of where the user would like to find parks by region and what the user wants to do at the park by category (such as recreational activities, sports, etc.), or by searching for the park by name. The app will in turn search for parks matching the selections the user has made and display the results in either a list or map view, where the user can select a specific park to see more details, such as the address, a picture, and features of that specific park.

11.3 Explorador design

The final app that will be discussed in this textbook is an interactive app designed to help users in Halifax, a Canadian city, find parks of interest based on location and amenities. As with the previous apps, we expect anyone will use this app so we cannot make assumptions about things like experience level. Users' main objectives for this app will be to search for local parks, using filters to refine their search criteria, and find parks that best match their interests. Users can filter their search results based on the location and type of park (i.e., parks that offer water activities, kid-friendly parks, smaller community parks, or large provincial parks).

After users submit their search, they will want a practical way to view the search results. As with any search results that are tied in some way to geography, it is useful to provide users with a map view, where the results are plotted by their location. Selecting a point on the map displays the detailed information for that park. Lists sometimes provide a better perspective to users, as information on each park can be displayed without users selecting every point plotted on the map. This app will allow users to switch dynamically between looking at the results on a map or in a list.

When applications become more complex and robust, it is useful (and arguably necessary) to use software engineering design practices to document user interactions with the app and create a mock-up of the proposed design. Two design practices that we'll focus on here are flow charts and wireframes.

Flowcharts map out how users move through the app at a high level. Flowcharts are useful because they allow the designer and developer to place themselves in the shoes of the user and consider the end-to-end experience, from the first screen the user sees on opening the app to the completion of a given task. A flowchart consists of each step or page in the app (represented by boxes), and each box is connected by arrows to signify the order of user operations. Once a flowchart is complete, you can trace the path of users moving through the app for various scenarios.

Wireframes are mock-ups of the app pages to help define the layout or structure of each page. Wireframes allow the designer to determine the best way to arrange the different page elements and how the navigation system will work together with the page. Wireframes are a good tool to ensure that all functionalities in the page are included. Flowcharts and wireframes were used in the creation of the Explorador app to map out the navigation of the app and create mock-ups of each page.

Some notable design decisions in the Explorador app include using a sliding menu that appears at the left of the screen when users slide their fingers to the right. This menu also appears when users click on the name of the app in the top left corner of the page. This user interface has become a very common way for users to navigate in mobile apps. The Explorador app adopts this design feature to reduce the learning curve for users who are likely to be familiar with this navigation feature from other apps. For the search criteria found in the menu, the default values are such that every single type of park and location will appear in the search results at first, until users deselect specific types of parks. This provides an easy starting point so that results will appear as soon as users start to interact with the app.

The search results are instantly recalculated when a user changes the search criteria. This feature ties into another point that we have made in the past. An app should respond quickly to provide users with feedback that their actions have been recognized. In the case of the Explorador app, the listing of parks (in either the map view or the list view) instantly loads and appears with each selection a user makes. For the map view, we also consider the appropriate zoom level for a mobile user. It is not useful to have the map zoomed out to show the whole world, when our users are only interested in one specific region. The zoom level is set so that the user can start tapping on parks right away and not have to adjust the map window.

Given the app's complexity and the fact that it is more likely to be used by someone who has not been introduced to it by an experienced user (such as in the case of the Thyroid app, which may be recommended by a doctor or fellow patient), we have created a landing page on the main screen of our app. A landing page helps explain to users what the app does and what value the app provides to them. Our landing page explains key features of the app and how to navigate to accomplish the users' goal of finding the best park for them.

11.4 Introduction to JavaScript templating using Handlebars

Figure 11.3 shows the body of the first Handlebars file, called header.handlebar in the subdirectory views/layout, that will be returned by the app.js when we send the request to http://140.184.132.239:8080.

```
views/layouts/header.handlebars:

<!DOCTYPE html>
<html lang="en">
<head>
 <meta name="viewport" content="width=device-width, initial-scale=1.0,
user-scalable=no">
 <link rel="stylesheet" href="/stylesheets/bootstrap.css">
 <link rel="stylesheet" href="/stylesheets/bootstrap-theme.css">
 <link rel="stylesheet"
href="http://blueimp.github.io/Gallery/css/blueimp-gallery.min.css">
 <link rel="stylesheet" href="/stylesheets/bootstrap-image-
gallery.min.css">
 <link rel="stylesheet" href="http://cdn.leafletjs.com/leaflet-
0.6.4/leaflet.css">
 <link rel="stylesheet" href="/stylesheets/esri/MarkerCluster.css">
 <link rel="stylesheet"
href="/stylesheets/esri/MarkerCluster.Default.css">
 <link rel="stylesheet" href="/metro-vibes/metro-vibes-css/style.css">
 <script src="http://cdn.leafletjs.com/leaflet-
0.6.4/leaflet.js"></script>
 <script src="js/leaflet/leaflet.markercluster.js"></script>
 <script src="js/leaflet/esri-leaflet.js"></script>
 <script src="js/leaflet/extras/clustered-feature-layer.js"></script>
 <script src="js/jquery.min.js"></script>
 <script src="js/bootstrap.js"></script>
 <script src="js/map.js"></script>
 <script src="js/menu.js"></script>
 <script src="js/bootstrap-image-gallery.min.js"></script>
```

```
<style>
  p{
    font-size :15px;
  }
  #text{
    position :relative;
    left :20px;
    color: 777777;
  }
  #top {
    background-color: #56B156;
  }
  .containr-class{
    background-color: #9acd32;
  }
  #content.snap-content{
    padding-top:70px;
  }
  span, span:hover {
    color: #FFFFFF;
  }
 </style>

<title>Explorador</title>
</head>
```

■ FIGURE 11.3 <head> Welcome screen for the iOS Explorador app File: header. handlebars

Most of the contents of this file look similar to the HTML5 code that we are already familiar with. We have a sequence of stylesheets from various sources including Bootstrap that we have used before. We also have the metro-vibes package from PixelKit that adds unique styling to the common Bootstrap elements, such as buttons and toolbars. We are also using some of the stylesheets from ESRI, www.esri.com, a company that provides geographic information services, including online mapping services for maps such as those embedded in this app. We will look at the facilities provided by these stylesheets when looking at the map. We also see a number of related JavaScript files that have been included. All except map.js and menu.js are obtained from other packages including Bootstrap and ESRI. Both map.js and menu.js are scripts that we have written and will discuss in detail in the following chapter. Following the JavaScript files are additional self-explanatory CSS3 elements for formatting our webpage.

Figure 11.4 shows the body of the header.handlebar file, which is relatively basic compared to what we have covered so far. On loading the page we invoke a JavaScript function called initSnapJS() that initializes the JavaScript for a sliding menu from the left of the screen. We will look at this function in detail in the following chapter. We are enclosing the body with a <div> section with class "row". The class "row" is defined by Bootstrap for managing groups of elements. We then have the element <ul> for an unordered list with an id "top". The unordered list uses a number of classes from the metro-vibes package. These classes are used to style the unordered list horizontally as shown at the top of Figure 11.2, where we see two items: Explorador and Map. These two items are

```
views/layouts/header.handlebars:

<body onload="initSnapJS()">
  <div class="row">
    <ul id="top" style="z-index:10" class="dropdown clearfix boxed">
      <li>
        <a href="file:#" id="open-left">
          <img src="images/icons/whiteMenu.png"
style="vertical-align:middle"></img>
          <span style="left=16px">Explorador</span>
        </a>
      </li>
      <li class="pull-right">
        <a href="file:#">
          <span id="listView">Map</span>
        </a>
      </li>
    </ul>
    {{> menu menuContext}}
  </div>
  {{{body}}}
</body>
```

FIGURE 11.4 <body> Element of the welcome screen of the Explorador app File: header.handlebars

added using the `<li>...</li>` list element tags. The first item has the id "open-left" and contains a small image with three horizontal lines using the picture whiteMenu.png and the text "Explorador". The link to # essentially points to the current page; however, tapping on Explorador will open the sliding menu, which is implemented through JavaScript in the initSnapJS function. The second link with id "listView", signifying the default way to view the results of a search for parks, also uses #, but initSnapJS() will bring up a map when it is tapped. The class for the second element is "pull-right", which makes it right justified.

The {{ }} syntax is used by Handlebars to indicate that it should incorporate other Handlebars files when rendering a page. The syntax '{{> }}' specifies a partial Handlebars file that should be included in this location when the page is requested. A partial is a piece of HTML code that either is going to be reused in many different files or is being extracted to simplify the main Handlebars file, as is being done here. The first parameter, menu, is the name of the Handlebars file, the second parameter is a JSON object that Handlebars will use to populate the file. This JSON object is provided by the server.

This file is located under the layout directory and will be configured as a layout file when Handlebars is started. Handlebars assumes that all layouts will be used for multiple pages, and thus each page will provide its own body or main content. The syntax '{{{ }}}' is used by Handlebars to mark where in the layout it should include the body code. The Handlebars file that will be used as the body is specified by the server. The implementation of both of these will be covered in detail in the following chapter, when we look at the corresponding JavaScript.

Figures 11.5 and 11.6 show the tutorial for the app that will come up when a user scrolls down the welcome screen. The HTML5 code for the tutorial is shown in Figures 11.7 and 11.8 from the file tutorial.handlebars. This is a partial HTML5 document that corresponds to the {{{body}}} in header.handlebars. The entire file is enclosed in a <div> section with id "content" and the class "snap-content". The class is defined in CSS by Snap.js, which we will look at later on.

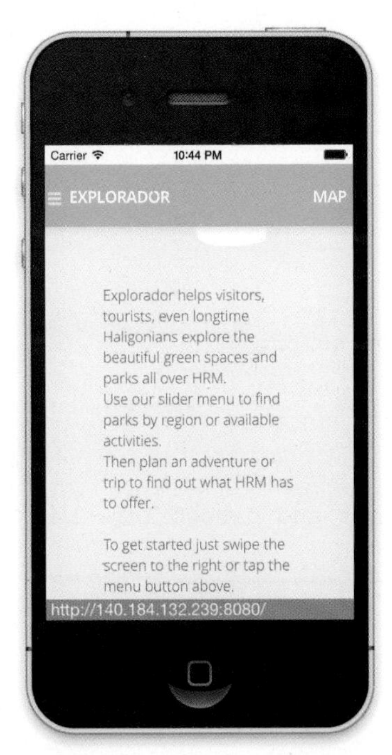

■ **FIGURE 11.5** Screenshot of the first screen in the tutorial

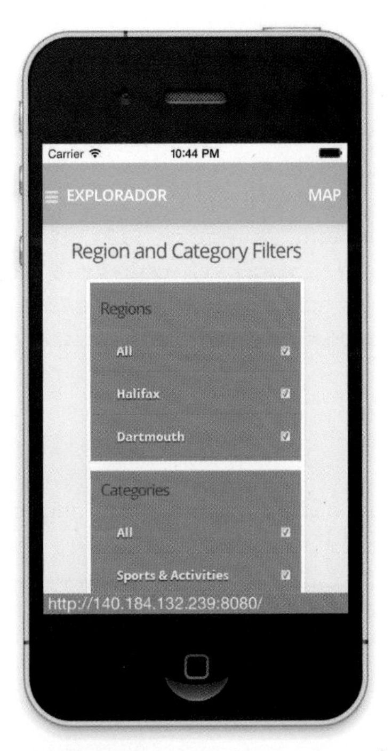

■ **FIGURE 11.6** Screenshot of the second screen in the tutorial

```
views/tutorial.handlebars:

<div id="content" class="snap-content">
 <div class="jumbotron">
   <div id="list" class="container">
     <h3>Welcome to Explorador</h3>
     <div class="col-xs-12 col-sm-6 col-md-3">
       <a href="file:#">
         <img src="images/icons/Swipe_Right256.png">
       </a>
     </div>
     <div class="col-xs-12 col-sm-6 col-md-3">
       <p>
         Explorador helps visitors, tourists, even longtime Haligonians
explore the beautiful green spaces and parks all over HRM.
         <br>
         Use our slider menu to find parks by region or available
activities.
         <br>
         Then plan an adventure or trip to find out what HRM has to
offer.
         <br>
         <br>
         To get started just swipe the screen to the right or tap the
menu button above.
       </p>
     </div>
   </div>
 </div>
```

■ **FIGURE 11.7** HTML5 code for the tutorial for the Explorador server—I

```
views/tutorial.handlebars:

 <div class="jumbotron">
   <div id="list" class="container">
   <div class="row">
     <h3>Region and Category Filters</h3>
     <div class="col-xs-12 col-sm-6 col-md-3">
       <a class="thumbnail" href="file:#">
         <img src="images/icons/tutRegionsCategories.png">
       </a>
     </div>
     <div class="col-xs-12 col-sm-6 col-md-6">
       <p>
         In the slider menu, you can select regions and categories that
you're interested in.
         <br>
         Categories with a '+' next to them can be expanded to allow
more refined choices
         <br>
         Browse your results in a list, or click the "Map" button in
the upper right corner to toggle to an interactive map.
       </p>
       <div class="col-xs-12 col-sm-6 col-md-6">
         <a class="thumbnail" href="file:#">
           <img src="images/icons/tutorialMap.jpg">
         </a>
       </div>
     </div>
   </div>
   </div>
 </div>
 <div class="jumbotron">
   <div id="list" class="container">
     <div class="row">
       <a
href="https://play.google.com/store/apps/details?id=com.smu.
cs.parkfinder">
         <img
src="https://developer.android.com/images/brand/en_app_rgb_wo_45.png">
       </a>
     </div>
   </div>
 </div>
</div>
```

FIGURE 11.8 HTML5 code for the tutorial for the Explorador server—II

We have three separate `<div>` sections for three parts of the tutorial. Each `<div>` section uses a Bootstrap class called jumbotron. The jumbotron class is used to separate each section of the tutorial. The first section has a picture that shows a palm swiping the screen to the right given by the figure in Swipe_Right256.png and some explanation about the functionality of Explorador. The second section has screenshots of the menu and maps and information on navigation. The third section provides a link to download the app from the Google Play store.

One interesting feature in these <div> sections is the use of a variety of "col" classes from the Bootstrap library. Bootstrap includes a grid-based column system to provide developers with an easy way to format content on the screen for any possible screen size. Bootstrap divides content on the screen with a grid of 12 invisible columns. For each div, we can specify how many visible columns should appear on the page with the syntax 'col-[screen size]-[number of bootstrap columns per visible column]'. For example, specifying 'col-xs-12' specifies that there will only be one visible column on xs, or extra small, screens (i.e. mobile phones), using 'col-sm-6' will specify that there will be two visible columns (each spanning 6 Bootstrap columns) for small screens such as tablets. Other possible screen sizes are md and lg.

11.5 Iteration and arrays in Handlebars

Figure 11.9 shows the screenshot of the sliding menu. Corresponding HTML5 and Handlebars code is shown in Figure 11.10. We are including the stylesheet snap.css and a JavaScript file snap.min.js from snap.js. Snap.js is an open source project available to developers to quickly and easily add sliding menus to any website. The primary developer for the Snap.js project is Jacob Kelley, and the project is available at (https://github.com/jakiestfu/Snap.js/).

The menu is enclosed in a <div> section with the class snap-drawers, which is defined in snap.css and is meant to contain a collection of <div> sections with class snap-drawer. A snap-drawer is a single menu. In our case, we only have one menu, but we still have to wrap it in a <div> with the snap-drawers class. The only menu we have is enclosed in the following <div> section with classes snap-drawer and snap-drawer-left. The snap-drawer-left means that the drawer comes out from the left. We further enclose it in a <div> section with id "snap-menu", which we will use when initializing the JavaScript for the menu. It encloses a <div> section with name "mainMenu", which contains all our menu items. The header for the menu is Explorador. Now we start a <div> section with id searchForm with an input element with id "search". This search box can be seen at the top of the menu in Figure 11.9. The following menu

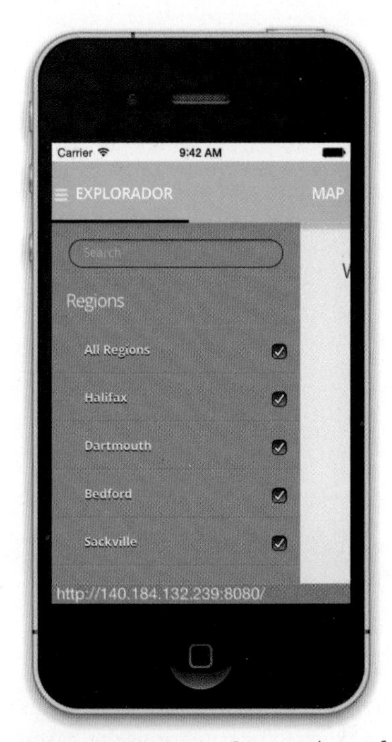

FIGURE 11.9 Screenshot of sliding menu in Explorador

```
views/partials/menu.handlebars:

<link rel='stylesheet' href='stylesheets/snap.css'>
<script src='js/snap.min.js'></script>

  <div class='snap-drawers' onload='initSnapJS()'>
    <div class='snap-drawer snap-drawer-left'>
      <div id='snap-menu'>
        <div name='mainMenu'>
        <h3>Explorador</h3>
          <div id='searchForm'>
          <input id='search' type='search' placeholder='Search'>
        </div>
        <h4>Regions</h4>
        <ul>
          <li>
            <label>All Regions</label>
            <input type='checkbox' name='regions' value='all' checked>
          </li>
          {{#each regions}}
          <li>
            <label>{{this}}</label>
            <input type='checkbox' name='region' value='{{@index}}'
checked>
          </li>
          {{/each}}
        </ul>
        <h4>Categories</h4>
        <ul>
          <li>
            <label>All Categories</label>
            <input type='checkbox' name='clusters' value='all'
checked>
          </li>
          {{#each categories}}
          <li>
            <label>{{this}}</label>
            <input type='checkbox' name='cluster' value={{addOne
@index}} checked>
          </li>
          {{/each}}
        </ul>
      </div>
    </div>
  </div>
</div>
```

FIGURE 11.10 The HTML5 and Handlebar code for the menu in Explorador

has two sections, the first with the header "Regions" and the other with the header "Categories". The parks in our database are divided using two criteria: region and categories. Region means a geographic region such as Halifax, Bedford, Dartmouth, and so on. The categories are based on the type of services available in the park, such as sports or water activities. Each of these sections has unordered list (<ul>) elements. The first item in each of the unordered lists is an input checkbox that allows you to select or deselect all the regions or categories. This first item is followed by a sequence of items specified using {{#each regions}}...{/each} and {{#each categories}}...{/each} respectively. One can look at this construct as a loop that iterates through all the regions and categories in the database. Let us look at the first loop {{#each regions}}...{{/each}}. We are iterating through an array of regions. Each element of the array is referred to as {{this}} within the {{each}} block of our code

```
<label>{{this}}</label}
```

and will put the value of the array elements as a label. Next to the label we have a checkbox element

```
<input type='checkbox' name='regions' value='{{@index}}' checked>
```

Here {{@index}} specifies the index of that array element and will end up having a value of 0, 1, 2, etc. The index in the menu corresponds to the values on our server that allow it to find out which element in the array is checked.

In the second menu section, we are iterating through an array of categories. Each element of the array is referred to as {{this}} in our code

```
<label>{{this}}</label}
```

and will put the value of the array elements as a label. Next to the label we have a checkbox element

```
<input type='checkbox' name='regions' value='{{addOne @index}}'
checked>
```

Here {{addOne @index}} is a helper function that adds one to the index of that array element. That will help our JavaScript find out which element in the array is checked. For regions, our array started at 0. However, the categories were created using a data mining software where category numbers start at 1, so we have created a helper function on the server to ensure that the values passed to the server match the values in the database.

Figure 11.11 shows the screenshot of the list of results obtained by a search. The HTML5 and Handlebars code can be seen in Figure 11.12. We have enclosed the entire page in a `<div>` section with id "content" and class "snap-content". The class "snap-content" comes from Snap.js. We then

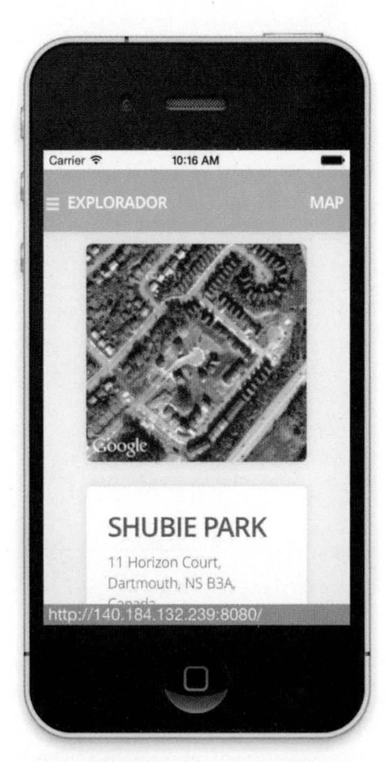

FIGURE 11.11 Screenshot of the list of results from Explorador

```
views/resultsList.handlebars:

<div id='content' class='snap-content'>
 <div class='jumbotron'>
   {{#each parks}}
   <div class='row col-sm-12'>
     <br>
     <div class='col-sm-4'>
       <div class='widget-container widget_gallery boxed'>
         <div class='inner'>
           <div class='carousel slide'>
             <div class='carousel-inner'>
               <div class='active item'>
                 <a href='/parkView?id={{_id}}'>
                   <img id='1'
src='http://maps.googleapis.com/maps/api/staticmap?center={{location.la
titude}},{{location.longitude}}&zoom=16&size=640x220&maptype=satellite&
sensor=false' alt='{{name}}'>
                 </a>
               </div>
             </div>
           </div>
         </div>
       </div>
     </div>
     <div class='col-sm-8'>
       <div class='tabs_framed styled'>
         <div class='tab-content boxed clearfix'>
           <div class='tab-pane fade in active'></div>
           <div class='inner clearfix'>
             <div class='tab-text'>
               <h3>{{name}}</h3>
               <p>
                 {{location.address}}
                 <br>
                 Features : {{formatFeatTypes featTypes}}
                 <br>
                 Distance : {{distance}} km
               </p>
             </div>
           </div>
         </div>
       </div>
     </div>
   </div>
   {{/each}}
 </div>
</div>
```

FIGURE 11.12 HTML5 and Handlebars code to display list of results from Explorador

use a <div> section with class "jumbotron" to display it on a gray block. We have seen the jumbotron class from the Bootstrap library before in the tutorial. Our search returns an array of parks, which we have appropriately named parks. As before, we iterate through the array using the Handlebars element

```
{{#each parks}...{{/each}}
```

Each result is enclosed in a <div> section with the row class from Bootstrap, which spans all 12 columns. It will then be divided into two sections, first with the <div> section spanning 4 columns and the second with a <div> section spanning 8 columns. The section with 4 columns contains an image of the park, and the one with 8 columns describes the features of the park. We will skip some of the <div> classes that are used for formatting these elements and look at the actual data itself enclosed in an <a href>...</a> element, which has an <img> element

```
<a href='/parkView?id={{_id}}'>
            <img id='1'
src='http://maps.googleapis.com/maps/api/staticmap?center=
{{location.latitude}},{{location.longitude}}&zoom=16&size=
640x220&maptype=satellite&sensor=false' alt='{{name}}'>
</a>
```

Each park contains an attribute called location, which in turn has an element for latitude and longitude. We access these attributes using {{location.latitude}} and {{location.longitude}} from Google's application programming interface (API) service

```
maps.googleapis.com/maps/api/staticmap
```

This service returns a satellite photo of the park at the coordinates we sent it. We provide the value of alt attribute for the <img> using {{name}}—that is, the value of the name attribute in the current array element.

In the remaining 8 columns, we list the features using the Handlebars elements {{name}}, {{location.address}}, {{formatFeatTypes featTypes}}, and {{distance}}.

- {{name}} is the name of the park.
- {{location.address}} is the address of the park, such as "923 Robie Street, Halifax".
- {{formatFeatTypes featTypes}} applies one of our Handlebars functions called formatFeatTypes and applies it to the attribute featTypes, which is an array of the types of features available at this park. We will study the function formatFeatType in the following chapter. At this time, all we need to know is that the values of feat-Types are all capital letters, which will be changed to title case (first letter of each word capitalized by our formatFeatTypes function.
- {{distance}} is distance in kilometers from the location of the user. The location of the user will be detected automatically by the app and calculated on the server by our JavaScript.

If users choose to use the map view (top right link, which toggles between the map and list views), we will see our results on a map as shown in Figure 11.13. We see two markers for two of the parks; tapping on one of the markers will bring up the details for that park as shown in Figure 11.14. The HTML5 and Handlebars code is shown in Figure 11.15. We see <div> sections with familiar classes such as snap-content and jumbotron as in the tutorial template. Let us focus on the <div> section with class "container". It consists of two <div> sections. The first <div> with id "parkResults" shows the information about the park using the Handlebars helper function {{stringify parks}}, which applies the stringify function to the JSON object parks. Stringify is a function we wrote that will be discussed

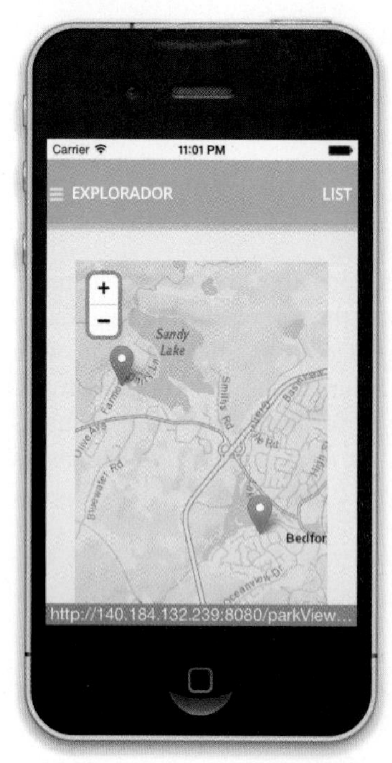

■ **FIGURE 11.13** Screenshot of results shown on a map in Explorador

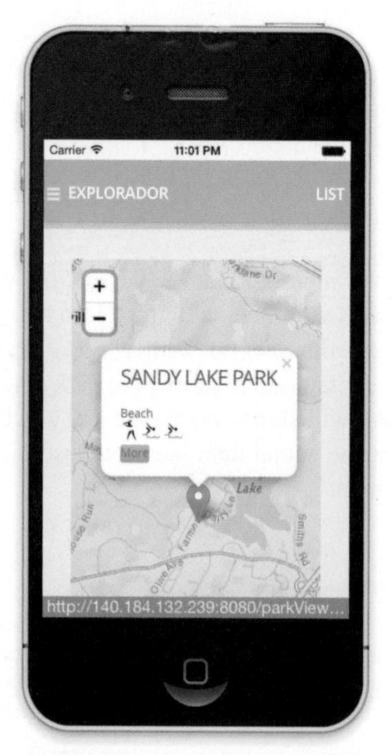

■ **FIGURE 11.14** Screenshot of detailed results shown on a map in Explorador

```
views/resultsMap.handlebars:

<div id='content' class='snap-content'>
 <div class='jumbotron'>
   <div class='container'>
     <div id='parkResults' class='hidden'>
         {{stringify parks}}
     </div>
     <div class='row'>
       <div id='map' style='height:400px;'></div>
     </div>
   </div>
 </div>
</div>
```

■ **FIGURE 11.15** HTML5 and Handlebars code for displaying results in a map

in the following chapter; it applies the JavaScript stringify function to the JSON object and returns it as a string. We have also set the class for the "parkResults" `<div>` as "hidden". This is an HTML5 class that means the information is not displayed on the screen, although it is still included when sent to the web browser. Our client-side JavaScript will grab the string of parks and use it to create the pop-up windows as shown in Figure 11.14. The second `<div>` section with class "row" from Bootstrap displays the map in the `<div>` section with id "map". The details of the map display are handled in the JavaScript that will be studied in the following chapter.

11.6 Conditional statements in Handlebars

Tapping on a particular park will give us detailed information about the park as shown in Figure 11.16. The HTML5 and Handlebars code is divided into two figures (Figures 11.17 and 11.18). We contain the entire page in a `<div>` section with id "content" and the class "snap-content" from Snap.js. It also contains a `<script>` element that runs the loadMap() function that will be discussed in the following chapter. We use the Bootstrap class called "jumbotron" for the entire page, further enclosed in a `<div>` section with the "row" class from Bootstrap that spans all 12 Bootstrap columns. Similar to the listView page, we have two `<div>` sections inside, the first spanning 4 columns and the second spanning 8 columns. The first `<div>` section spanning 4 columns is enclosed in another `<div>` section with class "carousel-inner" from PixelKit to format the layout of the `<div>`. The actual data is enclosed in an `<a href>...</a>` element, which has an `<img>` element

```
<a href='/parkView?id={{_id}}'>
              <img id='1'
src='http://maps.googleapis.com/maps/api/staticmap?center=
{{location.latitude}},{{location.longitude}}&zoom=16&size=
640x220&maptype=satellite&sensor=false' alt='{{name}}'>
</a>
```

Each park contains an attribute called location, which in turn has an element for latitude and longitude. We access these attributes using {{location.latitude}} and {{location.longitude}} from Google's application programming interface (API) service

```
maps.googleapis.com/maps/api/staticmap
```

This service returns a satellite photo of the park at the coordinates we sent it. We provide the value of alt attribute for the `<img>` using {{name}}—that is, the value of the name attribute in the current array element.

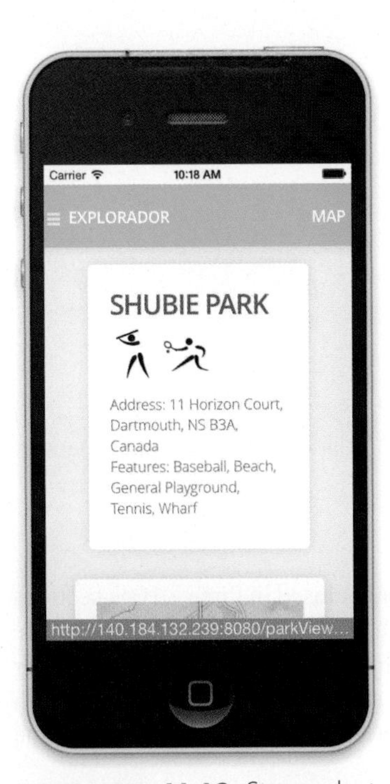

FIGURE 11.16 Screenshot of the detailed information of a park in Explorador

```
views/parkView.handlebars:

<div id='content' class='snap-content' onload='loadMap();'>
  <div class='jumbotron'>
    <div class='row col-sm-12'>
      <br>
      <div class='col-sm-4'>
        <div class='carousel-inner'>
          <a href='/parkView?id={{id}}'>
            <img id='1'
src='http://maps.googleapis.com/maps/api/staticmap?center={{park.locati
on.latitude}},{{park.location.longitude}}&zoom=16&size=640x220&maptype=
satellite&sensor=false' alt='{{park.name}}'>
          </a>
        </div>
      </div>
      <div class='col-sm-8'>
        <div class='tabs_framed styled'>
          <div class='tab-content boxed clearfix'>
            <div class='inner clearfix'>
<--! See Figure 11.?? for the code here -->
            </div>
          </div>
        </div>
      </div>
    </div>
    <div class='row'>
      <div class='col-sm-8'>
        <div class='tab-content boxed clearfix'>
          <div class='inner clearfix'>
            <div id='map' style='height: 400px'>
            </div>
            <div id='parkResults' class='hidden'>
              [{{stringify park}}]
            </div>
            <script type="text/javascript">
              loadMap();
            </script>
          </div>
        </div>
      </div>
    </div>
  </div>
```

FIGURE 11.17 HTML5 and Handlebars code for the detailed information of a park in Explorador—I

The second <div> section spanning 8 columns is further enclosed in three <div> sections with a number of classes from PixelKit and Bootstrap that we have seen previously. The contents of the <div> section with class "tab-text" are shown in Figure 11.18. Before looking at the figure let us finish the study of the rest of the template structure. The template ends with a <div> section with id "parkResults" that shows the information about the park using the

```
views/parkView.handlebars:

<h3>{{park.name}}</h3>
{{#each park.featTypes}}
  {{#ifEqual this 'RUNNING TRACK'}}
    <img src='images/icons/RUNNING TRACK' height='50' width='50'>
  {{/ifEqual}}
  {{#ifEqual this 'SOCCER'}}
    <img src='images/icons/soccerSvg.jpg' height='50' width='50'>
  {{/ifEqual}}
  {{#ifEqual this 'BASKETBALL HOOP'}}
    <img src='images/icons/basketball.jpg' height='50' width='50'>
  {{/ifEqual}}
  {{#ifEqual this 'TENNIS'}}
    <img src='images/icons/tennis.jpg' height='50' width='50'>
  {{/ifEqual}}
  {{#ifEqual this 'VOLLEYBALL'}}
    <img src='images/icons/Volleyball.svg' height='50' width='50'>
  {{/ifEqual}}
  {{#ifEqual this 'BASEBALL'}}
    <img src='images/icons/Baseball_pictogram.svg' height='50'
width='50'>
  {{/ifEqual}}
  {{#ifEqual this 'BASKETBALL COURT'}}
    <img src='images/icons/basketball.jpg' height='50' width='50'>
  {{/ifEqual}}
  {{#ifEqual this 'BOAT LAUNCH SMALL'}}
    <img src='images/icons/boat.svg' height='50' width='50'>
  {{/ifEqual}}
  {{#ifEqual this 'BOAT LAUNCH'}}
    <img src='images/icons/boat.svg' height='50' width='50'>
  {{/ifEqual}}
  {{#ifEqual this 'CRICKET'}}
    <img src='images/icons/Cricket.png' height='50' width='50'>
  {{/ifEqual}}
 {{/each}}
<p>
  Address: {{park.location.address}}
  <br>
  Features: {{formatFeatTypes park.featTypes}}
</p>
```

FIGURE 11.18 HTML5 and Handlebars code for the detailed information of a park in Explorador—II

Handlebars helper [{{stringify park}}], which applies the stringify function to the JSON object parks and presents it as an array using square brackets. Stringify is a function we wrote that will be discussed in the following chapter. It applies the JavaScript stringify function to the JSON object and returns it as a string. The array brackets are used on this page, as park is only one park. However, this section uses the same functions as the mapResults page, which expects an array of many parks to plot. The class for `<div>` section "parkResults" was "hidden" which is an HTML5 class that means the information is not displayed on the screen, although it is included in the page on the web browser. Our client-side JavaScript will grab it and display in a pop-up window similar to the one shown in Figure 11.14. We cannot see it in Figure 11.16 because the map appears below the visible area. The details of the map display are handled by the loadMap() function, which will be studied in the following chapter.

Now let us turn our attention to the sequence of conditional constructs shown in Figure 11.18. We have a loop {{#each park.featTypes}}...{{/each}} to iterate through an array of park's feature types (park.featTypes). For each element of the array, we are using {{#ifEqual condition}}...{{/ifEqual}} to selectively include images. For example,

```
{{#ifEqual this 'BOAT LAUNCH SMALL'}}
  <img src='images/icons/boat.svg' height='50' width='50'>
{{/ifEqual}}
```

will include the image boat.svg if the value of the current array element is 'BOAT LAUNCH SMALL'. This process is repeated for each element. Handlebars only provides a basic {{#if}}...{{/if}} helper function by default. Additional functions, such as ifEqual are defined by the server when the page is requested and rendered.

In this chapter, we looked at JavaScript templating using the Handlebars.js templating engine. We only looked at how the Handlebars elements are embedded in JavaScript and the corresponding results in the app screenshots for our Explorador app to explore parks in the Halifax region. We will be looking at the corresponding JavaScript in the following chapter.

Quick facts/buzzwords

JavaScript templating: Creating a consistent look and feel for a web app by allowing reuse and simplification of HTML5.

Handlebars: A popular package for JavaScript templating.

{{. . .}}: Double braces used to enclose Handlebars directives.

{{> }}: Handlebar directive to include other partial Handlebars files.

Snap.js: An open source library to add sliding menus.

PixelKit: A package to add more unique styling to the Bootstrap library.

{{#each array}}. . .{{/each}}: Handlebars construct to iterate through an array of objects.

{{this}}: Handlebars construct to denote entire object in the current context. Denotes the entire current object of the array when used in an {{#each}}...{{/each}} block.

{{@index}}: Handlebars construct to denote index of the current object in the array.

`maps.googleapis.com`: Website that provides an application programming interface to Google maps.

CHAPTER 11

Programming projects

1. This project continues over Chapters 11 and 12. We will enhance the database and create Handlebars templates in this chapter. The rest of the programming will be in the following chapter.

   Monitoring a collection of boilers: We will build on our previous app, which was used to monitor only one boiler. Our database server version allows us to store data for multiple boilers. Let us enhance the database to also include the map coordinates of each boiler. We will create an app modeled after Explorador that allows us to search for different boilers based on their IDs or location, using a map. Once we pick a boiler from this search, we will have the four options as before

   i. An option and corresponding page to allow you to change the basic information about the boiler.

   ii. An option and corresponding page to enter data—temperature and pressure.

   iii. An option and corresponding page to graph the data.

   iv. An option and corresponding page to make recommendations based on the values of temperature and pressure.

2. This project continues over Chapters 11 and 12. We will enhance the database and create Handlebars templates in this chapter. The rest of the programming will appear in the following chapter.

   Monitoring blood pressure of a collection of patients: We will build on our previous app, which was used to monitor only blood pressure of one person. Our database server version allows us to store data for multiple persons. Let us enhance the database to also include the map coordinates of each person. We will create an app modeled after Explorador that allows us to search for different persons based on their names, IDs, or location, using a map. Once we pick a person from this search we will have the four options as before

   i. An option and corresponding page to allow you to change the basic information about the person.

   ii. An option and corresponding page to enter data—blood pressure.

   iii. An option and corresponding page to graph the data.

   iv. An option and corresponding page to make recommendations based on the values of the blood pressure.

3. This project continues over Chapters 11 and 12. We will enhance the database and create Handlebars templates in this chapter. The rest of the programming will be in the following chapter.

   Monitoring a power consumption of a collection of plants: We will build on our previous app, which was used to monitor only one plant. Our database server version allows us to store data for multiple plants. Let us enhance the database to also include the map coordinates of each plant. We will create an app modeled after Explorador that allows us to search for different plants based on their IDs or location using a map. Once we pick a plant from this search we will have the four options as before

   i. An option and corresponding page to allow you to change the basic information about the plant.

ii. An option and corresponding page to enter data—power consumed.

iii. An option and corresponding page to graph the data.

iv. An option and corresponding page to make recommendations based on the values of power consumption.

4. This project continues over Chapters 11 and 12. We will enhance the database and create Handlebars templates in this chapter. The rest of the programming will appear in the following chapter.

   Monitoring body mass index (BMI) of a collection of patients: We will build on our previous app, which was used to monitor only BMI of one person. Our database server version allows us to store data for multiple persons. Let us enhance the database to also include the map coordinates of each person. We will create an app modeled after Explorador that allows us to search for different persons based on their names, IDs, or location using map. Once we pick a person from this search we will have the four options as before

   i. An option and corresponding page to allow you to change the basic information about the person.

   ii. An option and corresponding page to enter data—height (meter) and weight (kg).

   iii. An option and corresponding page to graph the data, height, and BMI.

   iv. An option and corresponding page to make recommendations based on the values of body mass index calculated using the weight and height entered.

5. This project continues over Chapters 11 and 12. We will enhance the database and create Handlebars templates in this chapter. The rest of the programming will appear in the following chapter.

   Monitoring line of credit for a collection of organizations: We will build on our previous app, which was used to monitor only one organization. Our database server version allows us to store data for multiple organizations. Let us enhance the database to also include the map coordinates of each organization. We will create an app modeled after Explorador that allows us to search for different organizations based on their IDs or location using a map. Once we pick an organization from this search we will have the four options as before

   i. An option and corresponding page to allow you to change the basic information about the line of credit.

   ii. An option and corresponding page to enter data—credit or debit.

   iii. An option and corresponding page to graph the balance.

   iv. An option and corresponding page to make recommendations based on the current balance.

CHAPTER 12

Maps, location, and multimedia databases

WHAT WE WILL LEARN IN THIS CHAPTER

1. How to manage Handlebars documents through JavaScript programs
2. How to use maps in our app
3. How to use location information from our app
4. How to use Snap.js, Leaflet, and ESRI modules in our server-side code

12.1 Introduction

This chapter discusses all the programming that goes into the Explorador app described in the previous chapter. We will use a number of more refined features of node.js from a number of modules including path, server-favicon, and morgan. We will also learn how to display maps and manage geographical data including the location of the user.

12.2 Setting up the environment in the Explorador app

As we saw in Chapter 11, the first step in getting our app going will be to launch the server with the app.js file shown in Figure 12.1. We run the file using the command node app.js

Let us have a quick look at the file. It is very similar to the app.js file we used to launch the Thyroid app server. We are including a number of modules using the require() function. These include

- express: It is a basic framework to launch a web server that we have seen previously.
- express3-handlebars: provides a Handlebars extension for the express server.
- path: is a module that is used to set the location of files such as images.
- serve-favicon: allows us to put an image in the URL window such as logo for our organization, when users come to our website.
- morgan: used to maintain log of activities on the server.
- parkMongoHandler.js: This is a module that we have written, to work with our MongoDB database. In particular we want to use the ParkProvider function from that file.

The statement

```
app = express()
```

creates the instance of the server, and we are ready to go.

```
/**
 *     explorador
 *     Created for the HRM apps4halifax contest
 *     Saint Mary's University, 2013
 *
**/

// Module dependencies
var express = require('express');
var expressHandlebars = require('express3-handlebars');
var errorHandler = require('errorhandler');
var path = require('path');
var favicon = require('serve-favicon');
var morgan = require('morgan'); // Logging
var ParkProvider = require('./parkMongoHandler.js').ParkProvider;
var app = express();
```

FIGURE 12.1 Launching the Explorador server File: app.js

Figure 12.2 shows the code that sets up the environment for the variable app (an instance of express server). On a UNIX platform, we have a number of environmental variables. On the first line in Figure 12.3, we check to see if the port for the server (process.env.PORT, the PORT environment variable in the terminal) is set. If it is not set, then we will use the value 8080. That means our port variable will get the value of either process.env.PORT or 8080 in that order.

The app.engine() method is used to specify the page rendering mechanism. In our case, we are going to use Handlebars. We are using the module that we included in Figure 12.1 as expressHandlebars. We are further specifying that the layout header.handlebars, as we saw in the previous chapter, will be used as the default for rendering all pages.

The app.set('view engine', 'handlebars') will make sure that when express checks the view engine the string 'handlebars' is returned.

```
app.use(favicon(__dirname + '/public/images/icons/favicon.ico'))
```

sets the icon that will appear next to the URL.

```
app.use(morgan('dev'))
```

```
// Environment setup
app.set('port', process.env.PORT || 8080);  //change port here
app.engine('handlebars',
expressHandlebars({defaultLayout:'header'}));
app.set('view engine', 'handlebars');
app.use(favicon(__dirname + '/public/images/icons/favicon.ico'));
app.use(morgan('dev'));
app.use(rawBody);
app.use(express.static(path.join(__dirname, 'public')));


// Setup the mongo handler. Specify local and port number.
var parkMongoGrab = new ParkProvider('127.0.0.1', 27017);
```

FIGURE 12.2 Setting up the environment in Explorador app File: app.js

```
// Custom middleware to replace express.bodyParser()
function rawBody(req, res, next) {
  req.setEncoding('utf8');
  req.rawBody = '';
  req.on('data', function(chunk) {
    req.rawBody += chunk;
  });
  req.on('end', function(){
    next();
  });
}
```

FIGURE 12.3 The function rawBody() to specify how to process incoming data
File: app.js

specifies the level of logging. We are using 'dev', which means we will get detailed logs of the activity on the server.

```
app.use(rawBody)
```

tells the server to use the function rawBody, which will appear in Figure 12.3.

```
app.use(express.static(path.join(__dirname, 'public')))
```

mentions that when express is looking for static files such as images, JavaScript, or stylesheets, they will be in the public directory.

We then set up a variable called parkMongoGrab, which creates an object of the type ParkProvider (which will appear in Figure 12.15) with the IP address of the localhost (127.0.0.1) and port 27017, where we can locate the MongoDB server using the following statement

```
var parkMongoGrab = new ParkProvider('127.0.0.1', 27017);
```

Figure 12.3 shows the function that is used to specify the fact that incoming data should be parsed in its raw form as opposed to doing any filtering or transformation.

12.3 Specifying express routes

Figure 12.4 and a number of the following figures will tell the server how to handle various http requests (express routes), starting with the request for the home page

```
http://140.184.132.239:8080
```

in Figure 12.4, specified using the app.get() method. The last line in the method calls the res.render() method, which is going to tell the server to render the page called tutorial (tutorial.handlebars file), and the relevant information is provided in a JSON object called context. The JSON object is created at the beginning of the app.get() method. It specifies two name/value pairs.

The first pair is a JSON object itself, called menuContext, with two attributes. The first attribute is an array called regions, and the second attribute is an array called categories. The values of the elements in the array are specified as well.

The second name/value pair, called helpers, specifies the function addOne that we use in the tutorial.handlebars to add one to an array index.

```
//Landing page routes to tutorial
app.get('/', function(req, res){
  var context = {
    menuContext: {
      regions: ['Halifax', 'Dartmouth', 'Bedford', 'Sackville', 'Eastern Shore',
'Outer HRM'],
      categories: ['Sports & Activities +', 'Kid Friendly +', 'Water Activities +',
'Big Parks', 'Small Community Parks']
    },
    helpers: {
      addOne: function(value) {
        return parseInt(value, 10)+1;
      }
    }
  };

  res.render('tutorial', context);
});
```

FIGURE 12.4 Processing the home page request in Explorador app File: app.js

Figure 12.5 tells the express server how to handle the request

```
http://140.184.132.239:8080/parkView?id=###
```

using the app.get() method, where ### is the ID of the park, as it is stored in the database. Typically, the request will come with an ID of the park. For example, Shubie park has the ID attribute of 133. We create a JSON object called query. It has one attribute called "_id", which is the attribute in our MongoDB database that stores the unique ID for the park, which stores the value of the park ID received as part of the request.

Next, we call the function parkMongoGrab.findAll(). We will be looking at the function in Figure 12.15. The function receives the stringified version of our JSON object called query, along with an anonymous callback function. The anonymous function receives two parameters, error and prks. The variable error can be used to indicate errors. The variable prks is an array of parks that we will render on the page. We first check to make sure that the array is not empty. If it is not empty, we call the res.render() method. The method receives two parameters. The first parameter asks the server to render the page called parkView (parkView.handlebars file), and the relevant information is provided in a JSON object that is supplied as the second parameter. It specifies three name/value pairs.

The first pair with the name parks stores the first element of the prks array, which is the JSON object with all the details about the park with the given id.

The second pair is a JSON object itself, called menuContext, with two attributes. The first attribute is an array called regions, and the second attribute is an array called categories. The values of the elements in the array are specified as well.

The third name/value pair, called helpers, specifies a number of functions that we will use in this rendering of the page. The first function is addOne that we saw previously, which will add one to a number; we use it to increase an array index.

The second function is called ifEqual and compares the two objects passed as the first two parameters. The function also receives a third parameter that we call option. Option is a parameter that Handlebars passes to every helper function it calls. This object contains two functions that we can use, fn and inverse, which specify either a successful (i.e., true outcome) or

```
//Route to individual park page
app.get('/parkView', function(req, res){

  var query="{\"_id\":"+req.query.id+"}";
  query=JSON.parse(query);

  parkMongoGrab.findAll(JSON.stringify(query), function(error, prks){

    if(prks.length>0)
    {
      res.render('parkView', {
          park: prks[0],
          menuContext: {
            regions: ['Halifax', 'Dartmouth', 'Bedford', 'Sackville', 'Eastern
Shore', 'Outer HRM'],
            categories: ['Sports & Activities +', 'Kid Friendly +', 'Water
Activities +', 'Big Parks', 'Small Community Parks']
          },
          helpers: {
            addOne: function(value) {
              return parseInt(value, 10)+1;
            },
            ifEqual: function(a, b, options) {
              if(a == b) {
                return options.fn(this);
              }
              return options.inverse(this);
            },
            stringify: function(object) {
              return JSON.stringify(object);
            },
            formatFeatTypes: function(featTypes) {
              var featuresString = new String(featTypes.join(', '));
              return featuresString.toLowerCase().replace(/^(.)|\s(.)/g,
function($1) {
                return $1.toUpperCase();
              });
            }
          }
      });
    }
  });

});
```

FIGURE 12.5 Processing parkView request in Explorador File: app.js

unsuccessful (i.e., false outcome) of running our helper function. When Handlebars encounters the syntax {{#function}} … `<!--HTML code-->` … {{else}} … `<!--HTML code-->` … {{/function}}, it will include in the output file the code before the "else" if fn is run, and the code after the "else" if inverse is run.

The third function, called stringify, receives one parameter and returns its string version created using the JSON.stringify() method.

The fourth function, called formatFeatTypes, takes an array of strings called featTypes that are in capital letters. For example ["CRICKET", "SOCCER", "GENERAL PLAYGROUND"] and sends back a single string that joins all the elements of the array with a comma (,) formatted such that the first character of every word is capitalized. In our example, the function will return

"Cricket, Soccer, General Playground"

We have two statements to do this: the first statement joins all the elements of the array using a comma followed by space (', '). This is stored in a variable called featureString. The second statement returns the formatted string with a little complicated manipulation, so let us analyze it in parts

- First we change the string to lowercase using the method toLowerCase().
- Then, we call the method replace that receives two parameters
    - The first parameter is a regular expression that picks the character at the beginning of the string identified by a special character ^ or space identified by a special character \s. We are using the vertical bar (|) to indicate or.
    - This character will then be sent to the second parameter, which is an anonymous function, as $1. The anonymous function returns its uppercase equivalent using the method toUpperCase().

This code is a little complicated and will not be easy to follow unless you know regular expressions. Readers who are not familiar with such sophisticated string manipulation may want to remember that we are taking an array of strings in capital letters such as ["CRICKET", "SOCCER", "GENERAL PLAYGROUND"] and sending back a single string that joins all the elements of the array with a comma (,) formatted such that the first character of every word is capitalized as

```
"Cricket, Soccer, General Playground"
```

Figures 12.6 to 12.10 describe how our app handles the request

```
http://140.184.132.239:8080/parkGrab
```

using the function app.post(). Note that all the previous requests were handled by app.get() methods. The method app.get() was used to get data from the server. The method app.post(), on the other hand, posts form data to the server. The app.post() function takes two parameters similar to app.get(). The first parameter is the request it will process. The second parameter is the anonymous function that processes the request. This anonymous function is rather long. However, it can be broken down into five logical pieces. That is why we have distributed it across five figures (Figures 12.6 to 12.10). The first part shown in Figure 12.6 creates two arrays that we have been using since Chapter 11. The first array is called regions and stores the names of all the regions. The second array, called featTypes, stores the values of all the feature types our parks can have.

Figure 12.7 shows the function starting to build up the MongoDB query, similar to what we saw in Chapter 9 for the Thyroid app. The criterion for finding parks is passed from the client to the server as a set of indices matching the indices of the array in Figure 12.6. If instead of numerical indices, an index with the value 'search' is sent, it means the user has used the search input to find a park by name. In that case, the query will search the database based on the name attribute and changes the input by the user to uppercase to match how we have stored the park names in MongoDB. An example of the MongoDB search query by name is

```
{"name":{"$regex":"COMMONS"}}
```

```
//Route for POSTing park queries
app.post('/parkGrab', function(req, res){

  var regions=new Array();
  regions[0]="Halifax";
  regions[1]="Dartmouth";
  regions[2]="Bedford";
  regions[3]="Sackville";
  regions[4]="EasternShore";
  regions[5]="OuterHRM";

  //All Sports and Activities values will be values 0-20
  //Kids values will be 21-
  var featTypes=new Array();
  featTypes[0]="Soccer";
  featTypes[1]="Ultimate Frisbee";
  featTypes[2]="Cricket";
  featTypes[3]="Rugby";
  featTypes[4]="Baseball";
  featTypes[5]="Basketball";
  featTypes[6]="Tennis";
  featTypes[7]="Ball Hockey";
  featTypes[8]="Volleyball";
  featTypes[9]="Lacrosse";
  featTypes[10]="Horseshoe";
  featTypes[11]="Lawn Bowling";
  featTypes[12]="Running Track";
  featTypes[13]="Outdoor Gym";
  featTypes[14]="Dirt Jump";
  featTypes[15]="Modular Ramps";
  featTypes[16]="Concrete Park";
  featTypes[17]="General";
  featTypes[18]="Shuffleboard";
  featTypes[19]="Outdoor Rink";
  featTypes[20]="Trails";

  featTypes[21]="Playfield";
  featTypes[22]="Playground";
  featTypes[23]="Outdoor Rink";
  featTypes[24]="Indoor Pool";
  featTypes[25]="Outdoor Pool";

  featTypes[26]="Beach";
  featTypes[27]="Spray Pool";
  featTypes[28]="Indoor Pool";
  featTypes[29]="Outdoor Pool";
  featTypes[30]="Wharf";
  featTypes[31]="Boat Launch Large";
  featTypes[32]="Boat Laundh Small";
```

FIGURE 12.6 Processing a parkGrab request in Explorador—I File: app.js

```
var indices=JSON.parse(req.rawBody);


  if('search' in indices)
  {
    var query="{\"name\":{\"$regex\":\""+indices.search.
toUpperCase()+"\"}}";
  }
  else
  {

  var regionsFind="";
  var clustersFind="";
  var featTypesFind="";

  //Minimum threshold for the fuzzy cluster
  var minClusterThresh=0.1;

  //Main Menu Query
  //Construct clusters query
  if(indices.clusters.length>0)
  {
    for(var i=0; i<indices.clusters.length; i++)
    {
      clustersFind+="\"clusters."+indices.
clusters[i]+"\":{\"$gt\":"+minClusterThresh+"}";
      if(i!=indices.clusters.length-1)
      {
        clustersFind+=",";
      }
    }
  }
```

■ **FIGURE 12.7** Processing a parkGrab request in Explorador—II File: app.js

Otherwise, the function will construct a find command based on where the indices passed by the client lie in the array, in region, category (or cluster), and feature type. In the code above we see that we have a minimum threshold for the fuzzy cluster the park belongs to. This is a data mining term and is outside the scope of this textbook. However, we can think of it as a rating of how similarly a park matches a specific category, on a rating of 0 to 1, where 1 means that the park matches the category completely. In our case, we want to exclude parks that have only a 0.1 or 10% match with a specific category, which is specified by the "$gt" (greater than) syntax.

Each cluster index that the user has requested has the required MongoDB query code created for it to be included in the find. The query code for each index will be similar to

```
"clusters.1":{"$gt":0.1}
```

Figure 12.8 shows the query string being further built to check if the region the park is located in matches one of the indices specified by the user. The syntax "$in" tells MongoDB to check for parks where the region attribute (which is the name of the region as seen in Figure 12.6) matches one of the regions specified by the user.

```
//Construct regions query
  if(indices.regions.length>0)
  {
    regionsFind="\"region\": {\"$in\":[\""
    for(var i=0; i<indices.regions.length; i++)
    {
      regionsFind+=regions[indices.regions[i]]+"\"";
      if(i!=indices.regions.length-1)
      {
        regionsFind+=",\"";
      }
    }

    regionsFind+="]}";

    if(indices.clusters.length>0)
    {
      regionsFind=", "+regionsFind;
    }
  }
```

■ **FIGURE 12.8** Processing a parkGrab request in Explorador—III File: app.js

The output of the regionFind string will be

```
"region":
{"$in":["Halifax","Dartmouth","Bedford","Sackville","OuterHRM"]}
```

Figure 12.9 shows the same process being completed for the feature types selected by the user. "$in" is used again here to check if any of the parks' feature types match those put in the array featArr. The names of the features requested are retrieved from the featTypes array, initialized in Figure 12.6, and transformed to uppercase to match the feature type format in the database.

Similar to the regionFind string, featTypesFind will appear as

```
"featTypes": {"$in":["BEACH","WHARF"]}
```

Once the featTypesFind string has been put together, the three attributes, clusters (categories), regions, and feature types find commands are grouped together into the final query string.

The final query string will look something like

```
{"clusters.1":{"$gt":0.1},"clusters.2":{"$gt":0.1},"clusters.3":
{"$gt":0.1},"clusters.4":{"$gt":0.1},"clusters.5":{"$gt":0.1},
"region": {"$in":["Halifax","Dartmouth","Bedford","Sackville",
"OuterHRM"]}, "featTypes": {"$in":["BEACH","WHARF"]}}
```

```
    //Construct featTypes query
    var featArr=new Array();
    var featQuery;
    if(indices.featTypes.length>0)
    {
      featTypesFind="\"featTypes\": {\"$in\":"
      for(var i=0; i<indices.featTypes.length; i++)
      {
        featArr.push(featTypes[indices.featTypes[i]].toUpperCase());
      }
      featTypesFind+=JSON.stringify(featArr);
      featTypesFind+="}";

      if(indices.clusters.length>0 || indices.regions.length>0)
      {
        featTypesFind=", "+featTypesFind;
      }
    }

    query="{"+clustersFind+regionsFind+featTypesFind+"}";

  }
```

FIGURE 12.9 Processing a parkGrab request in Explorador—IV File: app.js

```
parkMongoGrab.findAll(query, function(error, prks){
    if(prks.length>0)
    {
        for (var i=0; i<prks.length; i++) {
          var R = 6371; // Radius of the earth in km
          var lat2 = indices.geolocation[0];
          var lon2 = indices.geolocation[1];
          var lat1 = prks[i].location.latitude;
          var lon1 = prks[i].location.longitude;
          var dLat = (lat2-lat1)*Math.PI/180.0  // Javascript functions in radians
          var dLon = (lon2-lon1)*Math.PI/180.0
          var a = Math.sin(dLat/2) * Math.sin(dLat/2) +
                  Math.cos(lat1*Math.PI/180) * Math.cos(lat2*Math.PI/180) *
                  Math.sin(dLon/2) * Math.sin(dLon/2);
          var c = 2 * Math.atan2(Math.sqrt(a), Math.sqrt(1-a));
          var dist = R * c; // Distance in km
          prks[i].distance = dist.toFixed(2);
        }

        if(indices.view=="list")
        {
        res.render('resultsList', {
                      layout: false,
                            parks:prks,
                      helpers: {
```

```
                        formatFeatTypes: function(featTypes) {
                          var featuresString = new String(featTypes.join(', '));
                          return featuresString.toLowerCase().replace(/^(.)|\s(.)/g,
function($1) {
                            return $1.toUpperCase();
                          });
                        }
                      }
                  });
        }
        else if(indices.view=="map")
        {
                  console.log('Asking for map view');

                  res.render('resultsMap', {
         layout:false,
                             parks:prks,
         helpers: {
            stringify: function(object) {
              return JSON.stringify(object);
            }
         }
                  });
        }
    }
  return;
  });
});
```

■ **FIGURE 12.10** Processing a parkGrab request in Explorador—V File: app.js

Our MongoDB query string is then passed to parkMongoGrab.findAll, which will use it to query our MongoDB database for parks matching the users' request. The details of this function are presented in Section 12.4. When findAll has completed, it will run an anonymous function as its callback that will first check that there was at least one park found matching the users' search. The code that follows is a little bit complicated, but to summarize, it uses the longitude and latitude locations of the user, as passed to the server by the web browser to calculate the distance from the user to the park in kilometers. The more complex mathematics is used to make the calculation, taking into account factors such as the curvature of the Earth.

Also included in the request from the browser is the type of view it would like to receive, either a list or a map view. This information is passed using the indices.view attribute, and based on its value, our server decides which template to render.

Both resultsList and resultsMap Handlebars templates are passed context JSON objects with three attributes. Two of these attributes we have seen previously; parks includes the array of parks returned by the search results and will be used by the template to format these results for the user. Helpers is a set of functions to help Handlebars render the specific template. For resultsList, the only function required is formatFeatTypes, the full details of which we covered when discussing the /parkView route. For resultsMap, the only helper we provide is "stringify", which simply applies the JSON.stringify function to any JSON object passed to it as a parameter.

```
app.listen(app.get('port'), function(){
  console.log('Express server listening on port ' + app.get('port'));
});
```

■ **FIGURE 12.11** Method to launch the Explorador server

The third parameter specified is layout, which is set to false. You may recall when we set up the environment for this server that we set the defaultLayout for the app to be header.handlebars. In this case we will be passing only the body to the client/user and not the whole page, to prevent the web browser from retrieving and initializing the header and menu every time a user makes a new selection or search. By setting the layout parameter to false we prevent Handlebars from wrapping the resultsList.handlebars and resultsView.handlebars files in the header.handlebars body.

Finally, once we have specified all the express routes to service the different http requests made by the client to our Explorador app, we launch the server as shown in Figure 12.11 using the listen method from previous chapters.

12.4 Querying MongoDB

We are now going to study how our Explorador app interacts with the MongoDB server, which contains all the data about parks. Figure 12.12 shows us creating variables for the required components of the MongoDB module using a series of calls to the require() method. The modules we are importing are

- Db: Represents a MongoDB database as a JavaScript interface
- Server: Represents a Server connection to the MongoDB database

Figure 12.13 covers what happens when a new ParkProvider instance (the type of object that is assigned to the variable parkMongoGrab in our app.js file) is initialized. It is passed variables containing both the host and port that it should connect to. As seen in app.js

```
var parkMongoGrab = new ParkProvider('127.0.0.1', 27017);
```

```
var Db = require('mongodb').Db;
var Server = require('mongodb').Server;
```

■ **FIGURE 12.12** Specifying MongoDB connection parameters
File: parkMongoHandler.js

```
// Sets up connection to 'parkFinder' database
ParkProvider = function(host, port) {
  this.db= new Db('parkFinder', new Server(host, port, {safe: false},
{auto_reconnect: true}, {}));
  this.db.open(function(){});
};
```

■ **FIGURE 12.13** ParkProvider object constructor to connect to MongoDB
File: parkMongoHandler.js

Here the local machine IP address was specified, as well as the default port for MongoDB, 27017, where our database is waiting for connections.

Using the provided host IP address and port, ParkProvider initializes a server connection to the database. The option safe is set to false as we have not specified a function to run as a callback as the last parameter to this function (we have provided an empty object to signal that there is no callback) and the option auto_reconnect is set to true, so that in the case that our connection to the database is dropped for any reason, it should reconnect automatically.

Our database interface is provided with two parameters: the first is the name of the database containing all of our park data, called "parkFinder"; the second is the server connection to MongoDB.

The second line of the constructor function runs the Db function open() with an empty callback to open the connection to the parkFinder database to prepare it to accept queries.

Figure 12.14 shows the code required for the function getCollection(), which takes two parameters: the query to run on the database as a string and a callback to run with either the results or an error. This function is a wrapper around the MongoDB database function call to access a specific collection of documents. As ParkProvider will only ever require access to the collection of parks, that name is hard coded here. After retrieving the collection, this function runs the callback with either the error, if the database returned one, or the parks collection that it has successfully retrieved.

Figure 12.15 depicts the JavaScript code for the main functionality of ParkProvider. The function findAll accepts two parameters: the query to run on the database as a string, and a callback to run with either the results or an error. Using the keyword "this" to specify the

```
// Gets the 'parks' collection from database
ParkProvider.prototype.getCollection= function(query, callback) {
  this.db.collection('parks', function(error, park_collection){
    if( error ) callback(error);
    else callback(null, park_collection);
  });
};
```

FIGURE 12.14 ParkProvider.getCollection method to retrieve park collection from MongoDB File: parkMongoHandler.js

```
// Main grabbing function
ParkProvider.prototype.findAll = function(query, callback) {
    this.getCollection(query, function(error, park_collection) {
      if( error ) callback(error)
      else {
        // Parse the JSON formated string into a true JSON object
        parsedQuery=JSON.parse(query);
        // Pass JSON object to mongo database with find() call
        park_collection.find(parsedQuery).sort({area:-1}).
limit(13).toArray(function(error, results) {
          if( error ) callback(error)
          else callback(null, results)
        });
      }
    }); //end getCollection
};
```

FIGURE 12.15 ParkProvider.findAll method to retrieve the park collection from MongoDB File: parkMongoHandler.js

current instance, findAll uses the getCollection() wrapper function to obtain the collection of parks. If there was an error returned by getCollection(), then the callback provided to findAll() will also be called with the same error. Otherwise, the query string passed as a parameter will be parsed into a JSON object, which will then be run as a query on the parks collection using the find function of the collection. A couple of additional functions will be run on the results: the first, .sort({area:−1}) sorts the resulting array of parks found by the area attribute in descending order; the second, .limit(13) limits the number of search results to the 13 largest parks returned by our query.

After the results have been manipulated into a set that fits our specifications (matches our query, sorted in descending order by size, and limited to 13 search results as a maximum), they are converted from the MongoDB's native search result format to an array using the toArray() function. This find() command will return either an error, if the database ran into an error executing our query, or the results, containing an array of results that match our requirements. Our anonymous function will in turn use the callback provided to it to pass the list of parks found. An example of using this list of parks can be found in Figure 12.10.

Finally, in Figure 12.16 the syntax exports.ParkProvider is used to tell Node.js which object should be returned when another object requires this file, for example

```
var ParkProvider = require('./parkMongoHandler.js').ParkProvider;
```

which appears when we initialized our application server, in Figure 12.1.

```
// Allows ParkProvider function to be called outside this file (in app.js)
exports.ParkProvider = ParkProvider;
```

FIGURE 12.16 Exporting ParkProvider for access from other files File: parkMongoHandler.js

12.5 Client-side JavaScript

We will now turn our attention to the two JavaScript files running in the clients' web browser, menu.js and map.js, as included here in Figure 12.17 and those following.

- menu.js: Handles all the details of hiding and showing the sliding menu through the usage of Snap.js, as well as identifying the attributes the user requests when searching for a park, contacting the server with these requests, and managing the server responses.
- map.js: Interacts with the ESRI and Leaflet online mapping packages to create maps and plot markers for each park in the results.

```
var clustersQuery = new Array();
var regionsQuery = new Array();
var sportsActivitiesQuery = new Array();
var kidsQuery = new Array();
var waterQuery = new Array();

var menus = {
  main: 0,
  sportsActivities: 0,
  kids: 0,
  water: 0
};
```

```
//Geolocation interface global variable and options
var longitude = -63.5736;
var latitude = 44.6224;
var geoParams = {
  enableHighAccuracy: true,
  timeout: Infinity,
  maximumAge: 30000
};

var currentMenu = 0; //Start as main menu
var snapper;
```

FIGURE 12.17 Setting up the sliding menu File: public/js/menu.js

These two files, together with the external packages included in our app provide all the functionality necessary to navigate our app, request parks, and view results pages.

Figure 12.17 shows the initialization of many global variables that will be used by all functions in the menu.js file. These variables are

- clustersQuery: An array of category (also called cluster) integers that the user has selected on the main menu. Indices match the cluster IDs stored in our database, values between 1 and 5 inclusive.
- regionsQuery: An array of region indices that the user has selected from the main menu. Indices values match the regions array on the server, as seen in Figure 12.6.

So far, we have been discussing the two types of attributes users can select, categories (or clusters), which classify features of the park and regions, which specify the general location of the park. However, we also allow users to specify more specifically which feature they are looking for in three of our categories, Sports and Activities, Kid Friendly features, and Water features. As seen in Figures 12.18 and 12.19, these three categories have a plus sign next to them in the main menu and, when selected, open up a submenu for users to make further selections.

- sportsActivitiesQuery: An array of sports and activities indices. These indices match corresponding values on the server in the featTypes array, as seen in Figure 12.6.
- kidsQuery: An array of kids activities indices. These indices match corresponding values on the server in the featTypes array, as seen in Figure 12.6.
- waterQuery: An array of water activities indices. These indices match corresponding values on the server in the featTypes array, as seen in Figure 12.6.
- menus: An object that will contain the HTML for each of the submenus. Each time a user switches between a submenu and the main menu, the HTML of the menu currently on the screen will be saved in this object. This way, when a submenu is navigated to again, we can show the menu exactly as the user previously saw it.
- longitude, latitude: Variables to store the current location of the user, defaulted to the values for the center of Halifax.
- geoParams: A set of parameters that will be used by the HTML5 geolocation interface to obtain the users current location. The parameters of this object are self-explanatory, with the possible exception of maximumAge, which specifies how old a reading of the devices location will be accepted by our app.
- currentMenu: Value between 0 and 3, a number to identify which menu is currently being viewed by the user, where 0 is the main menu, and 1–3 specify one of the submenus
- snapper: The Snap.js menu instance

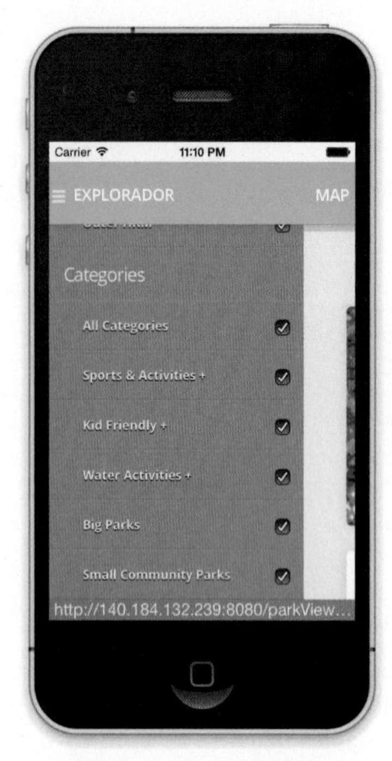

FIGURE 12.18 Main menu showing Categories that have submenus marked by a "+"

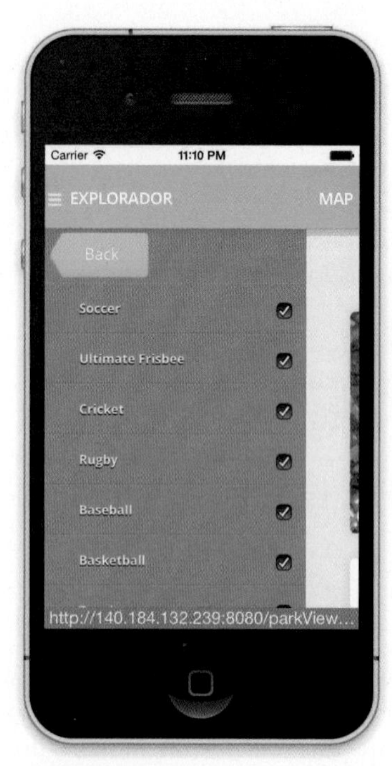

FIGURE 12.19 Submenu for the Sports and Activities category appears after a user selects "Sports and Activities +" in the main menu

Figure 12.20 displays a key function in the application, initSnapJS(). In Chapter 11 we saw in the Handlebars code that once the body has loaded, this function would run to set up our sliding menu. The first thing this function does is create a new instance of a Snap.js menu, where the main body, the object that will slide from side to side, is the element in our templates with the ID "content". Then, because Snap.js allows for menus that appear from the left, the right, or both, we disable the ability to swipe from right to left to open a menu. Following the menu initialization, we access the native HTML5 geolocation object, navigator.location and the function getCurrentPosition to find the latitude and longitude of the device, saving these values to our global variables in the anonymous callback function. If we are on a device or browser that does

```
function initSnapJS() {
  snapper = new Snap({
    element: document.getElementById('content'),
    disable: 'right'
  });

  if (navigator.geolocation) {
    navigator.geolocation.getCurrentPosition(function(pos) {
      longitude = pos.coords.longitude;
      latitude = pos.coords.latitude;
    });
  } else {
    longitude = -63.5736;
    latitude = 44.6224;
  }
```

```
    document.getElementById("listView").onclick = toggleViews;

    $("#search").keypress(function(event) {
      if (event.keyCode == 13) {
        doSearch();
      }
    });

    checkedTrigger();

    $("#open-left").click(function() {
      if (snapper.state().state == "left") {
        snapper.close('left');
      } else {
        snapper.open('left');
      }
    });
}
```

FIGURE 12.20 Initialize sliding menu, initSnapJS() File: public/js/map.js

not support geolocation, or the user has not granted us permission to access their location, we default to the center of Halifax. We then get the element with ID "listView", which is the second list element in the top menu bar and set an onclick listener to trigger the function toggleViews, which will switch between showing results in list and map formats. A keypress listener is added to the search element, so that whenever the enter key is pressed (identified by its ASCII value, 13), the doSearch() function will be run. The function checkedTrigger applies a function to handle a user's selection in the menu. Finally, we add a function to the other top navigation bar list element to open and close the sliding menu whenever it is pressed.

Figure 12.21 shows the function to be called by the span element, inside of the list element on the top right of our navigation bar to toggle between, showing the results in list or map format. Inside this function, "this" is the object that was clicked and triggered the function, which in this case is the span element. If it currently has the ID of "listView", that means that results are currently being displayed in the list format, as seen in Chapter 11 in the file

```
function toggleViews() {
  if (this.id == "listView") {
    var toggler = document.getElementById("listView");
    toggler.id = "mapView";
    toggler.innerHTML = "List";
  } else {
    var toggler = document.getElementById("mapView");
    toggler.id = "listView";
    toggler.innerHTML = "Map";
  }
  getQuery();
}
```

FIGURE 12.21 Toggle between list or map views of park results, toggleViews() File: public/js/menu.js

resultsList.handlebars. We then toggle the settings by grabbing the element on the page by its ID and changing the ID to "mapView" so that the results will be shown as a map. The text of the span is changed to "List" so that users will know that clicking on this element again will bring back the list results. The reverse is done if the current ID of the element is "mapView" and the results are currently being displayed on a map. Finally the getQuery() function is run to get the HTML5 code from the server in the newly requested format.

Figure 12.22 shows two functions that manage user selections on the menu. The first function, checkedTrigger(), adds a function to each checkbox to handle the required functionality for our app. The first thing that happens when a checkbox is clicked is that we change state from checked to unchecked or vice-versa. We need to do this ourselves as we will not be allowing the user to interact with checkbox directly, as discussed for the lineItemButton function. We then check whether the checkbox is labeled cluster, and if so, check whether it has an ID that matches one of our categories that have a submenu, which are Sports and Activities, Water Activities, and Kid Friendly features. Whenever the cluster matches a specific menu, we switch our sliding menu from the main menu to the corresponding submenu using the showMenu() function, found in Figure 12.23. Next we check the value of the checkbox; if it is assigned the value all, this checkbox is meant to select or deselect all of the checkboxes in its section. To do this, we get the current state of the checkbox, checked or unchecked, and grab all of checkboxes in the main menu. For each checkbox where the value matches our section, either "region" or "cluster", we set the value for that checkbox to match the state of the "all" checkbox. Finally, once the checkboxes are all in the correct state, the getQuery() function is run to get the new list of park results from the server.

```
function checkedTrigger() {
  $(":checkbox").click(function() {
    $(event.target).attr("checked", !$(event.target).attr("checked"));
    if (event.target.name == "cluster") {
      var currVal = event.target.value;
      //The category menus 1,2,3 have submenus
      if ((currVal == 1 || currVal == 2 || currVal == 3) && event.target.checked) {
        showMenu(currVal);
      }
    }
    var value = $(event.target).is(":checked");
    if (event.target.value == "all") {
      if (event.target.name == "regions") {
        $('#snap-menu').find(':checkbox').each(function() {
          if (this.name == "region") {
            $(this).prop('checked', value);
          }
        });
      }

      if (event.target.name == "clusters") {
        $('#snap-menu').find(':checkbox').each(function() {
          if (this.name == "cluster") {
            $(this).prop('checked', value);
          }
        });
      }
    }
    getQuery();
  });
```

```
    //Add on click to the menu list items
    //First remove any event listeners if they already exist
    $("li").off("click", lineItemButton);
    $("li").on("click", lineItemButton);
  }

  function lineItemButton(event) {
    if (event.toElement.localName != "input") {
      event.stopPropagation();
      $(this).find(":checkbox").trigger("click");
    }
  }
```

FIGURE 12.22 Checking menu selections, checkedTrigger() and lineItemButton() File: public/js/menu.js

```
function showMenu(index) {
  getQuery();
  saveCurrMenu(currentMenu);
  if (index == 0) {
    document.getElementById('snap-menu').innerHTML = menus.main;
    currentMenu = index;
  } else if (index == 1) {
    if (menus.sportsActivities == 0) {
      getCleanMenu(index);
    } else {
      document.getElementById('snap-menu').innerHTML = menus.sportsActivities;
    }
    currentMenu = index;
  } else if (index == 2) {
    if (menus.kids == 0) {
      getCleanMenu(index);
    } else {
      document.getElementById('snap-menu').innerHTML = menus.kids;
    }
    currentMenu = index;
  } else if (index == 3) {
    if (menus.water == 0) {
      getCleanMenu(index);
    } else {
      document.getElementById('snap-menu').innerHTML = menus.water;
    }
    currentMenu = index;
  }

  checkedTrigger();
}

function saveCurrMenu(index) {
  if (index == 0) {
```

```
      menus.main = document.getElementById('snap-menu').innerHTML;
    } else if (index == 1) {
      menus.sportsActivities = document.getElementById('snap-menu').
innerHTML;
    } else if (index == 2) {
      menus.kids = document.getElementById('snap-menu').innerHTML;
    } else if (index == 3) {
      menus.water = document.getElementById('snap-menu').innerHTML;
    }
}
```

FIGURE 12.23 Showing and saving the main menu and submenus, showMenu() and saveCurrMenu() File: public/js/menu.js

As well as setting the function to run whenever a checkbox is clicked, we also want to allow users to not have to click on just the checkbox but also to click on the entire list element in the menu, to make our app much more touch friendly. To do this we first set the click function for all `<li>` elements on our page to run the function lineItemButton(). This checks that the `<li>` element does not contain an `<input>` element (as is the case for the `<li>` containing our search input), then prevents other click functions from firing through the use of event.stopPropagation(), and then triggers the click function for that `<li>` elements checkbox. stopPropagation() will prevent JavaScript from running both the `<li>` elements click function and the checkboxes click function along with the default checkbox functionality for this event when a user presses on both the checkbox and the `<li>` element at the same time. Without this function call, the checkbox would first uncheck itself, and then check itself again, as its click function would be run twice.

The two functions in Figure 12.23, showMenu() and saveCurrMenu(), both relate to displaying the main menu or the submenus. Let us look at the smaller, second function first, saveCurrMenu(). This function is given the ID of the current menu being displayed as a parameter. It then checks the given ID to see which menu it matches and saves the inner HTML of the snap-menu `<div>` as a string to the "menus" JSON object global variable. This HTML string will be retrieved later, the next time the user accesses this menu.

We now turn attention to showMenu(), which takes the index of the checkbox that has just been selected as a parameter, or 0 if a user hits a "Back" button in a submenu. We first run getQuery() to save the current state of the menu. We then use saveCurrMenu(), passing in our global variable "currentMenu" as the index to be used to save the current menu appropriately.

Once the current menu is saved, the index of the requested menu to show is checked. The only way for the main menu to be requested is through the use of the "Back" button in a submenu, which means that the main menu must have been saved at least once previously, and it is loaded from the menus JSON object. The rest of the menus are first checked to see if they have the default value of 0, meaning they have not yet been shown to the user; then the HTML for a clean menu is set by getCleanMenu(), as seen in Figure 12.24. Otherwise the corresponding menu is retrieved from the menus object. Once the proper menu is placed in the "snap-menu" `<div>`, the index of the new current menu is stored in the global variable, and checkedTrigger() is called to apply the required click functions, as new submenus will not yet have these functions applied, given that they previously existed only as strings in JavaScript as described in Figure 12.24 and were not on the page.

Figure 12.24 shows the getCleanMenu() function, whose name is self-explanatory. We've seen this function used previously in the showMenu() function. The first time a given submenu is requested, this function sets the innerHTML of the menu to the string containing the HTML for a submenu in the default state, with all elements having been checked off. We have omitted some of the HTML code from the strings above, due to the repetitious nature of the menus.

The HTML contains a back button that will show the main menu when pressed, as well as a grouping of list elements, which list out all of the features related to the submenu for the user to select. The code here is similar to that of the output of the menu.handlebars file we saw in Chapter 11.

```
function getCleanMenu(index) {
  if (index == 1) {
    document.getElementById('snap-menu').innerHTML = '<h3 id="sportsMenu"> </h3>' +
      '<br><br><br><br>' +
      '<a class="btn btn-left" href="#" hidefocus="true" onclick="showMenu(0)"
style="outline: medium none; opacity: 1; padding-left: 5px; padding-bottom: 5px">' +
      '<span class="gradient">Back</span></a>' +
      '<ul>' +
      '<li><label>Soccer</label><input type="checkbox" name="soccer" checked
value="0"></li>' +
      '<li><label>Ultimate Frisbee</label><input type="checkbox" name="Ultimate Frisbee"
checked value="1"></li>' +
          // Parts of this string omitted due to repetition
'<li><label>Outdoor Rink</label><input type="checkbox" name="Outdoor Rink" checked
value="19"></li>' +
      '<li><label>Trails</label><input type="checkbox" name="Trails" checked
value="20"></li>' +
      '</ul>';
  } else if (index == 2) //kidsMenu index
  {
    document.getElementById('snap-menu').innerHTML = '<h3 id="kidsMenu"> </h3>' +
      '<br><br><br><<br>' +
      '<a class="btn btn-left" href="#" hidefocus="true" onclick="showMenu(0)"
style="outline: medium none; opacity: 1; padding-left: 5px; padding-bottom: 5px">' +
      '<span class="gradient">Back</span></a>' +
      '<ul>' +
      '<li><label>Playfield</label><input type="checkbox" name="Playfield" checked
value="21"></li>' +
          // Parts of this string omitted due to repetition
'<li><label>Outdoor Pool</label><input type="checkbox" name="Outdoor Pool" checked
value="25"></li>' +
      '</ul>';
  } else if (index == 3) //waterMenu index
  {
    document.getElementById('snap-menu').innerHTML = '<h3 id="waterMenu"> </h3>' +
      '<br><br><br><<br>' +
      '<a class="btn btn-left" href="#" hidefocus="true" onclick="showMenu(0)"
style="outline: medium none; opacity: 1; padding-left: 5px; padding-bottom: 5px">' +
      '<span class="gradient">Back</span></a>' +
      '<ul>' +
      '<li><label>Beach</label><input type="checkbox" name="Beach" checked
value="26"></li>' +
          // Parts of this string omitted due to repetition
'<li><label>Boat Launch Small</label><input type="checkbox" name="Boat Launch Small"
checked value="32"></li>' +
      '</ul>';
  }

}
```

FIGURE 12.24 Setting submenus, getCleanMenu() File: public/js/menu.js

Figures 12.25 and 12.26 show getQuery(), which is the key function for all the client-side code. This function gathers the selections the user has made from the menu and submenus and makes a request to the server for the relevant parks.

First, the function initializes new arrays that will be used to store the indices of the checkboxes that the user has selected. We then use a pair of jQuery selectors to get the snap-menu `<div>` and all of the checkboxes that it contains and for each of those checkboxes, it checks the name attribute to determine which array to insert the value into. We then determine which menu we're in based on which array(s) our checkboxes have fit into, by first checking to see if the main menu arrays have been filled; if not, we know the current menu is a submenu.

```
function getQuery() {
  //Arrays to carry all selected values of each type
  var regions = new Array();
  var clusters = new Array();
  //Submenus
  var featTypes = new Array();

  $('#snap-menu').find(':checkbox').each(function() {
    //For the categories
    if (this.checked && this.name == 'cluster') {
      clusters.push(this.value);
    } else if (this.checked && this.name == 'region') {
      regions.push(this.value);
    } else if (this.checked && this.name != 'location' && this.value != "all") {
      featTypes.push(this.value);
    }
  });

  /*When we change to a submenu, we save the query
   *So, if we're in a menu with featTypes, then we've saved the
   *main menu, otherwise, we've saved a submenu (if any)
   */
  if (featTypes.length > 0) {
    //We're in a submenu
    if (featTypes[0] > 20 && featTypes[0] > 26) {
      //In Kids Menu
      kidsQuery = featTypes;
    } else if (featTypes[0] > 25) {
      //In Water Menu
      waterQuery = featTypes;
    } else {
      //In sports menu
      sportsActivitiesQuery = featTypes;
    }
  } else {
    //In main menu
    regionsQuery = regions;
    clustersQuery = clusters;
  }
```

FIGURE 12.25 Constructing the query JSON object to send to the server, getQuery() File: public/js/menu.js

We then further determine which submenu is being displayed based on the indexes that have been stored in the featTypes array, which are set inside getCleanMenu() HTML strings. We save the values that we have found to the appropriate global variable matching that submenu, so that it will be stored the next time the menu is run—that is, after saving a submenus array of selected values. If the user returns to the main menu, the next time getQuery() is run, no featTypes will be found, but the previously selected featTypes will be saved and accessible from the global variable.

The function continues in Figure 12.26, where we start to build up the request JSON object to send to the server. This object will contain all of the desired parameters of the user's request. The first attribute set is the "view", which will be either "map" or "list" and depends on the ID currently set on the `<li>` element in the top right corner of the screen. Following setting the view, the global variables containing the query arrays for the regions and categories are selected, as well as an array containing the latitude and longitude of the user, to use when calculating the distance to the park. The three submenu queries are then checked to see if they have been set (i.e., the menu has been visited by the user). If so, the values from that submenu are concatenated into the "featTypes" attribute. Once we have built

```
    var request = {};
    if (document.getElementById("mapView")) {
      request.view = "map";
    } else {
      request.view = "list";
    }

    request.clusters = clustersQuery;
    request.geolocation = [latitude, longitude];
    request.regions = regionsQuery;
    request.featTypes = new Array();

    if (sportsActivitiesQuery.length > 0) {
      request.featTypes = request.featTypes.
  concat(sportsActivitiesQuery);
    }

    if (kidsQuery.length > 0) {
      request.featTypes = request.featTypes.concat(kidsQuery);
    }

    if (waterQuery.length > 0) {
      request.featTypes = request.featTypes.concat(waterQuery);
    }

    $.ajax({
      type: "POST",
      url: "/parkGrab",
      data: JSON.stringify(request),
      success: successful
    });
  }
```

FIGURE 12.26 Constructing query JSON object to send to the server, getQuery()
File: public/js/menu.js

up our request object, we send it to the server, making a post request to "/parkGrab". Since we only specify the route and not the domain, jQuery defaults to making a request to the current domain, in our case http://140.184.132.239:8080. The request data is stringified and set as the data portion of the request, to be parsed by the server. Finally, we set the function to run on a successful response from the server, which is called successful() and can be found in Figure 12.27.

Figure 12.27 contains the successful() function, which handles a successful response from the server. First the function checks to ensure that the server has responded with some response data. As described in Figure 12.10, the data received from the server will be the full template rendered by Handlebars. All our function needs to do with that data is to set the innerHTML attribute of the content `<div>` on the page. Once the data has been set, if the data received from the server contains an element with the id "map", or in other words, a map was requested, then the loadMap() function, which can be found in Figure 12.29, is run to set up the map on the screen.

The function doSearch(), as seen in Figure 12.28, is run when the user hits "Enter" while typing in the search field. The search input box is grabbed from the page, and if the user has entered something into the search bar, a request object is set up, very similar to how it was set up in getQuery(), with the exception being that instead of setting the various arrays containing indices of checkboxes the user has selected, the attribute "search" is set with the value from the input box. A POST request is made to the server with the request data as before. As well, we automatically close the Snap.JS menu by calling snapper.close() on the sliding menu to display the results of the query to the user.

```
function successful(data) {
  if (data != "") {
    var content = document.getElementById("content");
    content.innerHTML = data;

    if (data.indexOf("id='map'") !== -1) {
      loadMap();
    }
  }
}
```

FIGURE 12.27 Handling successful park results, successful() File: public/js/menu.js

```
function doSearch() {
  var term = document.getElementById("search");
  if (term.value != "") {

    var request = {};

    if (document.getElementById("mapView")) {
      request.view = "map";
    } else {
      request.view = "list";
    }
```

```
      request.search = term.value;
      request.geolocation = [latitude, longitude];

      //Jquery/AJAX POST of request
      $.ajax({
        type: "POST",
        url: "/parkGrab",
        data: JSON.stringify(request),
        success: successful
      });
      snapper.close();
    }
}
```

■ **FIGURE 12.28** Search by park name, doSearch() File: public/js/menu.js

The final file in our Explorador project is map.js. Map.js handles all of the logic required to load the map of Halifax and plot each of the parks on the map. In Figure 12.29 we have the global variable "map", which will be the map element on the page. Our map uses a combination of the open source Leaflet JavaScript library, http://leafletjs.com/ and the ESRI application programming interface (API), https://developers.arcgis.com/javascript/, for map image tiles. The loadMap() function first loads the map, using the Leaflet global variable "L" and getting the map element on the page, which has the ID "map" and sets the location of the map to be Halifax, with a default zoom level of 12, which allows us to see a reasonable area around Halifax. We then add a set of map images to the map that will provide a closer visual appearance to the rest of our application with ESRI's "NationalGeographic" set. Next we create a new

```
var map;
/*Initializes and loads the Esri/Leaflet Map*/
function loadMap() {
  /* Center on HRM, zoom level 12 */
  map = L.map('map').setView([44.67, -63.61], 12);
  L.esri.basemapLayer("NationalGeographic").addTo(map);
  parks = JSON.parse(document.getElementById("parkResults").innerHTML);
  var markerGroup = new L.MarkerClusterGroup();
  markerGroup.addLayers(getMarkers());
  markerGroup.addTo(map);
  function onLocationFound(e) {
    var radius = e.accuracy / 2;
    L.marker(e.latlng).addTo(map)
      .bindPopup("You are here").openPopup();

    L.circle(e.latlng, radius).addTo(map);
  }
  map.on('locationfound', onLocationFound);
}
```

■ **FIGURE 12.29** Load an embedded map of Halifax, loadMap() File: public/js/map.js

cluster group of markers. Cluster groups specify that when we zoom out far from where the markers are placed on the map, they will group together in a larger point listing the number of markers in that cluster, as can be seen in the screenshot in Figure 12.30. When the user zooms in, the cluster will break apart to show each marker.

We then make a call to getMarkers(), described in Figure 12.31, to obtain the HTML for all the markers to be plotted on the map and add them to our group of markers. That set of markers is then added to the map.

An inline function is created onLocationFound() that will be run once the map finds the location of the user. A pin will be plotted to show users their current location. Users are prompted automatically by Leaflet to obtain their location.

Leaflet accepts an array of markers to be plotted on the map. The job of the getMarkers() function, as displayed in Figure 12.31, is to convert the set of parks returned by the server as a JSON string to a set of markers. Markers have two key properties that we are using: the first is a location, in latitude and longitude, where the marker will be plotted. The second is HTML, which appears within the pop-up that appears when a user clicks on the marker.

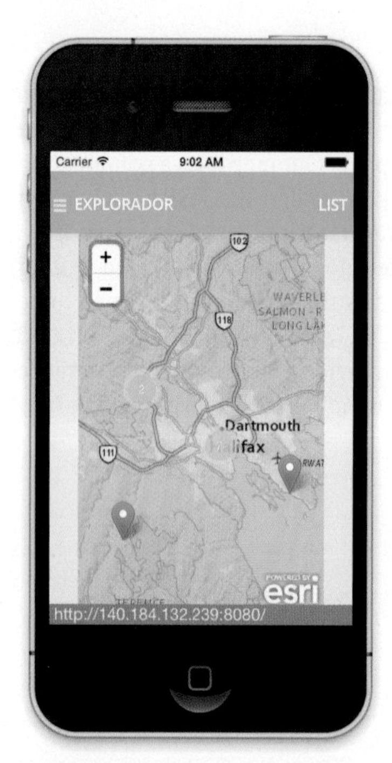

FIGURE 12.30 A cluster of park markers on the map

```
/*Evaluates the list of parks received from the server
  Creates a marker for each park
  Binds the appropriate details to the pop up*/
function getMarkers() {
  markerArray = new Array();
  for (var i = 0; i < parks.length; i++) {
    var prk = parks[i];
    var imageHtml = "";
    var sources = featFind(prk);
    for (var j = 0; j < sources.length; j++) {
      imageHtml += " <img src='" + sources[j] + "' width='20' height='20'/> ";
    }
```

```
    marker = new L.marker([prk.location.latitude, prk.location.longitude])
      .bindPopup("<h3>" + prk.name + "</h3>" + toProperCase(prk.featTypes.join(",
")) + "<br>" + imageHtml +
        "<br><a class=\"btn btn-success btn-sm\" href='/parkView?id=" + prk._id
+ "'>More</a>");

    markerArray.push(marker);
  }

  /*If only one search result, or in a parkView, center on that park*/
  if (parks.length == 1) {
    map.setView([parks[0].location.latitude, parks[0].location.longitude], 12);
  }

  return markerArray;
}

function toProperCase(s) {
  return s.toLowerCase().replace(/^(.)|\s(.)/g,
    function($1) {
      return $1.toUpperCase();
    });
}
```

FIGURE 12.31 Create map markers for each park, getMarkers() File: public/js/map.js

The getMarkers() function iterates through each of the parks, and for each park it uses the featFind() function, found in Figure 12.32, to retrieve the path to the images for each of the features. Each of those features is iterated through, and their HTML `<img>` tags are created and concatenated together. Once we have the imageHtml containing `<img>` tags for each of the features in the park, we create a new marker, with the park's location, and bind/attach the pop-up with HTML containing the park's name in an `<h3>` tag, the imageHtml, and a link to view the /parkView page for this park. Each map marker is added to the array. Finally, if there is only one park in our set, meaning that the search results only returned one park, or the map is in a /parkView page (which always only has one park), then we will center the map on that specific park. Otherwise, the map remains centered on the center of Halifax.

Lastly, the toProperCase() function performs functionality very similar to the formatFeatTypes Handlebars helper function used in the templates. It takes in a string of all capitals, as that is how the parks' names are stored in our database, and changes the string to title case, where the first letter of each word is capitalized. An in-depth description of the functionalities of this function can be found in Figure 12.5.

Figure 12.32 gives us the featFind function that iterates through all the features of a park, removes duplicates, and finds the corresponding path to icon images for those features. It is important to remove duplicates, as some parks have multiple baseball diamonds, but we only want to show one icon to users for this. The path to the images is saved in the source array; as we iterate through the park's features, we check whether the source has already been inserted into the array. If it hasn't, "indexOf" will return −1, and we know that the image path can be safely inserted without duplicates.

With this chapter, we conclude our app development efforts. In the following chapter, we will look at how to convert the HTML5/CSS3/JavaScript apps to native platforms, including iOS and Android, using a utility called Cordova by Apache corporation.

```
function featFind(prk) {
  var source = [];
  for (var k = 0; k < prk.featTypes.length; k++) {
    if (prk.featTypes[k] == "CRICKET") {
      if (source.indexOf("images/icons/Cricket.png") == -1) {
        source.push("images/icons/Cricket.png");
      }
    } else if (prk.featTypes[k] == "VOLLEYBALL") {
      if (source.indexOf("images/icons/Volleyball.svg") == -1) {
        source.push("images/icons/Volleyball.svg");
      }
    } else if (prk.featTypes[k] == "BASEBALL") {
      if (source.indexOf("images/icons/Baseball_pictogram.svg") == -1) {
        source.push("images/icons/Baseball_pictogram.svg");
      }
    } else if (prk.featTypes[k] == "BEACH") {
      if (source.indexOf("images/icons/beachOrSwimming.jpg") == -1) {
        source.push("images/icons/beachOrSwimming.jpg");
      }
    } else if (prk.featTypes[k] == "BOAT LAUNCH SMALL" || prk.featTypes[k] == "BOAT
LAUNCH LARGE") {
      if (source.indexOf("images/icons/boat.svg") == -1) {
        source.push("images/icons/boat.svg");
      }
    } else if (prk.featTypes[k] == "BASKETBALL HOOP" || prk.featTypes[k] ==
"BASKETBALL COURT") {
      if (source.indexOf("images/icons/basketball.jpg") == -1) {
        source.push("images/icons/basketball.jpg");
      }
    } else if (prk.featTypes[k] == "SOCCER") {
      if (source.indexOf("images/icons/soccerSvg.jpg") == -1) {
        source.push("images/icons/soccerSvg.jpg");
      }
    }
  }
  return source;
}
```

FIGURE 12.32 Find all different features related to a park, featFind() File: public/js/map.js

Quick facts/buzzwords

path: A node.js module to set the location of static files such as images.

serve-favicon: A node.js module to put an icon next to the URL in the URL box of the browser.

morgan: A node.js module to maintain a log of activities on the server.

express routes: http requests sent to the express server.

regular expression: A language for specifying text patterns.

leaflet: An open source JavaScript library for map management.

ESRI: A geographical information system that provides an API for maps.

CHAPTER 12

Programming Projects

1. We started with the Handlebars templating and database enhancement in the previous chapter. We complete the app in this chapter.

   Monitoring a collection of boilers: We will build on our previous app, which was used to monitor only one boiler. Our database server version allows us to store data for multiple boilers. Let us enhance the database to also include the map coordinates of each boiler. We will create an app modeled after Explorador that allows us to search for different boilers based on their IDs or location using a map. Once we pick a boiler from this search, we will have the four options as before

   i. An option and corresponding page to allow you to change the basic information about the boiler.

   ii. An option and corresponding page to enter data—temperature and pressure.

   iii. An option and corresponding page to graph the data.

   iv. An option and corresponding page to make recommendations based on the values of temperature and pressure.

2. We started with the Handlebars templating and database enhancement in the previous chapter. We will complete the app in this chapter.

   Monitoring blood pressure of a collection of patients: We will build on our previous app, which was used to monitor blood pressure of only one person. Our database server version allows us to store data for multiple persons. Let us enhance the database to also include the map coordinates of each person. We will create an app modeled after Explorador that allows us to search for different persons based on their names, IDs, or location using a map. Once we pick a person from this search, we will have the four options as before

   i. An option and corresponding page to allow you to change the basic information about the person.

   ii. An option and corresponding page to enter data—blood pressure.

   iii. An option and corresponding page to graph the data.

   iv. An option and corresponding page to make recommendations based on the values of blood pressure.

3. We started with the Handlebars templating and database enhancement in the previous chapter. We will complete the app in this chapter.

   Monitoring the power consumption of a collection of plants: We will build on our previous app, which was used to monitor only one plant. Our database server version allows us to store data for multiple plants. Let us enhance the database to also include the map coordinates of each plant. We will create an app modeled after Explorador that allows us to search for different plants based on their IDs or location using a map. Once we pick a plant from this search we will have the four options as before

   i. An option and corresponding page to allow you to change the basic information about the plant.

   ii. An option and corresponding page to enter data—power consumed.

iii. An option and corresponding page to graph the data.

iv. An option and corresponding page to make recommendations based on the values of power consumption.

4. We started with the Handlebars templating and database enhancement in the previous chapter. We will complete the app in this chapter.

   Monitoring body mass index (BMI) of a collection of patients: We will build on our previous app, which was used to monitor the BMI of only one person. Our database server version allows us to store data for multiple persons. Let us enhance the database to also include the map coordinates of each person. We will create an app modeled after Explorador that allows us to search for different persons based on their names, IDs, or location using a map. Once we pick a person from this search, we will have the four options as before

   i. An option and corresponding page to allow you to change the basic information about the person.

   ii. An option and corresponding page to enter data—height (meter) and weight (kg).

   iii. An option and corresponding page to graph the data, height, and BMI.

   iv. An option and corresponding page to make recommendations based on the values of body mass index calculated using the weight and height entered.

5. We started with the Handlebars templating and database enhancement in the previous chapter. We will complete the app in this chapter.

   Monitoring line of credit for a collection of organizations: We will build on our previous app, which was used to monitor only one organization. Our database server version allows us to store data for multiple organizations. Let us enhance the database to also include the map coordinates of each organization. We will create an app modeled after Explorador that allows us to search for different organizations based on their IDs or location using a map. Once we pick an organization from this search, we will have the four options as before

   i. An option and corresponding page to allow you to change the basic information about the line of credit.

   ii. An option and corresponding page to enter data—credit or debit.

   iii. An option and corresponding page to graph the balance.

   iv. An option and corresponding page to make recommendations based on the current balance.

CHAPTER 13

Cross-platform and native app development and testing

WHAT WE WILL LEARN IN THIS CHAPTER

1. What is native app development
2. What is Cordova
3. How to create and test a native iOS app
4. How to create and test a native Android app

So far we have developed apps that are essentially run through a browser. The browsers on mobile devices have evolved to provide an app experience that is almost as good as that provided by native apps. In this chapter, we will use a package called Cordova that hides the browser details to create standalone apps.

13.1 Native mobile apps

This book focuses on developing a mobile app that can work across a number of platforms including Apple's iOS (iPhone, iPad, and iPod touch), Android (phones, tablets, and phablets), and Windows (PCs, tablets, and phones). We used HTML5, CSS3, and JavaScript (jQuery Mobile and Node.js) as the underlying technologies for our app development. In order to create apps that exploit all the features of an app, we need to develop them in the native platforms. In the case of Apple devices that means programming in the language called Objective-C using an integrated development environment (IDE) called Xcode. For Android devices we need to program using Java (which can be done using a number of different IDEs, including Eclipse). For Microsoft Windows, we need Visual Studio, .Net, and any of the supported languages including C#.

While device-specific development will result in the most optimized app for that device, one can tap into many of the standard features of the devices such as the camera and accelerometer from JavaScript without going through the rigors of developing multiple apps for multiple platforms. Apache Cordova is a platform that can be used for building such native mobile applications using the three technologies (HTML5/CSS3/JavaScript) discussed in this book.

In this chapter, we will see how we can easily construct native apps for iOS, Android, and Windows 8 devices by converting our existing Explorador app using Apache Cordova, available for download at http://cordova.apache.org/. The chapter is meant to initiate the readers into the possibility of creating native apps from our cross-platform apps. Use of some of the fancier, native device functions, such as the camera and accelerometer, is beyond the scope of this book. However, we will list some of the resources for more sophisticated app development.

This chapter is meant for more adventurous and sophisticated readers who are comfortable with command line interface, as we will mostly use the command line interfaces in UNIX (Apple OS X). We will explain all the commands used to ensure that the chapter is self-contained. For iOS the apps can be developed only on an Apple OS X operating system, while Android apps can be developed on either UNIX (such as Mac OS X) or Windows platforms; therefore, we will use an Apple computer for both iOS and Android app development.

13.2 Setting up the Explorador app for Apple iOS and Android platforms

Figure 13.1 shows a command line session that begins the process of creating the native versions of the Explorador app for iOS and Android on an Apple computer. You should launch the Terminal app on the Apple computer to get to the command line interface. Navigate to the directory where you want to create the project. We assume a certain knowledge of file systems in this case.

The $ sign in Figure 13.1 is the prompt from the operating system for a command. We invoke Apache Cordova with the command cordova as

```
cordova create explorador com.smu.cs.explorador Explorador
```

The first parameter to the cordova command signals that we are creating a new app. The second parameter indicates the directory name where the code will be stored; in our case the code will be stored in the directory called Explorador. The third parameter is an ID in the form of a reverse web address. We are assuming a web address of explorador.cs.smu.com, so the ID will be com.smu.cs.explorador. We do not have to have the web address registered, but this ID is used by App Stores to uniquely identify your application; therefore, when building an application for the mass public, it is recommended that you are the owner, or have control over,

```
$ cordova create explorador com.smu.cs.explorador Explorador
Creating a new cordova project with name "Explorador" and id
"com.smu.cs.explorador" at location "/Users/mtriff/test/explorador"
$ ls
explorador
$ cd explorador/
$ ls
config.xml hooks      platforms  plugins    www
$ cordova platform add ios
Creating ios project...
... system messages deleted
... system messages deleted
Project successfully created.
$ cordova platform add android
Creating android project...
... system messages deleted
... system messages deleted
Project successfully created.
```

FIGURE 13.1 Commands to create native apps for iOS and Android platforms using Apache Cordova

at least the root of the web address, which in our case is cs.smu.ca. Finally, the fourth and last parameter is the name of the app, which is to be Explorador.

Once we hit enter, the system created the Cordova project as shown by the messages in Figure 13.1. The following command "ls" allows us to see the contents of the current directory. We see that there is only one listing "explorador", which is the directory where the cordova command has created all the relevant files to build our native app. We change to the Explorador directory with the "cd explorador" command. We use the "ls" command to see the contents of the directory created by the cordova command. There is a file called config.xml and four other subdirectories: hooks, platforms, plugins, and www. We will study these as and when it becomes necessary.

Now that we have created the project, and we are in the project directory, we can add the platforms iOS and Android as the two platforms we want to build our application for using the two commands

```
cordova platform add ios
cordova platform add android
```

The system gives rather verbose messages. We have deleted these messages for brevity. The final message from the system tells us that the project for each platform has been successfully created.

13.3 Building iOS app

Figure 13.2 shows the commands that are specific to building the iOS app. While still in the Explorador directory we type the command

```
cordova build ios
```

We will again get detailed system messages, which are omitted from the figure. Finally, we get a message that tells us that the build has succeeded.

We then change to the platforms directory using the command

```
cd platforms/
```

```
$ cordova build ios
... system messages
** BUILD SUCCEEDED **
$ cd platforms/
$ ls
android ios
$ cd ios/
$ ls
CordovaLib
Explorador
Explorador.xcodeproj
Build
Cordova
platform_www
www
$ open .
```

FIGURE 13.2 Commands to build iOS project using Apache Cordova

The directory listing using the command ls shows us directories for the two platforms as

```
android ios
```

Changing the directory to ios using the command cd ios/ and doing a directory listing using ls shows us the following files and folders:

```
CordovaLib
Explorador
Explorador.xcodeproj
Build
Cordova
platform_www
www
```

The file that we are most interested in is Explorador.xcodeproj. It is the Xcode project file for our app Explorador. We want to now switch to the graphical user interface mode for the next bit of computing. Therefore, we launch the Apple OS X finder with the command:

```
open .
```

The command open launches the folder. The parameter period (.) tells the command to open the current folder as shown in Figure 13.3. We open the project by double-clicking on the Explorador.xcodeproj. This will launch the Xcode IDE and load our app project as shown in Figure 13.4. If we press the play button on the top left, it will execute our app in an iPhone simulator as shown in Figure 13.5. Currently it is just an empty shell of our project with the default Hello World app; we still have to connect it to the real Explorador app.

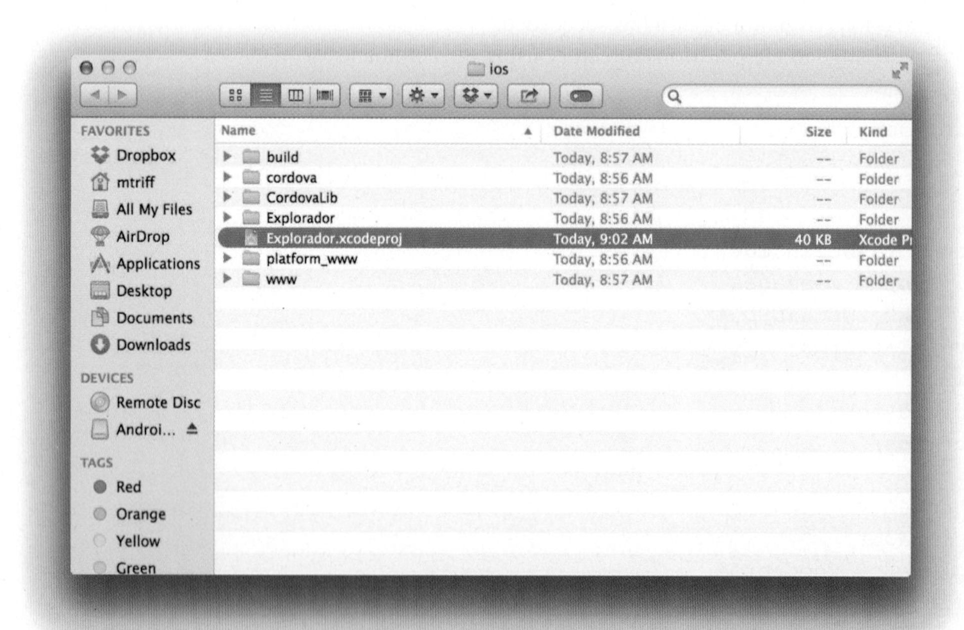

■ **FIGURE 13.3** Folder with iOS project for the Explorador app

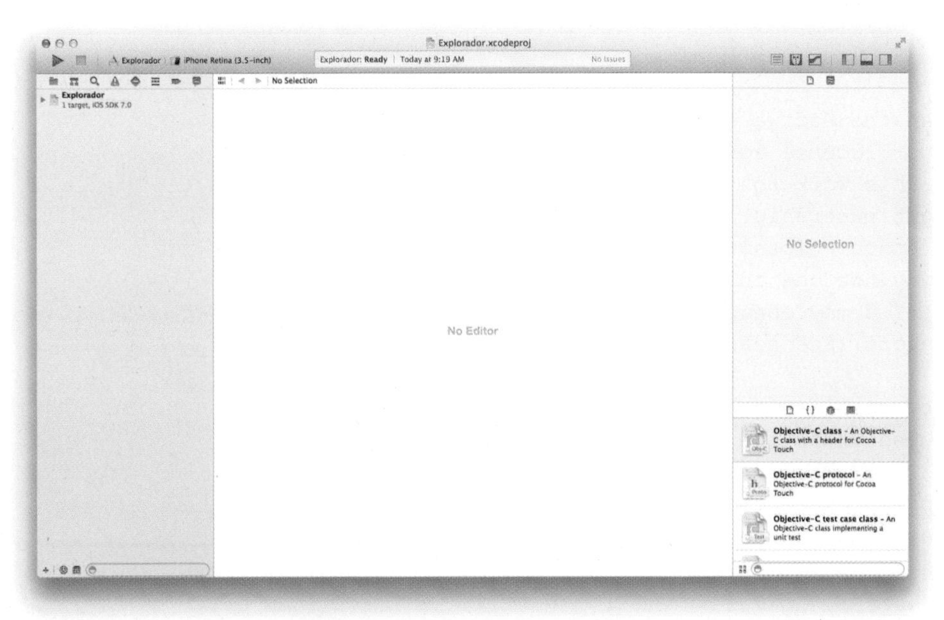

■ **FIGURE 13.4** iOS project for the Explorador app opened in the Xcode IDE

■ **FIGURE 13.5** Empty shell of the iOS Explorador app

Our iOS build operation created an empty shell of an app that uses an automatically generated index.html file, shown in Figure 13.6, in the www folder. We need to change it so that it will open our Explorador app in a window. Figure 13.7 shows the modified index.html file. We have set the title of the file to Explorador. The cordova.js script from the automatically generated file is necessary for all Cordova apps to start and communicate with the device. We add another JavaScript tag that adds an event listener using the method document.addEventListener() for the event "deviceready". When the app finishes loading and the device is ready to proceed, the event "deviceready" will be triggered by Cordova and the function launchExplorador() will be executed. The function has only one statement that calls

```
Original:
<!--
   Licensed to the Apache Software Foundation (ASF) under one
   or more contributor license agreements. See the NOTICE file
   distributed with this work for additional information
   regarding copyright ownership. The ASF licenses this file
   to you under the Apache License, Version 2.0 (the
   "License"); you may not use this file except in compliance
   with the License. You may obtain a copy of the License at
   http://www.apache.org/licenses/LICENSE-2.0

   Unless required by applicable law or agreed to in writing,
   software distributed under the License is distributed on an
   "AS IS" BASIS, WITHOUT WARRANTIES OR CONDITIONS OF ANY
    KIND, either express or implied. See the License for the
   specific language governing permissions and limitations
   under the License.
-->
<html>
   <head>
       <meta charset="utf-8" />
       <meta name="format-detection" content="telephone=no" />
       <!-- WARNING: for iOS 7, remove the width=device-width and height=device-
height attributes. See https://issues.apache.org/jira/browse/CB-4323 -->
       <meta name="viewport" content="user-scalable=no, initial-scale=1, maximum-
scale=1, minimum-scale=1, width=device-width, height=device-height, target-
densitydpi=device-dpi" />
       <link rel="stylesheet" type="text/css" href="css/index.css" />
       <meta name="msapplication-tap-highlight" content="no" />
       <title>Hello World</title>
   </head>
   <body>
       <div class="app">
           <h1>Apache Cordova</h1>
           <div id="deviceready" class="blink">
               <p class="event listening">Connecting to Device</p>
               <p class="event received">Device is Ready</p>
           </div>
       </div>
       <script type="text/javascript" src="cordova.js"></script>
       <script type="text/javascript" src="js/index.js"></script>
       <script type="text/javascript">
           app.initialize();
       </script>
   </body>
</html>
```

FIGURE 13.6 Automatically generated index.html file for the Explorador app

```
New:
<!DOCTYPE html>
<html>
   <head>
       <title>Explorador</title>
       <script type="text/javascript" src="cordova.js"></script>
       <script type="text/javascript">
           // Wait for the device API libraries to load
           document.addEventListener("deviceready", launchExplorador);

           function launchExplorador() {
               var ref = window.open('http://140.184.132.239:8080', '_blank',
'location=yes,toolbar=no');
           }
       </script>
   </head>
   <body>
   </body>
</html>
```

FIGURE 13.7 Modified index.html file for the Explorador app

the method window.open() to open our Explorador web page from port 8080 of our server 140.184.132.239. The additional parameters to the method include '_blank', which signals the window to open in a new screen and not directly within the Cordova application window. The options 'location=yes,toolbar=no', specify that the location should be shown, but that a toolbar, similar to that found in a web browser with back and forward navigation and a URL section should not be shown.

We need to make one more addition to our Cordova project before rebuilding our iOS app. The InAppBrowser functionality that we are using to access Explorador is not built into every project by default. We must run the command shown in Figure 13.8 to include the required plugin.

Once we have added the plugin, we are ready to rebuild the app. Before we do that we should make sure that our web server is serving the app as we have seen in Chapter 11. We are reproducing the step in this chapter in Figure 13.9. On our server 140.184.132.239, we go to the directory Explorador and type the command node app.js. We will not worry about the warning from the MongoDB as before. Coming back to our Apple computer, we go to the Explorador directory and rebuild the iOS app with the command shown in Figure 13.10. We then relaunch the app from the Finder window in Figure 13.3. We will now see the familiar Explorador app without all the paraphernalia that comes with the browser as shown in Figure 13.11. We can swipe to get the menu to make sure the app works as shown in Figure 13.12

```
$ cordova plugin add org.apache.cordova.inappbrowser
Fetching plugin "org.apache.cordova.inappbrowser" via plugin registry
```

FIGURE 13.8 Adding the InAppBrowser plugin to our project

```
Running the server:
pawan@HRM-ParkApps:~/explorador$ node app.js
========================================================================
=  Please ensure that you set the default write concern for the database by setting    =
=   one of the options                                                      =
=                                                                           =
=     w: (value of > -1 or the string 'majority'), where < 1 means          =
=        no write acknowledgement                                            =
=     journal: true/false, wait for flush to journal before acknowledgement  =
=     fsync: true/false, wait for flush to file system before acknowledgement =
=                                                                           =
=  For backward compatibility safe is still supported and                   =
=   allows values of [true | false | {j:true} | {w:n, wtimeout:n} | {fsync:true}]   =
=   the default value is false which means the driver receives does not     =
=   return the information of the success/error of the insert/update/remove =
=                                                                           =
=   ex: new Db(new Server('localhost', 27017), {safe:false})                =
=                                                                           =
=   http://www.mongodb.org/display/DOCS/getLastError+Command                =
=                                                                           =
=  The default of no acknowledgement will change in the very near future    =
=                                                                           =
=  This message will disappear when the default safe is set on the driver Db =
========================================================================
Express server listening on port 8080
```

FIGURE 13.9 Launching the Explorador server

```
$ cordova build ios
... system messages deleted
** BUILD SUCCEEDED **
```

FIGURE 13.10 Rebuilding the Explorador iOS app

FIGURE 13.11 Welcome screen for the iOS Explorador app

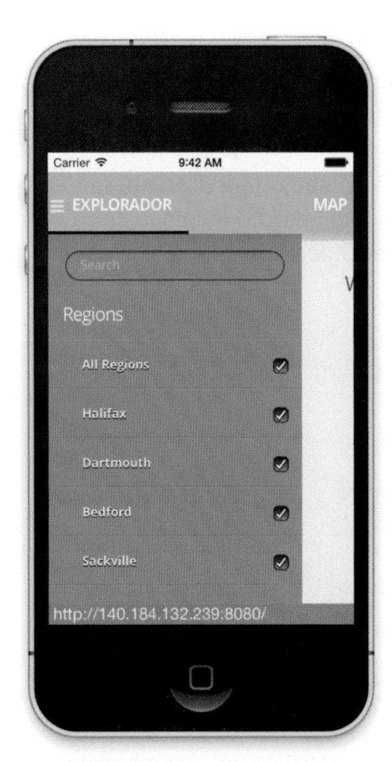

■ **FIGURE 13.12** Menu screen for the iOS Explorador app

13.4 Building the Android app

The build process for the Android app is a little simpler, especially since we have already modified the index.html file while building the iOS version. Figure 13.13 shows that all we have to do is build the Android version by typing the command

```
cordova build android
```

We launched the iOS version through an iPhone emulator by using the Xcode IDE. We can launch a simulator for Android as well. However, the Android simulator can take a very long time to load. Therefore, we connected an Android device, a Nexus 7 tablet, to launch the app on instead; the screenshot from the Nexus tablet can be seen in Figure 13.14.

```
$ cordova build android
... system messages deleted
BUILD SUCCESSFUL
Total time: 36 seconds

After plugging in device:
MacBook-Pro:explorador mtriff$ cordova run android
...
BUILD SUCCESSFUL
Total time: 5 seconds
WARNING : No target specified, deploying to device '08dbd5ab'.
Using apk: /Users/mtriff/test/explorador/platforms/android/ant-build/Explorador-debug-
unaligned.apk
Installing app on device...
Launching application...
LAUNCH SUCCESS
```

■ **FIGURE 13.13** Building the Android Explorador app

In this final chapter, we provided a glimpse for readers on how we can convert apps written in HTML5, CSS3 and JavaScript using Cordova on an Apple computer. It is possible to use a similar procedure for Windows 8 tablets and computers using a Windows computer that has Visual Studio installed.

Additionally, apps can reside directly on the device. With some modification the apps we looked at in the earlier chapters that have no server component can be included entirely in the www folder of the Cordova application to run natively, without connecting to web pages served over the Internet.

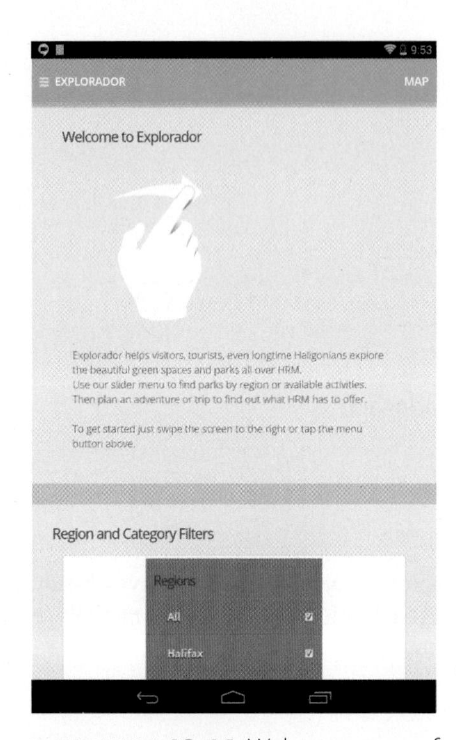

FIGURE 13.14 Welcome screen for the Android Explorador app

Quick facts/buzzwords

Apache Cordova: A platform for building native apps using HTML5/CSS3/JavaScript.

Java: Programming language used to build native Android apps.

Objective-C: Programming language used to build iOS apps.

IDE: Integrated development environment for building, compiling, and running programs.

Xcode: IDE used to develop iOS apps.

Visual Studio: IDE for building apps for Windows 8 tablets and phones.

C#: One of the languages that can be used to develop apps for Windows 8 tablets and phones.

CHAPTER 13

Self-test exercises

1. Name some of the devices that run Apple iOS.
2. Name some of the devices that run Android.
3. What is an IDE?
4. What is the language and IDE used for iOS app development?
5. What is the language and IDE used for Android app development?
6. What is the language and IDE used for Windows 8 app development?
7. What platform is used to convert a cross-platform app built with HTML5/CSS3/JavaScript to a native platform?

Programming projects

Use Apache Cordova along with an appropriate IDE to convert the following apps that you have developed before to your favorite mobile platform.

1. Compute the load distribution on a beam.
2. Binary operator
3. Electricity calculator
4. Amortization calculator
5. Temperature converter
6. The most refined version of the blood pressure monitor
7. The most refined version of the boiler monitor
8. The most refined version of the power consumption monitor
9. The most refined version of the body mass index (BMI) monitor
10. The most refined version of the line of credit manager

INDEX

M

N

T

V

W

X

INDEX OF HTML5 TAGS AND ELEMENTS

INDEX OF JAVASCRIPT AND OTHER PROGRAMMING ELEMENTS

INDEX OF CSS RESOURCES